T0310235

THE ECONOMICS OF HUMAN
SYSTEMS INTEGRATION

**WILEY SERIES IN SYSTEMS ENGINEERING
AND MANAGEMENT**

Andrew P. Sage, Editor

A complete list of the titles in this series appears at the end of this volume.

THE ECONOMICS OF HUMAN SYSTEMS INTEGRATION

Valuation of Investments in People's Training and Education, Safety and Health, and Work Productivity

Edited by

WILLIAM B. ROUSE

JOHN WILEY & SONS, INC., PUBLICATION

Published by John Wiley & Sons, Inc., Hoboken, New Jersey
Published simultaneously in Canada

For general information on our other products and services or for technical support, please contact our
Customer Care Department within the United States at (800) 762-2974, outside the United States at
(317) 572-3993 or fax (317) 572-4002.

Wiley also publishes its books in a variety of electronic formats. Some content that appears in print
may not be available in electronic formats. For more information about Wiley products, visit our web
site at www.wiley.com.

Library of Congress Cataloging-in-Publication Data:

Rouse, William B.
   The economics of human systems integration : valuation of investments in people's training and
education, safety and health, and work productivity / William B. Rouse.
       p. cm.
   Includes bibliographical references and index.
   ISBN 978-0-470-48676-4 (cloth)
   1. Human engineering.   2. Systems engineering.   3. Employees–Training of.   4.   Work
environment.   5. Human capital. I. Title.
   TA166.R68 2010
   658.3–dc22
                                                                            2010004245

Printed in Singapore

10 9 8 7 6 5 4 3 2 1

Contents

Preface

There is a long and rich history, with many successes, associated with effective integration of human behavior and performance into complex systems such as aircraft, automobiles, factories, process plants, and, more recently, service systems. Human systems integration (HSI)—as well as human-centered design—is now a well-articulated and supported endeavor. As discussed at great length in my recent book, *People and Organizations: Explorations of Human-Centered Design* (Wiley, 2007), we have accumulated much knowledge and the skills needed to enhance human abilities, overcome human limitations, and foster human acceptance.

However, as with any engineering activity, there are costs associated with HSI or human-centered design. Most would argue that these costs are actually investments in increased performance, higher quality, and lower operating costs. This book addresses the question of whether such investments are worth it. What are the likely monetary returns on such investments, and do these returns justify these investments?

I hasten to note that nonmonetary returns are often also of interest. However, this book is focused solely on getting the economics right. Admittedly, the numbers are not all that counts. But, you need to count the numbers correctly. Then you can trade off economic attributes versus noneconomic ones.

Understanding the economic attributes of HSI investments is not as straightforward as it may seem. First of all, there are several levels of costs. At the lowest level, there are the labor and material costs of the personnel who do HSI. Their efforts usually result in recommendations for improving the system of interest. These recommendations often involve second-level costs that are much larger than those associated with those doing HSI. At the third level, there are the costs associated with operating the system after the HSI-oriented recommendations have been implemented.

From an investment perspective, we would hope that the third-level costs are decreased by having incurred the first- and second-level costs. (Some HSI practitioners characterize these savings as "cost avoidance.") These reductions represent returns on having made the lower level investments. There may be additional returns associated with selling more units of a well-designed system, such as we have seen of late with Apple's iPhone (Apple Inc., Cupertino, CA). This increased demand can lead to greater production efficiencies and thereby increase profits per unit, creating a third source of return on investment.

The investment situation just outlined is summarized as follows. There are time series of upstream costs—or investments—and then time series of downstream returns. Standard discounted cash-flow analysis could be used to determine whether expected returns justify the proposed investments. However, it is not that straightforward. Were it so simple, this book would not be needed.

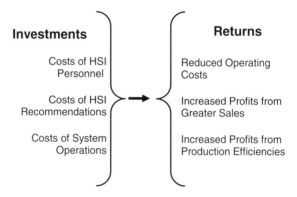

One problem is that is difficult to estimate the upstream and downstream time series of investments and costs. Point estimates will not suffice as there is much uncertainty. Thus, we need probability distributions, not just expected values. For all but the most sophisticated enterprises, this poses data-collection problems. Quite simply, although most enterprises understand their overall costs as seen on their income statements, most cannot attribute these costs to particular activities such as operations and maintenance of the systems they operate.

There are also uncertainties associated with what recommendations will emerge, which ones will be chosen for implementation, and whether the actual operating environment of the system once deployed will encounter operational demands that take advantage of the enhanced system functionality that was recommended by the HSI personnel. Consequently, the decision to invest in HSI is really a multistage decision. As shown in several chapters in this book, traditional discounted cash-flow analyses substantially underestimate the value of multistage investments. Although we have the analytical machinery to address these types of investments, many decision makers find this level of uncertainty daunting.

Beyond these technical and practical difficulties, there is often an enormous behavioral and social difficulty associated with the simple fact that different people and organizations make the investments and then see the returns. The organization developing or procuring a system is usually remote from the organization gaining the returns, both spatially and temporally. For example, engineering and manufacturing may incur the costs while marketing and sales see the returns. Furthermore, the costs may be incurred today while the returns are not seen until years from now.

This spatial and temporal separation is less difficult for highly integrated enterprises such as companies operating in the private sector. In contrast, for government agencies and companies operating in the public sector, there may be no one who "owns the future." In these situations, investments are treated as costs. Although these expenditures may yield assets that can provide future returns, government agencies—and Congress—have no balance sheet on which to tally the value of these assets. Thus, no value is explicitly attached to the future.

As formidable as this litany of difficulties may seem, we still make investments in training and education, health and safety, and performance enhancements. Culturally at least, we value a healthy, educated, productive, and competitive workforce. We have the right inclinations. However, we have not had the right data, methods, and tools to make stronger economic arguments for investing in people. This book is intended to improve this situation. The set of thought leaders recruited as contributors to this volume collectively provide a compelling set of data, methods, and tools for assessing the economic value of investing in people, not just in general but in specific investment situations. We hope that a broad cross section of policy makers and practitioners, as well as researchers, will benefit from this volume.

Pursuit of this topic has been a long journey, much of it in industry but also frequently in government, particularly the Department of Defense (DoD). Ken Boff, now a colleague at Georgia Tech, but for many years Chief Scientist of the Human Effectiveness Directorate at the Air Force Research Laboratory, has been a kindred spirit on this journey to understand the value created by investing in people and to develop methods and tools for economic assessment of value.

The Air Force Human Systems Integration Office within the Office of the Vice Chief of Staff played an initial and important role in bringing together key senior thought leaders interested in the economics of human systems integration. In particular, Senior Executives from DoD, the military services, and other government agencies, including John Young, Jack Gansler, Jim Finley, Mike Montelongo, Delores Etter, John Gilligan, Larry Spencer, Jay Jordan, Richard Gustafson, Mike Sullivan, Paul Chatelier, and John Retelle, provided important insights. Congressmen John Barrow, Mike Doyle, and Tim Murphy provided essential tutoring in the workings of Congress, both for defense and health care.

At first, the economics of HSI seem to be a straightforward issue of assessing costs and returns and then calculating economic value. However, as with most complex public–private systems, things quickly become complicated. This book is intended to make the issues fathomable and, thereby, to make assessment tractable. Better assessment will, in turn, hopefully lead to well-informed policy decisions. This, of course, is yet another level of human systems integration.

WILLIAM B. ROUSE

Atlanta, Georgia
October 2009

Contributors

W. Gary Allread, The Ohio State University, Institute for Ergonomics, Columbus, OH

Aruna Apte, Naval Postgraduate School, Graduate School of Business and Public Policy, Monterey, CA

Douglas A. Bodner, Tennenbaum Institute, Georgia Institute of Technology, Atlanta, GA

Ron Z. Goetzel, Emory University, Institute for Health and Productivity Studies, Rollins School of Public Health; Vice President, Consulting and Applied Research, Thomson Reuters

Keith Hartley, University of York, Huntington, York, United Kingdom

Ethan B. Kapstein, INSEAD, Business in Society Centre, Fontainebleau Cedex, France

Rivka Liss-Levinson, Emory University, Institute for Health and Productivity Studies, Rollins School of Public Health, Atlanta, GA

Kevin Liu, Engineering Systems Division, Massachusetts Institute of Technology, Cambridge, MA

William S. Marras, The Ohio State University, Institute for Ergonomics, Columbus, OH

Deborah J. Mayhew, Deborah J. Mayhew & Associates, West Tisbury, MA

Michael J. Pennock, Northrop Grumman Corporation, Fairfax, VA

Enid Chung Roenier, Emory University, Institute for Health and Productivity Studies, Rollins School of Public Health, Atlanta, GA

William B. Rouse, Tennenbaum Institute, Georgia Institute of Technology, Atlanta, GA

Daniel K. Samoly, Emory University, Institute for Health and Productivity Studies, Rollins School of Public Health, Atlanta, GA

Nachum Sicherman, Columbia University, Graduate School of Business, New York, NY

Maryam Tabrizi, Thomson Reuters (Healthcare), Washington, DC

Ricardo Valerdi, Engineering Systems Division, Massachusetts Institute of Technology, Cambridge, MA

PART I

Introduction

Chapter **1**

Introduction

WILLIAM B. ROUSE

1.1 INTRODUCTION

This book is concerned with the economic value of investing in people. A range of types of investments is of interest, for example:

- Investments in work technologies directly augment people's performance.
- Investments in education and training enhance people's potential to perform.
- Investments in health and safety enhance people's availability to perform.
- Investments in organizational processes enhance people's willingness to perform.

Such investments interact, as shown in Figure 1.1, to enable work performance that translates demands for products and services into supply of products and services.

Note that this line of reasoning applies to people who operate, maintain, and manage systems, as well as to those who research, design, and invest in systems. There are many stakeholders in the success of a system. It is likely that investing in the performance of several types of stakeholders can enhance this success. Therefore, for example, investing solely in enhancing the performance of

The Economics of Human Systems Integration: Valuation of Investments in People's Training and Education, Safety and Health, and Work Productivity. Edited By William B. Rouse
Copyright © 2010 John Wiley & Sons, Inc.

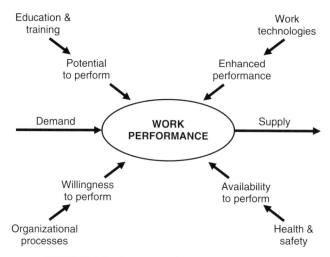

FIGURE 1.1 Examples of investments in people.

aircraft pilots will result in less success than achievable by also investing in aircraft mechanics and, perhaps, in aircraft designers.

Often the monies associated with these types of investments are simply viewed as operating costs. Investments in technologies such as computer workstations or manufacturing equipment usually show up as assets on an enterprise's balance sheet. However, monies spent on education, training, health, and safety usually only appear as expenses on the income statement. Thus, these expenditures may not be viewed as investments at all.

This book focuses on how to attach economic value to the returns provided by these expenditures. The goal is to provide an integrated view of how best to assess and project the economic value of people's performance, potential to perform, availability to perform, and willingness to perform. This involves considering both the costs of these investments and the subsequent economic returns, all over time. This also involves considering the uncertainties associated with these costs and returns.

It is useful to discuss why economic valuation of investments in people is difficult. One reason is the fact that such investments usually do not yield tangible assets—hence, they are absent from the balance sheet. One can inventory computers and equipment. However, it is difficult to "inventory" people's potential or availability to perform. An obvious reason is that one cannot own people, so they may deploy this potential elsewhere. Another complication is the fact that the circumstances may not call on people to perform (e.g., the demand in Figure 1.1 may be less than the work performance that could be supplied). Perhaps consumers will not want automobiles or refrigerators. Perhaps there will not be a war, a fire, or a crime.

Another reason for this difficulty is the typical lack of understanding of how work and work processes relate to value provided for customers or other constituencies. This makes it very difficult for enterprises to transform themselves when the nature of value fundamentally changes in a market (Rouse, 2006). Since the mids 1990s, (Hammer & Champy, 1994; Womack & Jones, 1996), there has been increased emphasis on understanding business processes and their relationships to value. Nevertheless, relatively few enterprises have mastered these skills.

Yet another reason underlying this difficulty is the lack of data upon which to base estimates of costs and returns, often in terms of cost savings. Perhaps surprisingly, organizations that have tendencies to document virtually every activity in their enterprise often have little ability to access this information for the purpose of estimating the parameters in economic models. The U.S. Department of Defense (DoD) is a notable example. The inability to estimate the costs of activities undermines the possibility of validating projected cost savings resulting from investing in people and, consequently, undermines the possibility of attaching value to these savings.

Despite such difficulties, it is very important that we have the methods and tools needed to attach economic value to investments in people. Many would agree with the general statement that a healthy, well-educated, and productive workforce is essential to a country's competitiveness. The question, however, is whether a particular health practice, educational program, or other investment will provide returns that justify the investment of scarce resources. Thus, we are less interested in the need to invest in general than we are in assessing and projecting the value of specific investments.

1.2 HUMAN SYSTEMS INTEGRATION

The issues and questions raised earlier could be addressed from a purely empirical perspective. One could collect data on the costs and returns of education, for example, and calculate an effective return on investment, for instance, for earning a college degree (see Chapters 4 and 5). Indeed, reports of such assessment frequently appear in newspapers and magazines. The general conclusion seems to be that investments in education do provide attractive returns in terms of enhanced incomes.

It this book, however, we are not concerned with investments in general. Instead, we would like to project the returns on investments in specific interventions for particular systems such as airplanes, ships, or factories. We would also like to address tradeoffs across alternative investments. For example, what are the relative returns from investments that directly augment human performance versus those that enhance the potential to perform (Rouse, 2007)? Should we invest scarce resources in a new flight management system or extended pilot training?

This book emphasizes the design, development, deployment, operation, and sustainment of complex systems. We want to engineer such systems so that the humans involved—operators, maintainers, and managers—are effective in performing their roles and in contributing to the value provided by these systems. The word "engineer" is used as a verb defining a set of activities rather than as a noun describing a discipline. Thus, the engineering of a system is perceived as involving many more disciplines than just engineering.

Systems engineering (SE) is the transdisciplinary set of activities that integrates across all the disciplines and activities involved in engineering complex systems. More specifically, "systems engineering is the management technology that controls a total system life-cycle process, which involves and which results in the definition, development, and deployment of a system that is of high quality, is trustworthy, and is cost-effective in meeting user needs" (Sage & Rouse, 2009). As might be imagined, SE involves a wide spectrum of methods, tools, and methodologies that address an enormous range of issues, many of which do not particularly relate to the humans associated with a complex system.

Human systems integration (HSI) is an element of SE that is concerned with understanding, designing, and supporting humans' roles and performance within a complex system. There are a variety of definitions of HSI. They fall into two broad classes (Booher, 1990, 2003; Pew & Mavor, 2007; Salvendy, 2006; INCOSE, 2008; Sage & Rouse, 1999):

One class of definitions focuses on integrating the knowledge, skills, and work outcomes of a range of human-related disciplines into the SE process:

- "HSI is a systems engineering process that integrates the seven technical domains of human factors engineering, manpower, personnel, training, habitability, personnel survivability, and safety/occupational health." (DoD, 2008)
- "HSI is synonymous with the traditional definition of human factors in the broadest sense. HSI adds to this traditional concept of human factors, the emphasis on integration of the individual HSI domains, and the integration of HSI into the acquisition process for emerging systems." (Malone, et al., 2007, p. 1)
- "HSI has come to be defined as the collection of development activities associated with providing the background and data needed for seamless integration of humans into the design process from various perspectives (human factors engineering, manpower, personnel, training, safety and health and, in the military, habitability and survivability) so that human capabilities and needs are considered early and throughout system design and development." (Pew & Mavor, 2007, p. 1)

The second class of definitions emphasizes HSI as a process within the SE process, with much less concern for articulating the distinct contributions of particular disciplines:

- "Human systems integration is primarily a technical and managerial concept, with specific emphasis on methods and technologies that can be utilized to apply the HSI concept to systems integration." (Booher, 1990, 2003, p. 4)
- "HSI is the interdisciplinary technical and management processes for integrating human considerations within and across all system elements." (INCOSE, 2008, p. 7)
- "Human-centered design is a process of assuring that the concerns, values, and perceptions of all stakeholders in a design effort are considered and balanced." (Rouse, 1991, 2007, p. 5)

This book is much more concerned with a process-oriented view of HSI. The seven or eight disciplines typically associated with the discipline-oriented view of HSI certainly represent important contributors to HSI, in particular, and to SE, in general. However, the human abilities, limitations, and inclinations of those who will operate, maintain, and manage the complex system of interest should be the central issues of interest rather than the extent to which these issues fall in the bailiwick of one discipline or another.

In summary, the focus is on the economics of investments in humans in the context of engineering complex systems. To many readers, this would seem to be synonymous with engineering economics (Newman et al., 2008; White et al., 2008). Chapter 7 addresses this field. However, this material is not sufficient when we are concerned with investments in people, in part because of the difficulties elaborated earlier. Furthermore, we need to understand the nature of investments in humans in the organizational contexts where these investments are considered.

1.3 ORGANIZATIONAL CONTEXTS

The two types of organizational context of interest are the system operational context (e.g., commercial airlines vs. military aircraft operations) and the systems engineering context, (e.g., commercial development vs. government procurement). These two types of context have enormous impacts on the ways in which operational value is defined and engineering value is created. Chapter 2, "Industry & Commercial Context," and Chapter 3, "Government & Defense Context," provide in-depth discussions of the ways in which operational and engineering considerations differ in these two domains. In this introductory chapter, the rationale for these distinctions is elaborated. Furthermore, a framework is outlined for characterizing the organizational contexts within which SE and HSI happens. This framework is carried forward into Chapters 2 and 3.

In studying the engineering of systems across many years, an overarching conclusion is that decision-making processes and decision outcomes occur in the context of the overall enterprise that, in turn, operates in a much broader context (Rouse, 2007). As shown in Figure 1.2, the success of an enterprise is both enabled and constrained by the nature of its markets and its internal capabilities to serve these markets, all of which occur in the broader context of the economy, which is increasingly global.

The value of investing in people depends on what an enterprise's markets value. Design, development, deployment (or distribution), and support of high-quality systems, products, and services requires investments that command higher prices than low-cost and/or commodity offerings. Some of these investments are in the people who design, develop, deploy, operate, maintain, and manage these offerings. Inadequate education and training, for example, result in less than high-quality service. Of course, investments could be made in technology that, for instance, enables Web-based self-service.

Thus, there are tradeoffs among alternative investments in people and across investments in other ways to succeed in the marketplace. The ways in which such

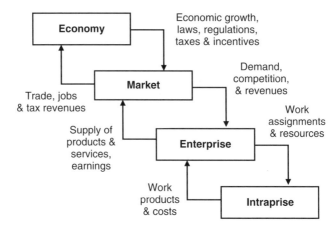

FIGURE 1.2   Context of a typical enterprise (Rouse, 2005).

tradeoffs are formulated and resolved depends, in part, on the nature of the market and, to a great extent, on the nature of the enterprise. As enterprises mature and become more successful, there is a strong tendency to develop "organizational delusions" that hinder approaching new problems in new ways (Rouse, 1998). For example, organizations where one discipline dominates (e.g., aerospace engineering in the aviation industry and electrical engineering in the semiconductor industry) tend to see all problems through these disciplinary lenses regardless of whether such a perspective is warranted.

The U.S. DoD provides a compelling example of how organizational context strongly affects how SE and HSI are pursued. As discussed in later chapters, the Government Accountability Office (GAO) has frequently criticized the DoD for acquiring weapon systems that do not meet performance requirements, far exceed original cost projections, and are deployed long after initial projections. This criticism cuts across ships, aircraft, and other large weapon systems. The GAO has concluded that these deficiencies result from not employing the best systems engineering practices. They also observe that this is not from a lack of knowledge but instead from a lack of will.

Figure 1.3 illustrates the U.S. enterprise for acquiring military ships. Note the large number of stakeholders in shipbuilding, many of whom have interests that go far beyond timely acquisition of high-performing, cost-effective ships. These stakeholders affect the ship-building enterprise in a variety of ways:

- Congressional interests and mandates (e.g., jobs and other economic interests)
- Service interests and oversights (e.g., procedures, documentation, and reviews)
- Incentives and rewards for contractors (e.g., cost-plus vs. firm fixed price)
- Lack of market-based competition (e.g., hiring and retention problems)
- Aging workforce and lack of attraction of jobs (e.g., outsourcing limitations and underutilization of capacity)

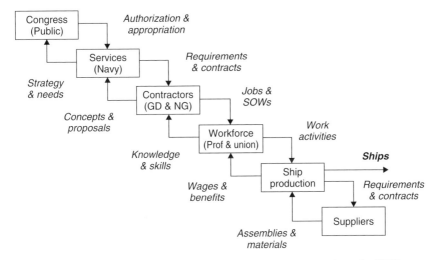

FIGURE 1.3 Enterprise for acquiring military ships (Pennock et al., 2007).

There have, in recent years, been many studies of the best commercial practices with a goal of reducing the costs and time required for military ship production. These initiatives have had positive impacts. However, there are important differences between military and commercial ships (Birkler et al., 2005). For example, the hull of a ship represents a much higher fraction of the value of a commercial ship compared with a military ship where the onboard systems are much more sophisticated.

Nevertheless, the interests of these stakeholders mitigate against acquiring ships faster and cheaper. Key stakeholders have staunchly defended the jobs and profits that would be lost were best practices adopted. Levinson's chronicle of adoption of containerized shipping by the commercial shipping industry provides an excellent illustration of key stakeholders doing their best to thwart fundamental changes (Levinson, 2006). Thus, the reactions of DoD stakeholders to attempts to adopt best SE practices and thereby transform the shipbuilding enterprise are far from unusual.

Organizational context plays an enormous role in the engineering of complex systems. The adoption of best SE and HSI practices is strongly affected by the context. Hence, the ways in which economic value is attributed to investments in people is likely to differ across contexts. The need to understand such differences has led to the following framework for characterizing organizational contexts. These ten questions provide a foundation for developing the economic models needed to assess and project economic value in a particular context:

1. What forces drive the acquisition of new systems?
2. What is the role of competition in providing new systems?
3. How large is the set of potential customers for new systems?
4. How are customers' requirements determined?

5. How are customers' budgets assessed or projected?
6. How large are production runs (i.e., number of units)?
7. How long are system life cycles (i.e., years)?
8. Who takes what risks in the system life cycle?
9. Who gains what rewards in the system life cycle?
10. How are future sales affected by past performance?

These questions are considered in depth in Chapters 2 and 3. Consequently, they are not elaborated here. Suffice it to say that comparing consumer electronics companies (e.g., Apple or Samsung) to defense contractors in shipbuilding (e.g., General Dynamics and Northup Grumman in Figure 1.3) results in very different answers to these ten questions.

Another organizational consideration of importance in this book is the scope of HSI. Adopting a human-centered perspective on this issue (Rouse, 1991, 2007), one quickly comes to the conclusion that HSI is pervasive. Beyond the humans associated with the complex system being acquired, there are the people and organizations that procure, design, and develop the system. Thus, beyond the teams or crews that operate, maintain, and manage the resulting system, there are customer teams that lead requirements definition and engineering teams that design and develop the system. There are also teams of companies that work together to engineer and manufacture the system and, increasingly, international alliances that come together to provide complex systems such as aircraft and automobiles. Human and organizational issues are pervasive among all these types of teams. Consequently, HSI can make contributions at several levels, as illustrated by the many authors who have contributed to this book.

1.4   OVERVIEW OF BOOK

This book is structured as follows. Chapters 1–3 provide the contexts of human systems integration and investment analysis. As is illustrated, context makes an enormous difference in how issues are best framed and analyzed. Chapters 4–7 provide a review of concepts, principles, models, methods, and tools drawn from economics. Chapters 8–11 discuss methods and tools of particular value for addressing the economics of human systems integration. Finally, Chapters 12–16 provide case studies of real-world economic valuations of investments in human systems integration. As emphasized in earlier discussions, all chapters in this book emphasize the monetization of the value of investments in humans. Although the intangible benefits of such investments are recognized, such benefits are not the concern of this book.

1.4.1   Introduction

This chapter sets the stage for this book. The importance of attaching economic value to investments in people is elaborated. Such investments range from work technologies, to education and training, to health and safety, to organizational processes. The reasons why such economic valuations are seldom done are elaborated. Approaches to economic valuation are addressed from the perspective of human systems integration, which is defined as a process within systems engineering. The impact of organizational context on how economic tradeoffs are formulated and resolved is considered, and a framework for characterizing and contrasting organizational contexts is introduced. This framework is employed in several subsequent chapters.

Chapter 2, "Industry and Commercial Context," considers human systems integration issues in the context of industrial organizations that primarily operate outside the aerospace and defense industry. Economic analyses of investment decisions in private versus public sectors are contrasted. A range of contemporary management practices are reviewed, including financial management, the innovation funnel, multistage decision processes, and portfolio management. Best practices are considered in terms of business process orientation, balanced scorecards, and how the "innovator's dilemma" is addressed.

Chapter 3, "Government and Defense Context," considers human systems integration issues in the context of enterprises that primarily operate in the aerospace and defense industry, as either government agencies or government contractors (e.g., defense contractors). The nature of public-sector acquisition is outlined with particular emphasis on defense acquisition. Past attempts at acquisition reform are reviewed. The acquisition enterprise is described in terms of the stakeholders and processes involved, including Congress, the military services, companies, unions, and suppliers. Management practices are reviewed and the extent to which best practices have been adopted is assessed.

1.4.2 Economics Overview

Chapter 4, "Human Capital Economics," provides a brief overview of human capital economics. The central notion is that monies used to train and educate people, as well as to ensure their health and safety, are capital investments rather than just expenditures. The measurement of returns on investments in human capital is discussed, using examples from training and health. The broader concept of human capital management is outlined, including the financial benefits of recognized management practices. Investment valuation is discussed, in terms of both attaching value to the capital created and direct valuation of investments and returns. Alternative investment metrics are reviewed. The notion of value mapping is introduced as a means for tracing process deficiencies and changes to value created. Finally, recognized best practices are summarized. These practices rely on financial models, but they extend well beyond purely monetary considerations.

Chapter 5, "Labor Economics," introduces some of the main questions studied by labor economists, their methodological approach, and some possible links between labor economics and human system integration. More specifically, the chapter discusses "human capital theory," the difference between "partial equilibrium" and "general equilibrium" approaches when considering the link between a system design and operators' skills and the difficulties in applying cost–benefit analysis in the absence of market transactions.

Chapter 6, "Defense Economics," addresses defense as a major user of scarce resources, including personnel and the human capital investments in training military personnel. The military employment contract is a distinctive feature of such training investments. Effectively, this contract "ties" labor to the Armed Forces for a specified period, allowing the Forces to obtain a return on their training investments. But defense spending raises a broader set of questions, namely, its impact on the economy and its labor markets: Are the effects harmful or beneficial? This chapter reviews the research and literature on this topic. The empirical results show all possible relationships, namely, positive, negative, and no relationships between defense spending and growth. A critique of the results and their limitations is presented.

Chapter 7, "Engineering Economics," begins with the observation that complex systems can be very expensive to research, design, develop, and deploy. They are often even more expensive to operate, maintain, sustain, and retire. Overall, 30% to 40% of the life-cycle costs can be associated with the human and organizational aspects of these systems. Upstream investments in human systems integration can yield substantial downstream savings in life-cycle costs. This chapter focuses on engineering economics and the concepts, principles, models, methods, and tools that can support analysis of HSI investments and operating costs. Engineering economics enables a much more rigorous approach to articulating the investment value of HSI than has traditionally been employed.

1.4.3 Models, Methods & Tools

Chapter 8, "Parametric Cost Estimation for Human Systems Integration," provides an approach for estimating the cost of human systems integration through the use of a cost model. As a backdrop, the authors discuss the history of HSI with respect to its role in the acquisition life cycle and its impact on systems engineering effort. They review several types of cost estimation approaches, focusing on the parametric model for systems engineering. To illustrate some of the most relevant cost drivers, they present a case study on HSI practice that highlights the importance of HSI requirements on system engineering effort. Finally, this chapter discusses how those requirements can serve as inputs into a parametric cost estimation model and provides recommendations for professional practice in HSI economics.

Chapter 9, "A Spreadsheet-Based Tool for Simple Cost–Benefit Analyses of HSI Contributions During Software Application Development," considers human systems integration applications that involve providing software applications to individual users to assist them in some aspect of performing their jobs. In this case, the HSI issue may be optimizing the productivity of the trained and experienced user (designing for "ease of use" or efficiency) or optimizing the ability of the new or casual user to get up to speed quickly with or without training (designing for "ease of learning"), or both. The focus of this chapter is to offer and explain a free spreadsheet-based tool to help estimate the potential return on investment

(ROI) in adding HSI resources and activities to an automation development effort or purchase.

Chapter 10, "Multistage Real Options," addresses the many real-world investment opportunities that are not instantaneous, now-or-never transactions but occur over time and present multiple opportunities for course corrections. This is particularly true for technology investments, which tend to be staged to mitigate risk. Traditional investment analysis fails to capture the value that staging provides, and consequently, real options analysis is required to assess appropriately multistage investments. This chapter presents both the theory required to understand real options and the methods used to solve them.

Chapter 11, "Organizational Simulation for Economic Assessment," argues that designing systems with significant levels of human integration involves substantial complexity and uncertainty. This makes economic analysis of such systems difficult. This chapter discusses the use of organizational simulation as a design tool that can aid in economic assessment. Simulation is a method of imitating a system's behavior for purposes of design and analysis. Traditional simulation methods allow for complex modeling and capture the effects of uncertainty and risk. However, they offer limited functionality for modeling human systems integration issues, especially as they relate to organizational phenomena. These phenomena are important especially in analyzing the economics of a system over its life cycle, which may be managed by multiple organizations. Organizational simulation is an emerging method that models organizational effects, processes, value creation, and the role of people. This chapter describes the current work and future directions of organizational simulation as applied to economic assessment of human systems integration.

1.4.4 Case Studies

Chapter 12, "HSI Practices in Program Management: Case Studies of Aegis," provides descriptive case studies that chronicle the operational and engineering processes used to reduce the total ownership cost for microwave tubes and radar phase shifters, components of the AEGIS Combat System, while dramatically improving their operational availability. They capture the program management practices, especially the integrated product teams, used in these processes. The processes used to achieve these results are important to understand in light of the current reductions in various acquisition support resources, including financial support, manpower, and in-house technical expertise. In particular, the cases highlight the role that Naval Warfare Centers and their resident technical staff can and do play in the acquisition process and their supporting engineering disciplines.

Chapter 13, "The Economic Impact of Integrating Ergonomics Within an Automotive Production Facility," notes that the advent of assembly-line systems has dramatically improved production efficiency. However, it often requires employees to perform similar physical activities throughout their workshifts. Because repetitive movements are linked to the development of work-related musculoskeletal disorders, care must be taken to design the human–system interface to minimize this injury risk. The case studies presented in this chapter illustrate how, by applying ergonomics principles to vehicle assembly work, companies cannot only improve employee safety but also significantly reduce production costs. Examples are given at the job level, across a facility's specific production department, and throughout a company as it launches new vehicle models.

Chapter 14, "How Behavioral and Biometric Health Risk Factors Can Predict Medical and Productivity Costs for Employers" begins with the observation that adults spend nearly a quarter of their lives at work. Businesses are beginning to recognize the potential economic benefits of investing in health-promotion and risk-reduction programs at the workplace. Researchers have developed economic models that establish the relationships between modifiable health risk factors among workers and their productivity and health-care use. Employers have developed real-world software applications that leverage research findings from these relationships. Predictive ROI models help employers build a business case for workplace health promotion programs and establish performance metrics for these programs. This chapter reviews existing evidence that supports investments in worksite health promotion and disease prevention programs and highlights two predictive models developed for the Dow Chemical Company and Novartis Pharmaceuticals.

Chapter 15, "Options for Surveillance and Reconnaissance," considers the value of defense investments in the context of a case study conducted for the Singapore Ministry of Defense. The framing of defense investment decisions is elaborated. Alternative investments for surveillance and reconnaissance missions are discussed. Economic valuations of these investments are presented, using both traditional and real options methods and tools. Economic results are integrated into a multiat-tribute analysis to develop an overall investment strategy. The resulting investment decisions are discussed.

Chapter 16, "Governing Opportunism in International Armaments Collabora-tion: The Role of Trust," notes that international joint ventures suffer high failure rates, and academic research has placed much of the blame on ungovernable prob-lems of opportunism. Not only do partners sometimes shirk their responsibilities and hold up a given venture—for example, by failing to deliver quality products on time and within budget—but they also engage in "technology poaching" or illicit efforts to procure proprietary knowledge from the other firm(s). Although these problems are difficult enough to manage within a purely domestic setting, they become much more intractable when it comes to operating across borders, where laws and cultural norms may differ between the partner companies, ren-dering contracts inefficient. A particularly "hard case" where such opportunism is likely to be rife is provided by international armaments cooperation, which is the focus of Chapter 16. The argument made is that if the partners in an inter-national armaments project are able to structure their relationship in such a way as to codevelop a complex weapons system, this suggests important lessons for the governance and management of cross-border joint ventures in high technology more generally. Drawing from behavioral economics, one finds that crucial to an effective partnership is the development of trust mechanisms. It is also suggested, however, that there are some limits associated with trust building in the context of complex international projects and indeed that trust building usually serves as a compliment to a set of strategic policies aimed at reducing the risk of opportunism.

1.5 CONCLUSIONS

The state of the art is such that we know how to attach economic value to investments in people. The concepts, principles, models, methods, and tools are readily available. However, these practices are not frequently employed. There are three overarching reasons.

First, operational costs are seldom tracked at a level that an organization can assess the benefits of enhancing human behavior and performance. Pervasive and integrated enterprise information systems are changing this situation. Thus, the lack of cost data may be less and less a barrier.

Second, investments in people do not appear as assets on balance sheets. As one cannot own people, one cannot own employees' knowledge and skills despite the fact that one may have invested in creating this knowledge and skills these employees now possess. Nevertheless, increasingly the "knowledge capital" of enterprises is their dominant asset. It will, however, probably be quite some time before including such assets on balance sheets will be a generally accepted accounting principle (also known as "GAAP").

Third, we have difficulty accounting for the value of investments where returns will accrue 5, 10, or 20 years in the future. This is particularly the case for public–private enterprises. The U.S. Congress has no balance sheet. All expenditures are operating costs, accounted for, in effect, on the income statement. Substantial savings on the future operating costs of military airplanes, for example, has no value with this approach to financial management. Overcoming this barrier to best practice would involve an enormous cultural change.

So, the good news is that we know how to address the economics of human systems integration. The bad news is that there are significant impedances to employing these means. Consequently, we are likely to continue to underinvest in people's health, education, and productivity. The implications for our long-term competitiveness are, as a result, less than rosy.

REFERENCES

Birkler, J., Rushworth, D., Chiesa, J., Pung, H., Arena, M.V., & Shank, J.F. (2005). *Differences Between Military and Commercial Shipbuilding*. Santa Monica, CA: RAND Corporation.

Booher, H.R. (Ed.). (1990). *MANPRINT: An Approach to Systems Integration*. New York: Van Nostrand Reinhold.

Booher, H.R. (2003). *Handbook of Human Systems Integration*. New York: Wiley.

DoD. (2008). Operation of the Defense Acquisition System. *Defense Directive Number 5000.02*. Washington, DC: Department of Defense, December 2.

Hammer, M., & Champy, J. (1994). *Reengineering the Corporation: A Manifesto for Business Revolution*. New York: Harper Business.

INCOSE. (2008). *Human Systems Integration Working Group Meeting*. Albuquerque, NM.

Levinson, M. (2006). *The Box: How the Shipping Container Made the World Smaller and the World Economy Bigger*. Princeton, NJ: Princeton University Press.

Malone, T., Savage-Knepshield, P., & Avery, L. (2007). Human-systems integration: Human factors in a systems context. *Human Factors and Ergonomics Society Bulletin*, December, 1-3.

Newman, D.G., Eschenbach, T.G., & Lavelle, J.P. (2008). *Engineering Economic Analysis* (10th Edition). New York: Oxford University Press.

Pennock, M.J., Rouse, W.B., & Kollar, D.L. (2007). Transforming the acquistion enterprise: A framework for analysis and a case study of ship acquisition. *Systems Engineering*, *10*(2), 99–117.

Pew, R.W., & Mavor, A.S. (2007). *Human-System Integration in the System Development Process*. Washington, DC: National Academies Press.

Rouse, W.B. (1991). *Design for Success: A Human-Centered Approach to Designing Successful Products and Systems*. New York: Wiley.

Rouse, W.B. (1998). *Don't Jump to Solutions: Thirteen Delusions That Undermine Strategic Thinking*. San Francisco, CA Jossey-Bass.

Rouse, W.B. (2005). A theory of enterprise transformation. *Systems Engineering*, *8*(4), 279–295.

Rouse, W.B. (Ed.). (2006). *Enterprise Transformation: Understanding and Enabling Fundamental Change*. New York: Wiley.

Rouse, W.B. (2007). *People and Organizations: Explorations of Human-Centered Design*. New York: Wiley.

Sage, A.P., & Rouse, W.B. (1999). *Handbook of Systems Engineering and Management* (1st Edition). New York: Wiley.

Sage, A.P., & Rouse, W.B. (2009). *Handbook of Systems Engineering and Management* (2nd Edition). New York: Wiley.

Salvendy, G. (2006). *Handbook of Human Factors and Ergonomics*. New York: Wiley.

White, J.A., Case, K.E., & Pratt, D.B. (2008). *Principles of Engineering Economic Analysis* (5th Edition). New York: Wiley.

Womack, J.P., & Jones, D.T. (1996). *Lean Thinking: Banish Waste and Create Wealth in Your Corporation*. New York: Simon & Schuster.

Chapter **2**

Industry and Commercial Context

WILLIAM B. ROUSE

2.1 INTRODUCTION

This chapter focuses on the economics of human systems integration (HSI) within the private sector, with emphasis on industrial organizations outside the aerospace and defense industry. Such organizations have more latitude in the systems engineering and management practices they adopt as well as greater accountability for the consequences of these practices in terms of both rewards and risks. This chapter summarizes the best practices developed and used by companies in a wide range of markets.

2.1.1 Contexts of Experience

The discussions in this chapter draw on experiences in domains ranging from aviation to appliances, computers to communications, and drugs to data warehouses. Hundreds of planning engagements—focused on overall strategy and new product planning—form this experience base. It is essential at the outset to note that few, if any, of these experiences involved explicit use of the phrase "human systems integration." Although human-related issues are often primary in system design, development, deployment, and sustainment, these issues and how they are addressed are rarely labeled "HSI" outside of the aerospace and defense industries. However,

The Economics of Human Systems Integration: Valuation of Investments in People's Training and Education, Safety and Health, and Work Productivity. Edited By William B. Rouse
Copyright © 2010 John Wiley & Sons, Inc.

as this chapter delineates in some detail, the HSI philosophy permeates most of the best practices industry wide.

2.1.2 Overview of Chapter

This chapter proceeds as follows. The next section contrasts private- and public-sector practices in the process of researching, designing, developing, and deploying products and systems. The following section reviews contemporary management practices, including financial management, the innovation funnel, multistage decision processes, and portfolio management. The final section reviews a variety of best practices, including business process orientation, balanced scorecards, and how the "innovator's dilemma" is addressed. Throughout these discussions, private- and public-sector perspectives are contrasted, with particular emphasis on impacts on economic decisions.

2.2 PRIVATE VERSUS PUBLIC SECTORS

The primary differences between system engineering and management in the private and public sectors are indicated in Table 2.1. In this section, these differences are first elaborated, and then an overarching explanation is offered for why these differences are manifested.

Both the private and public sectors are driven by customers' needs and desires, but the level of specificity relative to these needs and desires differs substantially. In the public sector, procurement plans and budgets are publicly available information[1]—customers' intentions are explicit, often many years in advance, although political factors can derail these plans. In the private sector, needs and opportunities must be inferred from buying patterns and possibly technology trends—customers' intentions may be only vaguely articulated and very open to change. Hence, the market uncertainties in the private sector are substantially greater than in the public sector.

It is, however, very important to recognize the possibility that well-documented and communicated requirements for public systems can nevertheless be ill-conceived. Such a starting point provides ample opportunities for cost overruns, schedule slippages, and much finger pointing (GAO, 2008a, 2008b). Ensuring that requirements are well founded is important to successful HSI in all domains, whether private or public (Sage & Rouse, 1999, 2009).

Winning a public-sector contract to provide products and systems often insures a long stream of revenues, perhaps for decades in the case of defense systems. In the private sector, new offerings and new players emerge more frequently and customers may switch to these providers if the costs/benefits are better. Thus, in

---

[1] An obvious exception to this generalization concerns systems associated with national security where a "need to know" may be a prerequisite to gaining access to information. Those involved with proposing solutions inherently have this need.

TABLE 2.1 Comparison of Private- and Public-Sector Characteristics

Comparison	Private Sector	Public Sector
Driving Force	Market needs and opportunities	Procurement plans and budgets
Competition	Continued new offerings and players	Little once production contract won
Customers	Many potential sales opportunities	Few potential sales opportunities
Customers' Requirements	Typically researched and inferred	Publicly specified
Customers' Budgets	Seldom openly available	Publicly available information
Production Runs	Often very large	Usually relatively small
Product Life Cycles	Usually relatively short	Often very long
Risks	Usually borne by developer	Usually borne by customer
Rewards	Determined by Marketplace	Controlled by Customer
Sales of New Offerings	Brand and Relationship Loyalty Key	Usually Competitively Bid

the private sector, winning provides only a temporary advantage. On the other hand, losing results in only a temporary disadvantage. Thus, overall private-sector market relationships are much more dynamic and responsive to change than in the public sector.

There are usually only a few possible customers for public systems. If, for instance, the Federal Aviation Administration does not buy your air traffic control system, then to whom else can you sell this system? As another example, if the U.S. military does not buy your defense system, there are unlikely to be many foreign military sales. In contrast, the private sector typically has many potential customers, ranging from 20 to 30 airlines that buy commercial aircraft and to millions of people who buy automobiles and computers.

Public-sector customers' requirements are usually specific and publicly available. All potential providers compete to offer the most cost-effective way to meet these requirements. Private-sector vendors have to research and infer requirements, often because customers do not really know what they want—hence, market research such as described by Blattberg et al. (1994) is pursued. As a result, alternative solutions tend not to have the same functions and features. Eventual winning solutions may have significant competitive advantages thanks to proprietary technology, functionality, performance characteristics, and so on.

Budgetary information is publicly available for public-sector customers. Thus, providers know what is expected to be spent and when this spending is projected. Information on private-sector customers' budgets is seldom openly available. In fact, there may be no budget items in many areas. Sales of products and systems in the private sector may, therefore, depend on arguing the costs/benefits of possible solutions and, in effect, on creating needs that were not previously perceived.

Public-sector production runs tend to be relatively small, typically ranging from hundreds (e.g., aircraft) to thousands (e.g., small defense systems). In contrast, production runs of hundreds of thousands to millions are not uncommon in the private sector. This enables amortization of research and development (R&D) costs over many more units. It also results in significantly greater cost savings as providers move down production learning curves. Consequently, prices for private-sector products tend to decrease, often as quality also improves.

The product life cycles for public-sector products tend to be long, with many defense systems, for example, remaining in use for several decades. The private sector often experiences product life cycles as short as a few months to as long as a few years. The provider who gets to market first, and makes it down the production learning curve the fastest, tends to capture market share and realize the best margins. Of course, there is also the risk of getting to market too quickly or getting there with the wrong sets of functions and features.

The risks associated with the creation of public-sector products and systems are usually assumed by the customer—who often is the only customer. In the private sector, such risks are assumed by the developers of the product. If they are too early, too expensive, or off target in terms of functions and features, then they must accept the consequences.

Those who accept risks often earn the greatest rewards. Thus, public-sector profit margins are often modest and explicitly controlled by customers. Private-sector profits are determined by the marketplace and typically are not visible to customers, at least not in terms of profit per unit. Of course, was this not the case, private-sector product developers would be unlikely to accept the inherent risks.

Sales of new offerings in the public sector usually involve an open competitive bid, despite superior past performance, service, and so on. Brand and relationship loyalty provide much more advantage in the private sector, often resulting in sales of new offerings without competition. In fact, the extent of loyalty may be such that customers no longer even consider the possibility of alternative ways to meet their needs.

The impact of the differences summarized in Table 2.1 can be considered in the context of a typical product/system life-cycle model. System design in the private sector involves much more early uncertainty and risk, particularly in terms of market and competitive factors rather than technology. Thus, more effort and time is invested in concept definition, requirements analysis, and conceptual design, in part because concepts and requirements must be evaluated to ensure likely market acceptance. If, for example, a product has major usability deficiencies, private-sector markets cannot be, in effect, forced to buy the product until these deficiencies are remedied.

Across the whole product life cycle, private-sector system design and development usually involves many more hypotheses and tests, regularly trying to catch bad ideas quickly—"bad" meaning things that will not sell or may sell but lead to warranty or product liability problems. Consequently, there is much more reliance on "spiral" models of development, rather than the "waterfall" model common in aerospace and defense. The requirements analysis, conceptual design, detailed design, and production and testing phases of the life cycle are repeated relatively quickly, each time learning and refining the product or system concept to improve the design and spiral in toward a good design.

Best practices for public-sector system development do include evolutionary waterfall models and spiral models (DoD, 1996; Sage, 1995, 1999), often discussed in terms of integrated product and process development (IPPD). Thus, public system developers are often well aware of many, and perhaps most, of system design and development best practices. Unfortunately, complications of procurement processes, as well as political processes, can extend and distort life cycles in ways that undermine the benefits of these practices (GAO, 2008a, 2008b).

Faster development processes, as well as less customer control, results in private-sector products evolving much more quickly. This enables faster technology upgrades and insertion of new technologies. Lessons learned in operations and maintenance are quickly fed back to the evolution of new releases of the product or, more appropriately, the evolving product family. Consequently, private-sector products and systems are much more likely to include the latest, leading-edge technologies.

This blessing for private-sector customers comes with the curse—for developers—of greatly reduced sustainability of competitive advantage. Substantial revenues for spare parts and maintenance of decades-old products and systems are rare in the private sector. Competitive displacement of older technologies tends to be merciless in all but near-monopoly private-sector markets (e.g., commercial aviation). Thus, products usually cannot be viewed as "loss leaders" for downstream recurring revenues.

One very major exception to this assertion involves products in which there are substantial ongoing service components. Automobile maintenance is a good example where innovations in products are, to a great extent, intended to attract and retain buyers who will avail themselves of the ongoing services for the products or systems. Thus, the automobile dealer, in this case, does not have to make very large margins on the initial product sale.

Summarizing the distinctions drawn in this section, private-sector system design differs from public-sector product/system development in terms of uncertainties, risks, and rewards. Private-sector system design involves much greater uncertainty and many more risks. However, the *potential* rewards are much greater thanks to both large unit volumes and much greater profits per unit.

These differences strongly affect how HSI issues are pursued. Human-related concerns—such as the possibility of creating products or systems with wrong function/feature sets or products with major usability problems—receive substantial attention because the consequences to the developer are so significant. On the other hand, concerns such as long-term health and safety considerations receive much less attention, mostly because they do not affect near-term sales but also because of the substantial discounting of long-term impacts in general.

To a great extent, differences in how HSI issues are addressed are determined by who suffers the consequences of being wrong. Public-sector product/system development projects are typically sold before development begins; thus, the sale does not depend on how HSI issues are addressed, although it may depend on having a plan for addressing these issues. In contrast, private-sector products/systems are sold after they are developed—customers who are not satisfied with how HSI concerns have been addressed can simply choose to not make purchases. In general, things get done when sales are dependent on them!

The differences summarized thus far are related to both the nature of products and systems created in the private and public sectors and to the inherent dissimilarities between these sectors. A large percentage of system design in the private sector involves creating standard products (solutions) for a relatively homogeneous market (stakeholders) that only has scrutiny over the end product. In contrast, a large portion of systems developed in the public sector, for which HSI has received the most attention, involve tailored solutions for which there are a wide range of stakeholders—users, customers, employees, politicians, and so on—who have considerable scrutiny over both system characteristics and the process whereby the system is designed and developed.

As a consequence, many system engineering and management best practices are difficult to implement in the public sector, despite that practitioners are well aware of these practices. For example, it is often the case that the military end user—the warfighter—will dictate design decisions, including design changes that adversely affect budget and schedule. Congress may, for instance, preempt design changes that would adversely affect developers in favored congressional districts. And, of course, the whole federal procurement process can complicate and extend the acquisition process in ways in direct opposition to system design and development best practices.

It is also useful to note that private-sector system design and development efforts involving tailored solutions for heterogeneous stakeholders who have considerable scrutiny of both product and process can encounter some, if not all, of the same difficulties encountered in the public sector. For example, Fryer (1999) reports that at least 90% of enterprise resource planning (ERP) projects end up late or over budget, often taking six to seven years or more to realize positive returns. This tends to result in large "expectations gaps," often created by vendors of ERP systems. Thus, dissimilarities between the private and public sectors provide only a partial explanation of the differences summarized in Table 2.1.

2.3   MANAGEMENT PRACTICES

In this section, four central elements of contemporary management practices in the private sector are reviewed, including financial management, the innovation funnel, multistage decision processes, and portfolio management. These practices have an enormous impact on what gets counted and how it gets counted, which in turn, pervasively affect private-sector enterprises.

2.3.1   Financial Management

Financial management is a very substantial topic (Brigham & Gapenski, 1988; Smithson, 1998), and a full review is far beyond the scope of this chapter.

Fortunately, a few key issues can be summarized by just considering the nature of financial statements in the private sector, in contrast with financial management in the public sector. In particular, the income statement and balance sheet are of interest.

The income statement summarizes the revenues, costs, and net income or profits of an enterprise. Private-sector enterprises attempt to increase revenues faster than costs—better yet if they can increase revenues while decreasing costs via, for example, production learning. The goal is to increase profit margins, for which these enterprises will be rewarded by the stock market with higher share valuations, assuming the enterprise is listed in the stock market. The income statement also includes depreciation costs attributable to assets on the balance sheet (definition to follow).

In contrast, enterprises selling to public-sector customers typically find that their profit margins are dictated by their customers, often as a percentage of costs. Thus, decreasing costs decreases profits. Typically, these enterprises' costs primarily consist of labor and materials. In fact, the costs of the materials are often driven by the labor costs of suppliers. The overall impact of this is the simple fact that these enterprises are selling the government labor hours, not airplanes, ships, or tanks. Therefore, the more labor hours provided, the higher the costs, and the greater the profits.

The goal in the public sector is to have the highest costs possible but remain sufficiently competitive to not lose in the competitive bidding process. Once a contract is won, cost overruns are very profitable as long as they are driven by customers' requests for modifications and extensions, and they are not so huge that the contract is canceled. Consequently, the income statements for private- and public-sector enterprises should be interpreted differently.

The balance sheet of an enterprise lists its assets, liabilities, and ownership equity. Assets can be tangible (e.g., facilities, equipment, and inventory) or financial (e.g., cash, securities, and accounts receivable). Liabilities include accounts payable, debts, deferred taxes, provisions for warranties, and so on. Equity is the difference between assets and liabilities.

As discussed in Chapters 1, 4, and 7, balance sheets do not currently include investments in human capital assets. Investments in productivity-enhancing technology, for example, may be capitalized on the balance sheet. In contrast, investments in human knowledge and skills are simply viewed as operating costs on the income statement. This means that HSI is viewed as an expense rather than as an investment.

Broadly speaking, there are three different types of HSI expenditures. An enterprise can make HSI investments in improving its own productivity. It can also invest in HSI-related aspects of its products and services; in this case, it may be improving its customers' productivity. Third, it can invest in HSI competencies that enable the other two types of HSI investments. To the extent that these investments result in tangible assets, these assets appear on the balance sheet. Otherwise, these expenditures are recorded as costs on the income statement.

This financial management practice creates difficulties for HSI. Private-sector companies can improve profits by lowering costs by decreasing HSI expenditures. Market forces counteract this tendency, justifying all three types of HSI expenditures in terms of competitiveness, market shares, and profit margins. Nevertheless, for knowledge-intensive companies such as pharmaceuticals and software companies, the balance sheet does not include a major portion of the assets that underlie their competitive positions in their markets.

Companies serving customers in the public sector face the dilemma of their major customer not having a balance sheet. For the U.S. Congress, everything is an operating expense. There are no investments, and future returns on current HSI expenditures (not viewed as investments) are not tracked or captured. The total focus is on the expenditures in the current fiscal year. Consequently, these companies focus on cost management rather than on making investments to yield future returns. The market forces to counteract this tendency are limited.

In summary, companies operating in the private sector pursue HSI because their markets demand it. Companies operating in the public sector pursue HSI if their customers require it and pay for it. In both cases, the costs of HSI appear on the income statement unless the HSI endeavors result in assets that can be capitalized on the balance sheet. In general, HSI competencies and HSI work products can be important competitive assets to an enterprise, especially in the private sector, but generally accepted accounting principles (GAAP), do yet support this practice.

2.3.2 The Innovation Funnel

The uncertainties and risks associated with product and system design and development in private-sector markets result in many more concepts being pursued than eventually make it to the marketplace. The ratio of initial ideas to market innovations through the product development "funnel" ranges from 3,000:1 (Stevens & Burley, 1997) to 10,000:1 (Nichols, 1994). These daunting numbers are generalized in Figure 2.1.

The numbers become more imaginable when one thinks about the number of projects that must be funded to yield one innovation. The Air Force Scientific Advisory Board conducted a study of best practices associated with science and technology investments (Ballhaus, 2000). Two companies, one a large chemical company and the other a large telecom equipment manufacturer, reported investing in 300 projects to get one or two innovations in the market.

When these numbers were mentioned, someone in the audience asked, perhaps in jest, "Why didn't you just invest in the right ones in the first place?" After a bit of laughter, the Chief Technology Officer of the chemical company said, "R&D is a very uncertain business. You cannot get rid of the uncertainty and we don't try to. Our goal is to be the best at managing uncertainty and thereby gain competitive advantage."

Another way to think about this phenomenon is to imagine there being no uncertainty. The future and the nature of value in the future would be crisply defined. In every market segment, all competitors would produce exactly the same products,

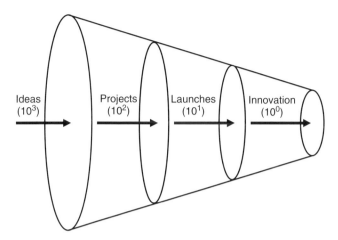

FIGURE 2.1 The innovation funnel.

systems, and services. Everything would be a commodity, and all profit margins would be very slim or perhaps zero. This does not sound like an appealing world.

Nevertheless, many managers in the public sector, both in government and in contractor organizations, emphasize risk reduction. They argue, for example, that every R&D investment should "transition" to fielded capability. They want to transform the funnel in Figure 2.1 into a pipe. If they succeed, the pipe gets clogged with a plethora of mundane ideas that are sure to work, (i.e., transition), and are also sure to provide little, if any, significant enhancement to current capabilities.

2.3.3 Multistage Decision Processes

Companies who accept the funnel and manage the inherent winnowing process with firm go/no-go decision points perform much better than those who are more ad hoc in their approach. Cooper (1998) has pioneered the formalization of multistage decision processes for making these decisions. An example of a multistage decision process is depicted in Figure 2.2.

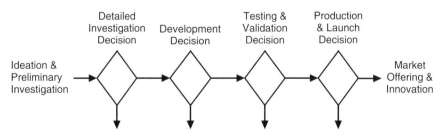

FIGURE 2.2 Typical multistage decision process.

At each stage, projects must pass specified criteria to move on to the next stage. Not passing these criteria results in projects being killed, shelved, or possibly retained at the earlier stage. Of great importance, project managers know exactly what the criteria are for each stage long before having to satisfy these criteria.

Investment levels increase substantially with each stage, adding significant potential value to the concepts being pursued by projects. The gates are designed to eliminate projects where the value added downstream will not justify the costs. The criteria change with each stage, shifting emphasis from strategic relevance and technical feasibility to economic return and risk management. For attributes that are relevant across stages, "hurdle rates" or decision thresholds increase with each stage.

The nature of multistage decision making just depicted results in many more "failures," than experienced for public systems. Most ideas do not make it to applications. However, these nonsuccesses are viewed as simply part of the process of bringing new products and systems to market. In contrast, as noted, one often hears public-sector leaders decry the lack of transition of ideas from laboratories to deployed systems.

Put differently, product and system design and development in the private sector is much less risk averse than in the public sector. Emphasis is on assessing and managing risks rather than on eliminating them. Of course, the consequences of risks are spread across many, many companies. In the public sector, the government absorbs most risks and nonsuccesses can result in endless scrutiny. Furthermore, companies who develop products and systems for the public sector face the risk that their one and only customer will no longer be able to justify an acquisition, resulting in substantial loss of revenue although seldom significant loss of capital.

Possible criteria for transitions between the stages in Figure 2.2 are shown in Table 2.2. Note how the attributes of interest remain unchanged across stages but the criteria become more stringent as projects make their way through the innovation funnel and require greater commitments of resources. Criteria may also change with stages thanks to varying stakeholders across stages (i.e., the next user usually differs from the end user for all but the last stage).

There is a natural tendency to assume that a central premise underlying such multistage processes is that projects can and will smoothly move through the stages. However, some projects in exploratory development, for example, may stay at that stage for an extended period of time without explicit intentions to transition. Furthermore, not all ideas and projects enter the funnel from the left. More mature ideas and projects (e.g., originated by partners), may enter at exploratory or advanced development. Thus, the metaphor of a funnel should be loosely interpreted.

It is important to realize that the nature of what transitions between stages may vary, ranging from people to information to prototypes to requirements but rarely in the form of off-the shelf technology. This can be thought of in terms of data versus information versus knowledge, transitioned either in formal documents or informally in people's heads. In fact, it often appears that the speed with which projects move through the funnel is highly related to the flexibility with which people move throughout the enterprise.

TABLE 2.2 Criteria for Decisions at Each Stage of Process

			Decision		
Decision Criteria	Idea → Concept Paper	Concept Paper → Initial Project	Initial Project → Exploratory Development	Exploratory Development → Advanced Development	Advanced Development → Technology Transition
Strategic Fit	NA	Possible	Definite	Priority	Programmed
Payoff	NA	Imaginable	Articulated	Projected	Demonstrated
Schedule	NA	One-year deliverables	Multiyear sequence of deliverables	Multiyear sequence of demonstrations	Technology transition plan
Resources	No budget	Discretionary budget available	Budget scoped appropriately	Costs/benefits projected	Costs/benefits assessed
Technical Risk	NA	NA	Anticipated	Managed	Minimized
Application Risk	NA	NA	NA	Anticipated	Managed
Personnel	Interest and commitment	Commitment and credibility	Commitment and credibility	Credibility and availability	Credibility and availability
Competencies	Desirable and obtainable	Desirable and developing	Available internally and externally	Available internally and externally	Demonstrated and available

2.3.4   Portfolio Management

Investments in all the projects in the funnel are managed as portfolios (Cooper et al., 1998, 1999, 2000). Plots of returns versus risks are typically used to represent portfolios (see Figure 2.3). Such plots enable comparisons of alternative investments.

Return is expressed in terms of net present value (NPV) or net option value (NOV). The former is used for those investments where the lion's share of the commitment occurs upstream and subsequent downstream "exercise" decisions involve small amounts compared with the upstream investments. NPV calculations are close enough in those cases. Chapter 7 discusses and provides the equations for NPV and NOV. Chapter 10 discusses multistage NOV.

Risk (or confidence) is expressed as the probability that returns are below (risk) or above (confidence) some desired level—zero being the common choice (Rouse, 2001, 2007). Assessment of these metrics requires estimation of the probability distribution of returns, not just expected values. In some situations, this distribution can be derived analytically, but more often Monte Carlo analysis or equivalent is used to generate the needed measures.

A common goal is to create an "efficient" portfolio such that each project in the portfolio has the minimum risk for a given level of return or the maximum return for a given level of risk. The line connecting several projects (P_A, P_B, P_H, and P_Z) in Figure 2.3 is termed the "efficient frontier." Each project on the efficient frontier is such that no other project dominates it in terms of *both* return and confidence. In contrast, projects interior (below and/or left) to the efficient frontier are all dominated by other projects in terms of both metrics. Ideally, from an economic perspective at least, the projects in which one chooses to invest should lie on the efficient frontier. Choices from the interior are usually justified by other, typically noneconomic attributes (e.g., "strategic criticality") and may result in elements of portfolios that are inefficient from a purely financial perspective.

A primary purpose of a portfolio is risk diversification. Some investments will likely yield returns below their expected values, but it is very unlikely that all of

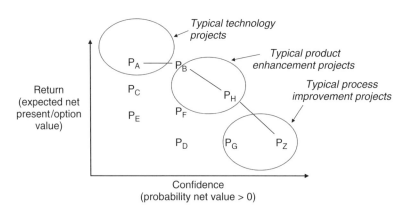

FIGURE 2.3   Portfolio of investments.

them will—unless, of course, the underlying risks are correlated. For example, if the success of all HSI projects depends on a common scientific breakthrough, then despite a large number of project investments, risk has not been diversified. Thus, one usually designs investment portfolios to avoid correlated risks.

Although this makes sense, it is not always feasible—or desirable—for HSI investments. Often multiple investments are made because of potential synergies among these investments in terms of technologies, markets, people, and so on. Such synergies can be beneficial, but they must be balanced against the likely correlated risks.

2.4 BEST PRACTICES

Multistage decision processes and portfolio management are both recognized as best practices in most private-sector markets. Both practices have been widely adopted in the private sector. These practices are reasonably well known in the public sector as well. However, adoption is often in name only. The discipline necessary for successful adoption of these practices is often difficult to maintain with a wide range of stakeholders with competing and conflicting interests.

2.4.1 Business Process Orientation

Businesses have traditionally thought of themselves in terms of functions such as marketing, engineering, manufacturing, finance and accounting, human resources, and so on. These functions have been budgeted to conduct a set of activities associated with their roles in the enterprise. Over time, as enterprises have grown and flourished, these functions have become like principalities with inherited resources and perquisites.

This organizational model met with serious difficulties in the 1970s and 1980s as traditional U.S. industries encountered strong competition, particularly from Japan. The resulting anguish and gnashing of teeth precipitated a new way of thinking. The new organizational model involved processes that create value for customers, supported by functions. Process "owners" became the functions' customers, providing resources (money) for the services (people) provided by the functions.

Hammer and Champy (1993) pioneered the notion of "business process reengineering," with emphasis on rethinking the business rather than just automating the current ways of doing things. Womack and Jones (1996) championed the idea of "value streams" and "lean production" where all activities are scrutinized to assess the extent to which they add value and, if they do not, are eliminated if at all possible.

These ideas raised questions of how best to account for costs. Cooper and Kaplan advocated the notion of "activity-based costing" as a means to minimize undifferentiated "overhead" and to understand the true costs of business activities (Kaplan & Cooper, 2008). This concept is elaborated in Chapter 7. The key idea is to know *both* what value activities add to processes and the true costs of these activities. This, in turn, enables well-informed reengineering.

A business process orientation has been adopted by many leading enterprises in the private sector to enable transformation of their enterprises to new levels of success (Rouse, 2006). The public sector has undertaken many initiatives labeled with the transformation rubric. However, fundamental change has been slow in coming, with much resistance from many stakeholder groups whose basic business models presume "business as usual" for public-sector agencies.

2.4.2   Balanced Scorecard

Beyond managing the portfolio of investments—HSI and otherwise—one needs to manage the overall enterprise. Traditionally, the enterprise's financial statements, as discussed earlier in this chapter, were considered sufficient for assessing the health of the enterprise. However, this approach to strategic management was found to be wanting during the crises of the 1970s and 1980s.

In response to these difficulties, Kaplan and Norton (1996) developed the notion of a "balanced scorecard." The idea was to develop a means for assessing the health of the enterprise that provides balance across the enterprise's key processes. Creation of a balanced scorecard involves defining two to four critical measures in each of four strategic areas:

- Customer
- Financial
- Internal Business Processes
- Learning and Growth

For each measure in each area, the balanced scorecard specifies objectives, measures, targets, and initiatives.

As an illustration, Table 2.3 indicates a possible balanced scorecard for an R&D organization. Rather than attempting to develop a single index of value, this balanced scorecard recognizes the multi-attribute nature of value and the need to define attribute-specific targets that, in turn, drive the initiatives undertaken. Note that only a few of the objectives shown relate directly to an enterprise's near-term financial statements. However, most people would agree that all of these objectives will affect an enterprise's long-term financial success.

In practice, creation of a balanced scorecard involves key stakeholders debating, discussing, and agreeing to the objectives, measures, and targets included in the balanced scorecard. Initiatives then become the responsibility of individuals, managers, and staff personnel. Progress is typically reviewed quarterly. Results are reviewed annually, which prompts reconsideration of objectives, measures, and targets. During this annual review, the organization collectively takes responsibility for the whole scorecard. The tenor of the discussion then focuses on where "we" succeeded or failed rather than on where those responsible for individual initiatives succeeded or failed.

A wide range of enterprises have developed balanced scorecards and employ them in strategic management. Professional groups have developed around the

TABLE 2.3 Balanced Scorecard for R&D Organization

Customer

Objectives	Measures	Targets	Initiatives
Customer Satisfaction			
Technology Transferred			
Operational Cost Savings			

Financial

Objectives	Measures	Targets	Initiatives
Internal Budget External Resources			
Collaboration/ Joint Ventures/ Partnerships			

Internal Business Processes

Objectives	Measures	Targets	Initiatives
Budget/Service Variances			
Goal Variances			
Facilities/Equipment Quality			

Learning and Growth

Objectives	Measures	Targets	Initiatives
Publications/Patents			
Recognition of People & Technologies			
Training Accomplished			

practice and refinement of this construct. Two aspects of this practice deserve particular mention. First, the discussions and debates associated with creation and use of the balanced scorecard provide a very useful means for developing a shared vision and vocabulary for what matters and how to measure it. Second, the breadth of the balanced scorecard provides greatly increased opportunities for members of the organization to understand how what they do is integral to the enterprise's success.

2.4.3 The Innovator's Dilemma

Most enterprises want to be innovative, creating change in the marketplace with their ideas and inventions. In fact, many enterprises start this way, with a new idea that they pursue with passion and persistence. If they are fortunate, they grow and prosper, refining their idea and capturing more and more customers.

A few of these enterprises become large and very successful, perhaps dominating the market in which their product or service is now the leader. Most of their resources are devoted to maintaining their competitive position. However, some resources may be invested in R&D to create new ideas.

Eventually, someone comes up with an idea that, if successful, will make the existing product or service obsolete. This results in the "innovator's dilemma" (Christensen, 1997). Should the enterprise invest in an idea that will undermine their main source of revenue and profits? How will they maintain growth while maturing this nascent idea that customers may or may not embrace?

As Christensen has clearly documented, many companies keep their resources focused on the existing product lines and ignore the new opportunity. In this way, Digital Equipment Corporation (DEC) did not embrace personal computing; Apple and then IBM took this market, and DEC no longer exists. Ford and General Motors ignored the market for small, reliable economy cars; Toyota and Honda now dominate this market with steadily increasing market shares.

More recently, Christensen and Raynor (2003) have developed "the innovator's solution" to help enterprises escape the dilemma. In particular, the goal is to be able to take advantage of new ideas while maintaining existing lines of business. Put simply, the goal is to dovetail disruptive innovations with incremental improvements of existing products in ways that sustain growth by smoothly replacing the decline of "old" revenues with the growth of "new" revenues.

Based on theories of innovation, they recommend a few key practices:

- Focus on what key customers need to get done, and then look for unmet needs that can be turned into the seeds of disruptive or sustaining innovations. Look for nonconsumption from lack of availability rather than from lack of desire.

- Innovate around circumstances—why someone buys something—not around products or customers. Put another way: Focus on the benefits sought rather than on the thing that enables the benefits. View products as things that provide services.

- Do not expect the current organization and leadership, which has sustained existing business, to be well matched to disruptive opportunities. Do not require emergent new lines of business to conform to "business as usual" practices.

This kind of thinking has lead to online bookstores, GPS in automobiles, iPods (Apple Inc, Cupertino, CA), and social computing in the private sector. It can have value in the public sector also. For example, does the military really want aircraft (a product) or the ability to airlift people and equipment (a service)? The latter is clearly more important than the former. Consequently, the Ministry of Defense in the United Kingdom buys airlift capabilities by the hour rather than own the airplanes. This type of change is very disruptive for companies that want to sell airplanes rather than provide services. On the other hand, innovators who determine how to provide such services at good margins could end up dominating the airlift business.

2.5 CONCLUSIONS

This chapter has considered the economics of HSI in the context of the private sector where market considerations and profit motives drive investment decisions. The characteristics of private-versus public-sector research, design, development, and deployment were contrasted. Contemporary management practices were discussed in terms of financial management, the innovation funnel, multistage decision processes, and portfolio management. Other best practices discussed included business process orientation, the balanced scorecard, and approaches to overcoming the innovator's dilemma.

Competitive pressures, as well as providers having to absorb most risks, result in private-sector enterprises being very sensitive to HSI issues. The possibilities of products or systems not selling, having to be recalled, and leading to legal suits provide strong motivations for paying attention to and resolving HSI issues. On the other hand, for new, innovative products and systems, the marketplace is often forgiving with regard to HSI limitations.

To the extent that public-sector products and systems are similar to private-sector offerings, one may be able to rely on similar competitive forces to ensure responsiveness to HSI issues. On the other hand, for public system procurements that are sufficiently unique to require significant deviations from off-the-shelf, private-sector solutions, HSI compliance may have to be a regulatory requirement. Further, as with all aspects of the acquisition of public systems, the customer will have to pay for HSI.

Overall, the issues in private- and public-sector system design are similar. However, the motivations for pursuing and resolving these issues tend to be different. Multistage decision processes apply equally well in both domains, but the nature of the stakeholders—beyond users—is very different. Consequently, *common* practices tend to differ significantly, although one could reasonably argue for similar *best* practices.

REFERENCES

Ballhaus, W., Jr., (Ed.). (2000). *Science and Technology and the Air Force Vision*, Washington, DC: U.S. Air Force Scientific Advisory Board.

Blattberg, R.C., Glazer, R., & Little, J.D.C. (Eds.). (1994). *The Marketing Information Revolution*. Boston, MA: Harvard Business School Press.

Brigham, E.F., & Gapenski, L.C. (1988). *Financial Management: Theory and Practice*. Chicago, IL: Dryden.

Christensen, C.M. (1997). *The Innovator's Dilemma: When New Technologies Cause Great Firms to Fail*. Boston, MA: Harvard Business School Press.

Christensen, C.M., & Raynor, M.E. (2003). *The Innovator's Solution: Creating and Sustaining Successful Growth*. Boston, MA: Harvard Business School Press.

Cooper, R.G. (1998). *Product Leadership: Creating and Launching Superior New Products*. Reading, MA: Perseus Books.

Cooper, R.G., Edgett, S.J., & Kleinschmidt, E.J., (1998). *Portfolio Management for New Products*. Reading, MA: Addison-Wesley.

Cooper, R.G., Edgett, S.J., & Kleinschmidt, E.J. (1999). New product portfolio management: Practices and performance. *Journal of Product Innovation Management*, *16*(4), 333–351.

Cooper, R.G., Edgett, S.J., & Kleinschmidt, E.J. (2000). New problems, new solutions: Making portfolio management more effective. *Research Technology Management*, *43*(2), 18–33.

DoD. (1996). *Guide to Integrated Product and Process Development*. Washington, DC: Office of the Undersecretary of Defense for Acquisition and Technology.

Fryer, B. (1999). The ROI challenge: Can you produce a positive return on investment from ERP? *CFO*, September, 85–90.

GAO. (2008a). *Defense Acquisitions: Assessments of Selected Weapons Programs* (GAO-08-467SP). Washington, DC: Government Accountability Office.

GAO. (2008b). *Defense Acquisitions: Results of Annual Assessment of DoD Weapons Programs* (GAO-08-674T). Washington, DC: Government Accountability Office.

Hammer, M., & Champy, J. (1993). *Reengineering the Corporation: A Manifesto for Business Revolution*. New York: Harper Business.

Kaplan, R.S., & Cooper, R. (2008). *Activity-Based Costing*. Boston, MA: Harvard Business School Press.

Kaplan, R.S., & Norton, D.P. (1996). Using the balanced scorecard as a strategic management tool. *Harvard Business Review*, January-February, 75–85.

Nichols, N.A. (1994). Scientific management at Merck: An interview with CFO Judy Lewent. *Harvard Business Review*, January-February, 88–99.

Rouse, W.B. (2001). *Essential Challenges of Strategic Management*. New York: Wiley.

Rouse, W.B. (Ed.). (2006). *Enterprise Transformation: Understanding and Enabling Fundamental Change*. New York: Wiley.

Rouse, W.B. (2007). *People and Organizations: Explorations of Human-Centered Design*. New York: Wiley.

Sage, A.P. (1995). *Systems Management*. New York: Wiley.

Sage, A.P. (1999). Systems reengineering. In A.P. Sage and W.B. Rouse, Eds., *Handbook of Systems Engineering and Management* (Chapter 23). New York: Wiley.

Sage, A.P., & Rouse, W.B. (Eds.). (1999). *Handbook of Systems Engineering and Management*. New York: Wiley.

Sage, A.P., & Rouse, W.B. (Eds.). (2009). *Handbook of Systems Engineering and Management* (2nd Edition). New York: Wiley.

Smithson, C.W. (1998). *Managing Financial Risk: A Guide to Derivative Products, Financial Engineering, and Value Maximization*. New York: McGraw-Hill.

Stevens, G.A., & Burley, J. (1997). 3000 raw ideas = 1 commercial success! *Research Technology Management*, *40*(3), 16–27.

Womack, J.P., & Jones, D.T. (1996). *Lean Thinking: Banish Waste and Create Wealth in Your Corporation*. New York: Simon & Schuster.

Chapter 3

Government and Defense Context

WILLIAM B. ROUSE AND DOUGLAS A. BODNER

3.1 INTRODUCTION

This chapter focuses on the economics of human systems integration (HSI) within public–private enterprises that acquire and operate complex systems. In this context, HSI can be viewed from two perspectives. First and more traditionally, HSI addresses the effectiveness of human performance in the systems developed by these public–private enterprises (e.g., aircraft or other vehicles). However, HSI also addresses the effectiveness of human performance, primarily decision making, in the enterprise "system" itself. Here, humans interact with enterprise system elements, including business processes, information systems, and networks of quasi-independent organizations that collaborate (and sometimes compete) in the public–private enterprise. In both cases, effectiveness is judged by numerous, sometimes conflicting metrics, including cost, schedule, and performance. The range of enterprises of interest includes the Department of Defense (DoD), Federal Aviation Administration (FAA), National Aeronautics and Space Administration (NASA), and many more. We focus in particular on the DoD, although many of our observations pertain to the FAA, NASA, and other agencies.

As is later recounted, the possibility of transforming public-sector acquisition of complex systems has received decades of attention and investment, with minimal success. Charette (2008) reviewed a range of troubled acquisition programs for

The Economics of Human Systems Integration: Valuation of Investments in People's Training and Education, Safety and Health, and Work Productivity. Edited By William B. Rouse
Copyright © 2010 John Wiley & Sons, Inc.

aircraft (e.g.,F-22 and F-35), combat information systems, (e.g., Future Combat System or FCS), and other systems. He concluded that the problems associated with these systems—over budget, late delivery, and inadequate performance—can be attributed to several factors:

- Politicization of a technical process
- Reliance on unproven, exotic technologies
- Enormous complexity and interconnectedness of new military systems
- Lowballed cost projections that allow too many programs to be approved
- Shortage of skilled engineers, program managers, and contract oversight staff

These are primarily organizational and social phenomena rather than purely technology issues. As such, HSI methods are applicable in addressing them, albeit at the level of business processes. The focus on defense jobs in congressional districts, tendencies to attempt to mature unproven technologies during development, lowballing projections to make it through approval wickets, and the graying of the government and contractor workforce are not amenable to solely engineering solutions. In contrast, the increased complexity and connectedness are driven by technology opportunities and needs.

Charette suggested that these problems reflect a "collective conspiracy," despite decades of blue ribbon panels repeatedly recommending the same reforms (see the next section). We need a collective will not to accept these limitations any longer and create high-performing acquisition enterprises. These difficulties and the nature of the acquisition processes later outlined provide the context for considering models for economic valuation of HSI investments.

This chapter addresses such valuations for public-sector systems, with particular emphasis on defense systems. This chapter proceeds as follows. First, we discuss the history of acquisition reform and identify past failures. Next, we describe the acquisition enterprise that serves as the context within which to consider economic valuation of investments in HSI. Finally, we consider the extent to which government and defense have adopted—and actually employ—the range of best practices discussed in Chapter 2.

3.2 PUBLIC-SECTOR ACQUISITION

The acquisition of public-sector complex systems is time consuming, very expensive, and rife with uncertainties. Enterprises that acquire these systems face serious cost challenges. The costs of military platforms (e.g., ships), space platforms (e.g., space stations), and transportation systems (e.g., airports) have increased enormously in the past few decades, far beyond inflation during this period. Consequently, the public-sector enterprises that acquire these systems anticipate buying fewer of them. This tends to sacrifice needed capabilities as well as exacerbate the cost challenges.

This situation has resulted in many proposals for the transformation of public-sector acquisition, particularly defense acquisition as it consumes an enormous fraction of the discretionary portion of an increasingly constrained Federal budget. Such transformation proposals can be viewed, in many respects, as proposals to transform the human–system nature of the acquisition enterprise. For instance, such proposals may seek to transform the processes by which decision makers interact with one another to make decisions or the criteria by which decisions are made. Certain costs associated with transformation can then be viewed as HSI investments, whereas the returns from this investment may be enhanced performance of deployed systems or reduced enterprise cost. Such transformation proposals may also change the way in which HSI investments are made in the development of individual systems, resulting in reduced cost or enhanced performance of that system. Therefore, it is valuable to understand the effects of these previous reform efforts and the current state of research on the acquisition enterprise itself.

3.2.1 Acquisition Reform

The defense acquisition enterprise is unique; it operates with public funds, with primarily one buyer, little competition, contracts signed years in advance based on cost estimates, and decisions made in complex stages by multiple organizations. The process is infused with disparate goals and objectives, to have the highest performing technology at the lowest price possible in the fastest amount of time, to ensure the defense industry and related economies remain solvent, and to encourage small businesses, minority contractors, and women-owned businesses. (Cancian, 1995). The number of participants in the acquisition enterprise is large, and they have different goals and measures of success. There seems to be little agreement on what needs to be reformed, let alone on how to fix it. The result has been 60 years of failed attempts to reform the system (Charette, 2008).

Historically, reforms have been enacted for primarily two reasons: increasing complexity of the technologies involved and individual corruption and abuse for monetary gain. Excesses in time and cost, or deficits in performance, are some of the more obvious outward signs that reform is warranted. But these are just symptoms, and it is instructive to elucidate the contributing factors. First, is the government acquiring the right systems to meet its needs, and second, is it acquiring those systems well?

The first question addresses the agility of the acquisition enterprise. With an ever-changing world, the actions of both adversaries and allies can alter the efficacy of military systems, both deployed and under development, with little warning. Consequently, a program could be run with perfect efficiency and achieve all of its performance objectives; yet the resulting systems could be useless after completion. Although this does not constitute a failure in the traditional sense, a lack of agility in the acquisition system means that resources continue to be expended on a program even after it is recognized that it is no longer viable.

The second question addresses the efficiency of the acquisition process. That is, assuming that the mission is sound, does the acquisition enterprise deliver systems in the most cost-effective way possible? This category includes most of the issues one typically associates with acquisition failings, including excessive oversight, lack of competition, political interference, requirements creep, and the inclusion of immature technologies. Issues with acquisition efficiency are linked to the structure of the acquisition process as well as with the discipline with which the process is implemented.

With acquisition it is sometimes difficult to define a failure because even troubled programs often result in the acquisition of something. However, in hindsight at least, it is not always the case that the right weapon was acquired to address the right threat. Furthermore, the costs of acquired systems often far exceed original projections, and the desired capability is often provided much later than originally planned. These factors determine the effectiveness of acquisition. History has shown that not all acquisition efforts are successful with regard to these factors. These phenomena can be better illustrated by providing some examples.

Loss of Mission occurs when the threat that was to have been addressed by the system is no longer viable or when a new type of threat emerges. One such example is the B-70 Valkyrie. The Valkyrie was intended to be a high-altitude, Mach 3+ strategic bomber. However, concerns over the aircraft's vulnerability to surface-to-air missiles as well as the increasing dominance of Inter Continental Ballistic Missiles (ICBMs) in the nuclear strike role led both the Eisenhower and Kennedy administrations to question its military viability. Eventually, the program was transformed into a research program, the XB-70. Another example of loss of mission is the Drone Anti Submarine Helicopter (DASH). It was originally developed as an expendable antisubmarine platform. However, since submarines were not a significant threat during the Vietnam War, the DASH program was canceled in 1969. Both of these examples illustrate a lack of agility in the acquisition process in that resources were redeployed long after the changing threat had been identified.

Process Failure can cause the cancellation of programs as well. For example, the M247 Sergeant York DIVAD (Division Air Defense gun) was born out of the Army's need for a replacement for the aging M163 20 mm Vulcan A/A gun and M48 Chaparral missile system. Despite that the system used as much off-the-shelf technology as possible, when the first production vehicles were delivered in late 1983, there were many performance deficits, including issues with the fire control system, clutter handling, turret traverse rate, and Electronic Counter-Counter Measures (ECCM) suite. Consequently, in December 1986 after approximately 50 vehicles had been produced, the entire program was terminated. Of course, most acquisition process problems do not lead to cancellation. Many acquisition programs deliver highly capable systems but only after delays and cost overruns. An example of such is the F-22 Raptor. Considered one of the most technologically advanced aircraft in the world, it is also one of the most expensive. The program began with the award of the Advanced Tactical Fighter Demonstration/Validation contract in 1986 and achieved initial operational capability in 2005. The inclusion of many advanced technologies such as advanced avionics and low-observable materials helped contribute to the long duration and high cost of the program.

These and many other instances have driven desires for acquisition reform. However, past reform efforts have been less than fully successful, as shown by Drezner et al. (1993). Drezner reported that reform initiatives from 1960 to 1990 did not reduce cost growth on 197 defense programs. In fact, the average cost growth on these programs was 20% and did not change significantly for 30 years. Christensen et al. (1999) reaffirmed this conclusion and found that initiatives based on the specific recommendations of the Packard Commission did not reduce the average cost overrun experienced (as a percent of costs) on 269 completed defense acquisition contracts evaluated across an eight-year period (1988–1995). Actually, cost performance experienced on development contracts and on contracts managed by the Air Force worsened significantly.

In part, this lack of reform success can be attributed to one or more of the causes discussed. Since the 1980s, the military threat has changed from full-scale thermonuclear war to domestic terrorism, insurgencies, information warfare, and asymmetric warfare. Not only are weapons programs designed for a Cold War threat not always appropriate, but the entire system of acquisition has become too slow to adapt to emerging threats. Performance and politics still have an impact, but the rate of technological change has advanced so rapidly that weapon systems can become obsolete before they leave the design stage. In response, the Department of Defense has attempted large-scale, fundamental change in all facets of its operation.

Former Secretary of Defense Donald Rumsfeld made the comprehensive transformation of the Department of Defense a priority during his tenure. Although he departed in November 2006, the current Defense Secretary Robert Gates indicated that he shares Rumsfeld's vision of military transformation, centered around a lighter, more mobile force that heavily relies on technology.

Beyond transforming how it pursues military engagement, the DoD has begun transforming its acquisition process to create more efficient and effective ways to acquire goods and services faster, better, and cheaper (DAU, 2005). The exponential rate of technological advance combined with the availability of new technologies on the commercial market has added a sense of urgency to the acquisition environment. The DoD would like to access these advances before adversaries can use them against the United States.

A good example of the types of changes sought is the pursuit of evolutionary acquisition strategies that rely on spiral development processes. This approach focuses on providing the warfighter with an initial capability (that may not be the final capability) as a tradeoff for earlier delivery, flexibility, affordability, and risk reduction. The capabilities delivered are provided across a shorter period of time, followed by subsequent increments of capability over time that incorporate the latest technology and flexibility to reach the full capability of the system (Apte, 2005). Pennock has shown that the overhead costs associated with each increment may have a significant impact of total costs across all increments (Pennock, 2008; Pennock & Rouse, 2008). Potentially, such an approach can facilitate better HSI for individual systems, because interaction between the system and its human users happens earlier in deployment of initial capability, thereby providing increased opportunities for lessons learned. The efficacy of HSI investments in evolutionary acquisition versus traditional acquisition has not yet been studied rigorously.

In the recent Defense Science Board summer study on transformation (DSB, 2006), it was recommended that the Undersecretary of Defense (AT&L) "should renew efforts to remove barriers that prevent the entry of non-traditional companies to the Defense business and Defense access to commercial technology, attacking the myriad rules, regulations, and practices that limit the use of OTA, Part 12, and other programs to reach beyond traditional defense companies." The study goes on to recommend intense integration with global and commercial supply chains, as well as transforming the export license process.

This brief review shows that acquisition reform has long been sought and that the results are mixed, at best. Although the call for reform is persistent, these findings raise the question of whether it is possible to transform the acquisition enterprise and to have the varied stakeholders agree to any extent that the process has actually improved. This issue leads to the question of what is known about the fundamental nature of acquisition.

3.2.2   Acquisition Research

A quick review of recent acquisition research topics indicates a tendency to concentrate on single-issue concepts such as outsourcing, contractors, leasing, privatization, contingency contracting, performance measurement, and financial management. Considering the 2004–2006 Annual Conferences on Acquisition Research, topics covered included:

- Acquisition avenues such as market-based acquisition, capabilities-based acquisition, competitive sourcing, and outsourcing
- Acquisition issues such as program management, performance management, and business process reengineering
- Financially oriented topics such as financial management, total cost of ownership, and real option models

Furthermore, acquisition policy in general was, of course, a recurring theme. Although improving the performance and/or judging the effectiveness of each of these topics is worthwhile, it is also important to study the overall acquisition enterprise as an integrated and interactive complex system, amenable to HSI methods for design and management.

Currently, however, only limited acquisition research is being conducted— primarily by internal DoD organizations, such as the Naval Postgraduate School, Defense Acquisition University, Air Force Institute of Technology, and DoD federally funded research and development centers (FFRDCs) (e.g., RAND and LMI). Although these research projects offer valuable assessments of current practices and suggestions for improvements, the results are often limited in scope and may only address one specific problem at a time; often replicate previous or parallel work; and generally have limited general application. These efforts constitute only a fraction of the effort that is warranted by the size, complexity, and changing nature of DoD's acquisition challenges. They are not a substitute for disciplined, replicable academic research (Gansler & Lucyshyn, 2005).

No significant reform effort has addressed this issue at the broad enterprise level. Viewing the challenges of the acquisition process from a broad systems-oriented perspective provides an opportunity to understand where change can be leveraged and the economic value of such change. This will not obviate the impacts of changing threats and technologies, or political forces. However, modest changes of the acquisition process can have enormous economic benefits (Pennock et al., 2007).

3.3 HSI INVESTMENTS IN THE ACQUISITION ENTERPRISE

There are a variety of economic, political, and social arguments for why acquisition reform has not provided the benefits envisioned. As noted, our sense is that the lack of a broad view of acquisition has made it difficult to articulate and estimate the economic benefits of acquisition process changes. Furthermore, the lack of a broader view has limited the ability to estimate the increased value of the systems acquired using new processes. This section provides the needed broader view.

3.3.1 Acquisition Life Cycle

Figure 3.1 depicts the Defense Acquisition Management Framework provided in the Defense Directive 5000.1 (DoD, 2006). This process provides the context for economic valuation of acquisition investments. The ways in which the many stakeholders in the acquisition enterprise exercise this process strongly affect the time, costs, and uncertainties associated with the acquisition of complex systems. In light of the past and current Secretary of Defense's stated transformation priorities, this process would seem to be a good candidate for fundamental change.

3.3.2 Acquisition Process

Figure 3.2 summarizes the acquisition process as provided in Defense Directive 5000.2 (DoD, 2008). This process occurs within the context of the life cycle shown in Figure 3.1.

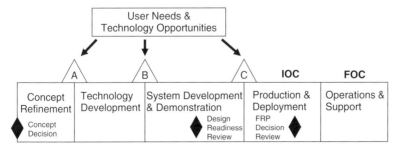

FIGURE 3.1 Defense acquisition management framework.

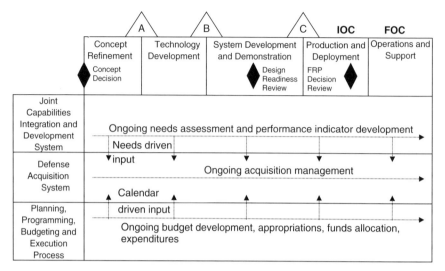

FIGURE 3.2 Defense acquisition processes.

Milestones A, B, and C are critical decision points, at which point a system may be discontinued. A Milestone Decision Authority (MDA) judges these decisions. Three different organizational systems interact within the context of system acquisition (Cochrane & Hagan, 2005). The Joint Capabilities Integration and Development System (JCIDS) identifies and documents war-fighting needs based on mission deficiencies and technological opportunities. The Defense Acquisition System is the set of processes whereby these needs are translated into affordable and sustainable systems. The Planning, Programming, Budgeting and Execution (PPBE) Process establishes the processes whereby funding decisions for acquisition are made. Clearly, these three systems must interact effectively for there to be successful acquisition. This provides a rich set of human–system integration issues at the enterprise level.

In determining needed capabilities, the JCIDS reviews doctrine, organization, training, materiel, leadership and education, personnel, and facilities for capability gaps. Once such a gap has been identified, the next step is the development of an initial capability document (ICD), which defines the capability need and guides concept definition (CD) and pre-Milestone A decisions, as well as progress through the Concept Development and Technology Development phases of the acquisition process. Similarly, during the Technology Development phase, a capability development document is developed to guide system development and demonstration by providing thresholds and objectives for a proposed system to meet. Finally, a capability production document (CPD) is developed during the System Development and Demonstration to guide Milestone C decisions.

In guiding a system through the acquisition process in terms of meeting capability needs, JCIDS relies on a variety of supporting processes that account for strategic guidance from the President and Joint Chiefs and for planned operational concepts 10 to 15 years in the future as well as for analyses that address functional requirements, the ability of current and programmed systems to meet those requirements, and solutions for capability gaps. Of particular concern is interoperability of systems across different DoD services, different types of platforms, and different systems operated by U.S. allies. Interoperability is characterized as the ability to exchange data, information, material, and services across different systems.

For a particular program, the Defense Acquisition System provides processes and milestones for a program manager to monitor the progress of the program, which includes contracting with companies to perform work associated with system design, development, and production. Normally, a program is initiated at Milestone B. However, multiple entry points into the acquisition process are possible, depending on concept and technology maturity. Preacquisition processes govern concept refinement and technology development. Concept refinement processes use innovation and competition from various outside sources (e.g., companies) and evaluate potential commercial off-the-shelf approaches. Technology development processes rely on technology demonstrations to monitor and review progress.

Once Milestone B is passed, the System Development and Demonstration phase is entered. Generally, to enter this phase, systems must have a system architecture (i.e., set of subsystems and their interactions) defined as well as an operational architecture (i.e., specification of the interaction of the system with other systems). The first step is then system integration, whereby the designed architecture is translated into a prototype of integrated subsystems with demonstrated functionality. Along with the demonstrated functionality, a design readiness review (DRR) assesses the design maturity and corrections to any deficiencies. Successful completion of the integration phase moves the system into system development, whereby the design and demonstrated functionality are developed into a prototype that demonstrates useful capability according to performance indicators from JCIDS. In traditional acquisition, a system encounters Milestone B only once. However, in evolutionary acquisition, each increment of capability requires its own Milestone B.

A successful system development phase moves the system to Milestone C, at which point a decision is made whether to commit to production. The first subphase of the Deployment and Production phase is low-rate initial production, during which manufacturing capability is developed such that the system can be efficiently produced. The first system units that can be used in combat-ready situations provide initial operating capability (IOC). Successful low-rate production can result in a favorable full-rate production review, which also considers system life-cycle issues before granting full-rate production to the system. The Operations and Support phase is next. It concentrates on designing and operating logistics and sustainment systems to support system operation as well as on end-of-life procedures for system retirement.

The PPBE Process provides the DoD's budget request to the President for inclusion in his or her budget. Once this request is developed, it moves to an enactment phase whereby it is included in the President's budget, which is in turn sent to Congress for hearings, amendments, and eventual authorization and appropriation. Once appropriated, funds are apportioned to the DoD by the executive branch

(Office of Management and Budget), and the DoD in turn apportions funds to its various agencies. Finally, execution is the process whereby funds are obligated for defense programs via contracts and then expended as the terms of the contracts are met.

This whole process is driven by a biennial calendar for congressional appropriations. Even years are characterized as "on-years," in which largely new budgets are formulated for approval by Congress. Odd years, or "off-years," then focus on the execution and modification of the "on-year" budgets. Because funding is driven by a calendar, and acquisition is event-driven, it is critical that the Defense Acquisition System align its activities so as to support successful budgeting and funding for programs according to the appropriations calendar.

3.3.3   Opportunities for Change via HSI Investment

The acquisition process, as described, offers several opportunities for potential improvements resulting from HSI investments, both at the individual system level and at the enterprise level.

At the level of individual systems, investments in HSI upstream in the acquisition process often have the effect of preventing the need for downstream "fixes" to HSI problems that cause delays, additional costs, and degraded capability. For instance, cockpit usability issues might be more economically addressed in system integration, rather than in low rate initial production. Formal models of how such upstream investments can play a role in preventing downstream problems would be important if such models could quantify the economic benefits of prevention. To be fully useful, such estimates of downstream economic benefits would have to be factored into upstream decision making.

At the enterprise level, HSI investments that improve enterprise decision making are of particular interest. Possible opportunities include system life-cycle cost estimation tools, risk understanding and mitigation tools, and scenario analysis tools. Realization of such benefits will require that key stakeholders embrace the assertion that effectively addressing HSI issues at the enterprise level is central to achieving the full benefits of addressing HSI issues at the individual system level. Put simply, it is difficult for a dysfunctional organizational system to deliver the best physical systems possible.

3.3.4   Future Acquisition Issues

Human systems integration issues are likely to become more important as acquisition becomes even more complex. Consider the example of the Joint Strike Fighter system currently under development in low rate initial production. This system is being developed as three variants on a common platform, for three service applications (Air Force fighter, Navy carrier-based fighter, and Marine vertical takeoff and landing fighter). Thus, in the United States, there are three customers, rather than the customary single-service customer.

As is customary, there will also be international customers. However, the innovation taken here is that to support international sales, design and production will be undertaken by an international set of collaborating companies, so as to encourage international sales (Kapstein, 2004). This includes final assembly, which for the first time for a U.S. fighter will be partially done outside the United States. In addition, sustainment will be performed using an international network of suppliers and service stations. Sustainment and production, in fact, are slated to be performed using the same underlying logistics system for parts delivery. Given the large-scale nature of the program in terms of the number of systems planned and the number of participating suppliers and countries and their export policies, taxes, and tariffs, the scope of the logistics system in terms of size and points of potential risk is large and complex (Smith et al., 2006).

The intent is to transform the traditional contract-based, hierarchical relationship that a program lead company for a particular system has with its suppliers to a collaborative network relationship between the lead and its partners. This is a major and difficult culture change in and of itself, notwithstanding the transformations associated with international development and integrated production sustainment. This is a possible future model for acquisition, if it is successful. There are a wide variety of HSI problems, ranging from managing complexity to addressing cultural norm issues that must be resolved for successful transformation.

3.4 MANAGEMENT PRACTICES

Management practices with government and defense tend to be different from industry and commercial practices. As is elaborated in the following discussion, government and defense are well aware of the contemporary management practices—and best practices—elaborated in Chapter 2. However, adoption of these practices by government and defense has been limited. Consequently, the ways in which HSI investments are addressed vary from those encountered in industry and commercial enterprises.

We hasten to note at the outset that this discussion pertains to government agencies and their contractors. These contractors also often function as private-sector businesses. However, in this chapter, the concern is with how they function as contractors and work for their government customers—rather than with how they attempt to provide value to their shareholders. The contractors' work is very strongly affected by their customers' dictates and requirements. This has an enormous impact on the extent to which contractors can adopt the best practices discussed in Chapter 2.

3.4.1 Financial Management

The nature of how financial resources are managed has an enormous impact on how investments are addressed, both for HSI and other types of investments. The DoD Instruction, "Economic Analysis for Decision Making" (DoD, 1995), specifies a methodology for financial modeling based on discounted cash flow with net present value (NPV) as the dominant metric. However, most DoD investments are staged.

Thus, net option value (NOV) is a more appropriate metric (discussed in Chapters 7, 10, and 15) because it attaches economic value to the management flexibility of terminating poorly performing investments. The NPV, in contrast, assumes investments will continue despite poor performance and, thereby, penalizes risky investments (i.e., the NPV is always lower, and often much lower, than the NOV).

The result is that investments seem to be much less valuable than they actually are considering the management flexibility to terminate failing investments early. Because Congress only appropriates money one to two years at a time, they have frequent opportunities to terminate poorly performing investments. However, economic and political forces often thwart such terminations and poor investments are continually sustained. Thus, the financial value of flexibility may—in the reality of public systems—be overstated. Furthermore, financial value does not drive Congressional decision making. The "profit" possible from savings in providing a given level of public service (e.g., defense) is much less a driver than the capabilities being procured and the jobs created, or sustained, in the process.

Overarching all this is the fact that current-year spending is viewed as operating costs regardless of how the monies are being spent. Expenditures made to decrease future operating costs can be characterized as investments whose returns will be the future savings. However, such expenditures are viewed in the same way as monies spent on expendable commodities. Thus, the focus is on the income statement, not on the balance sheet, which does not exist for Congress or government agencies.

Finally, the lack of necessary data undermines adoption of best financial management practices. The government often seems to monitor and record every transaction. However, the purpose is to avoid fraud, waste, and abuse, not to manage the enterprise better. Consequently, although an enormous amount of data is collected and archived, little information is available with which to apply modern financial management methods and tools. For the most part, the government knows the checks it writes and to whom they are sent. However, it usually cannot attribute these expenditures to particular activities and the specific value provided (Kaplan & Cooper, 2008).

3.4.2   Role of R&D

The U.S. Government is an enormous investor in research and development (R&D) through the departments of Defense, Energy, Health & Human Services, and so on, as well as through the National Aeronautics and Space Administration and National Science Foundation. These investments yield knowledge and skills important to the future of the country. Enterprises make similar investments in their futures.

A great deal is known about how best to make and manage R&D investments (Rouse, 2001). Multistage investment processes, portfolio approaches to balancing returns and risks, and option-based project valuation are among the best practices. Despite knowledge of these practices, their adoption within government and defense is not widespread (Ballhaus, 2000).

There are several reasons for this. For basic research, where most of the funding goes to academia, there is a strong sense that research should not be "managed."

Freedom of exploration, as judged by peers rather than by managers, is a strong element of the academic landscape. The sense is that the outcomes of basic research will be valuable even if they are unpredictable.

For exploratory research and advanced development, considerable efforts are made to manage these expenditures to ensure that valuable contributions result. Great emphasis is placed on fielded contributions, which often places a premium on incremental contributions that are less risky and more likely to be accepted in the field. Overall, the portfolio of these types of expenditures is highly fragmented and the stakeholders are risk averse. The overarching perspective is one of securing and preserving R&D budgets.

An alternative view is that the purpose of these investments is to create science and technology "options" for meeting the contingent future needs of the country. Creating high-value options is the role of R&D (Rouse & Boff, 2004). This view works well for private enterprises that have a strong sense of future opportunities and threats, as well as the contingencies needed for taking advantage of opportunities and addressing threats. Options-based approaches, using NOV as a decision criterion, to making and managing these investments have been shown to create more value than approaches based on NPV (Bodner & Rouse, 2007).

This approach has been successfully applied to government R&D investments, as illustrated in Chapter 15. However, there are two primary impediments to adoption of this practice. First and foremost, there needs to be agreement on future opportunities and threats, as well as on the contingencies needed to address them. Such consensus can be difficult to achieve in an environment where preservation of budgets is paramount.

The other difficulty is the aforementioned lack of financial data in a form that can be employed for developing and evaluating financial models. Given the lack of benchmarks with regard to past financial returns and volatility, it is easy for negatively affected stakeholders to dismiss models and analyses. Similarly, it can be easy for potentially positively affected stakeholders to make modeling assumptions that cast their investment opportunities in the most positive light.

3.4.3   Decision Processes

Government agencies tend to be thorough in developing decision process maps, particularly for the acquisition of complex systems. Multistage acquisition processes such as illustrated earlier in this chapter are used to procure ships, planes, bridges, and hospitals. Such maps are created, published, and maintained but not always fully followed.

Several factors underlie this situation. First, the number of stakeholders associated with the acquisition of complex systems is large. The public's interests are represented by Congress who, not surprisingly, can be parochial with regard to the interests of their local constituencies. The acquiring agency (e.g., military service) is also a key stakeholder. The companies associated with designing and developing systems exert strong influence. Labor, perhaps as represented by unions, is another key stakeholder. All these stakeholders strongly affect decisions.

Amidst these many competing interests, leaders in government tend to have more influence than power. They cannot command all the stakeholders to behave in particular ways. Unlike a corporate CEO, they cannot remove members of Congress, contractor executives, or union officials. In particular, they cannot preempt stakeholders from exerting influence on decision processes. Consequently, the actual decision processes may be far from the pristine process maps frequently seen on agency walls.

Finally, these decision processes can be undermined by the length of time (often decades) associated with the acquisition of complex systems. This results in much turnover of people and chances for environmental change, such as mission change or administration priorities. In contrast, in enterprises with a short product life cycle (e.g., those driven by Moore's Law), the discipline to follow agreed-upon decision processes is much easier to maintain.

3.4.4   Portfolio Management

The idea of managing investments as portfolios is well known in government and defense. However, it is difficult to adopt in practice. Once a program gains Congressional authorization and appropriation, it operates independently of other programs in response to its mission and the agency tasked with this mission. Making decisions and moving resources across programs is difficult.

This results, in part, from highly fragmented lines of authority between programs and other elements of the enterprise. The Program Executive Officer on the government side and the Program Manager on the contractor side have strong incentives to ensure the program's success, independent of all other programs and everything else happening in their enterprises. Local optimization without regard to the broader system—termed "suboptimization"—is common and, considering the incentive and reward system, is rational on the part of these two stakeholders.

It is important to note that government and defense managers are much more oriented to thinking in terms of operational portfolios to achieve missions, whether these missions are in pursuit of military objectives or other agency goals. Consideration of the performance and costs of mixes of assets—or recently, capabilities—is a common practice. Thus, the lack of portfolio management for investments may stem more from the lack of an investment orientation in general than from an inability to think in terms of portfolios.

3.5   BEST PRACTICES

In this section, we consider the extent to which government and defense have adopted the best practices discussed in Chapter 2. We also discuss some best practices that have been identified within the government and defense context.

3.5.1   Business Process Orientation

The federal government is very process oriented when it comes to acquisition of complex systems, as exemplified in the 5000 Series of DoD Directives and other guidelines (DoD, 2006, 2008). However, although the private sector focuses on how processes create business value, government focuses on processes for getting required work done. The ways in which this work creates value is seldom explicitly addressed, in part because of a lack of consensus on what value means as well as because of the heterogeneity of stakeholders' perception of value.

The General Accountability Office frequently assesses the effectiveness of defense programs (GAO, 2008a, 2008b). One study focused on best practices in the design and development of complex systems, based on examples of success such as the Caterpillar 797 (Peoria, IL) mining truck and the Bombardier BRJ-X (Montreal, Canada) regional jet (GAO, 2001). The best practices they identified include:

- Matching customer expectations to available resources prior to setting requirements and launching development programs.
- Creating an initial design of the system that ensures that only proven technologies, design features, and productions processes are used.
- Making investments to address uncertainties such as new technologies or by reducing the system's initial performance requirements

The GAO found that government acquisition programs that did not follow these practices were more likely to not meet performance requirements while overrunning budgets and schedules. This results, in part, from "lowballing" of estimates of budgets and schedules to decrease risks of not making procurement cuts. Another factor was the natural tendency to include immature technologies in long lead time projects, perhaps reflecting a "now or never" psychology (Pennock, 2008; Pennock & Rouse, 2008). As discussed in Chapter 15, developers also tend to invest incremental resources in enhancing system functionality rather than in reducing time and risk.

3.5.2   Balanced Scorecard

Government and defense managers are aware of this construct (Kaplan & Norton, 1996). Attempts have been made to adopt this idea (e.g., for running military laboratories). However, the complexity of the social system with its many stakeholders and interests makes adoption extremely difficult. As indicated in earlier discussions, this lack of adoption results from the difficulty of getting agreement on the outcomes that matter.

It also reflects the fact that considerable attention is focused on securing and sustaining budgets. Such budgets represent a small portion of an overall balanced scorecard. In fact, budgets as inputs to the organization may not seem at all as one of the organization's outcomes. The notion that one of the primary purposes of an organization is to consume resources does not fit well within a value creation perspective of organizations. Thus, to an extent, this best practice does not fit in with organizational cultures typical in government and defense.

3.5.3 The Innovator's Dilemma

As with many large enterprises (Christensen, 1997, Christensen & Raynor, 2003), government and defense often have difficulty converting their own inventions into fielded innovations. Although the broader economy does take advantage of government R&D investments, government agencies have great difficulty coordinating current investments with future public needs. The broader social returns of government spending are discussed in Chapter 6. The concern here is with direct returns to agencies from their investments.

A primary difference between the contexts of Chapter 2 and this chapter is the relationship between business units and R&D. For government and defense, programs are, in effect, semi-autonomous business units. Such business units usually do not yet exist at the time that R&D investment would be needed to yield knowledge and technologies to feed the subsequent programs. Consequently, once a new program is authorized, the program manager will tend to fund R&D during development, resulting in less mature technologies being adopted with the intention to mature these technologies during the program. This practice is in complete conflict with the GAO best practices discussed earlier.

If the business units directly funded a portion of the R&D, like IBM and other companies, this might work differently. Business units would have a say in what R&D projects were pursued. They also would be sufficiently involved to shape the R&D, which would greatly increase the opportunities for the results of the R&D to contribute to programs' successes. This would require, however, that business units span multiple programs and that programs be considered as portfolios rather than independently. The current level of independence stems from Congressional processes for advocacy, authorization, and appropriation by program.

The innovator's dilemma often emerges because new technologies cannot garner sufficient market share and revenue to gain the attention of senior management. These factors are not inherent drivers in government and defense. Hence, new technologies should have a greater chance of making it from invention to innovation. However, organizational impediments, rather than market impediments, can make this difficult.

3.5.4 Lessons Learned

Government and defense frequently invest in study contracts and advisory panels who attempt to glean lessons learned from ongoing and past programs. A notable example is the comparison of the F/A-22 and F/A-18E/F development programs by Younossi and his colleagues (2005). Selected lessons learned include:

- Early, realistic cost and schedule estimates set programs on the right paths, for the rest of the program, but they must be adjusted over time.

- Stable team structure, proper team expertise, clear lines of responsibility and authority, and leadership responsible for success are critical.

- An experienced management team and contractors with prior business relationships help eliminate early management problems.

- Concurrent development of new technology and the system adds significant risk for both the individual technology and the integration of the system.

- Reducing the costs and risks of system development should be a key focus of the concept development phase rather than later in the program.

- Preplanned, evolutionary modernization of a high-risk system can reduce risk and help control costs and schedules, especially for rapidly changing technologies.

Clearly, government and defense repeatedly learn and relearn the right lessons. However, the complexity of the social system affects what they do and how they do it. This makes it difficult to adopt and sustain best practices despite knowledge of them. There are far too many stakeholders, with too many issues and interests, to allow for straightforward adoption of agreed-upon best practices.

3.6 CONCLUSIONS

This chapter has discussed the difficulties that government and defense have in adopting best practices from industry and commercial enterprises. This is not from a lack of knowledge of best practices. Instead, it is from, in part, some practices not being applicable across contexts (Birkler, et al., 2005). More significant, however, is the nature of the government and defense environment.

The complexity of the social system, with a wide range of stakeholders and frequently conflicting interests, makes it difficult to adopt practices such as business process reengineering (Hammer & Champy, 1993) and lean thinking (Womack & Jones, 1996). Companies who provide services to the government have been more successful in adopting such practices themselves compared with government agencies. However, in their roles as contractors, such companies usually have to comply with government dictates.

How does this affect human systems integration? At the level of individual systems, it can mean that HSI issues do not get addressed early when the costs of addressing these issues would be modest compared with dealing with them later during production and deployment. Furthermore, unaddressed HSI issues that negatively impact operational costs in future years may be the victims of the enterprise being unable to attach value to future savings.

At the level of the enterprise, HSI issues are typically not recognized. All energies are focused on the system. The fact that the enterprise is not fully prepared to address system-level HSI issues is difficult to both recognize and accept.

Furthermore, the changes that such recognition and acceptance might prompt may negatively affect one or more key stakeholders. This can certainly impede change.

Fortunately, as several later chapters illustrate, there are many stories of successful framing and of resolving major HSI issues at both system and enterprise levels. Despite the complexity of the government and defense context, the ability to economically value HSI investments is having an increasing impact.

REFERENCES

Apte, A. (2005). *Spiral Development: A Perspective*. Monterey, CA: Naval Postgraduate School, Graduate School of Business and Public Policy.

Ballhaus, W., Jr., (Ed.). (2000). *Science and Technology and the Air Force Vision*, Washington, DC: U.S. Air Force Scientific Advisory Board.

Birkler, J., Rushworth, D., Chiesa, J., Pung, H., Arena, M.V., & Chank, J.F. (2005). *Differences Between Military and Commercial Shipbuilding*. Santa Monica, CA: RAND Corporation.

Bodner, D., & Rouse, W.B. (2007). Understanding R&D value creation with organizational simulation. *Systems Engineering*, *10*(1), 64–82.

Cancian, M. (1995). Acquisition reform: It's not as easy as it seems. *Acquisition Review Quarterly*, Summer, 190–192.

Charette, R.N. (2008). What's wrong with acquisition? *IEEE Spectrum*, *45*(10), 32–39.

Christensen, C.M. (1997). *The Innovator's Dilemma: When New Technologies Cause Great Firms to Fail*. Boston, MA: Harvard Business School Press.

Christensen, C.M., & Raynor, M.E. (2003). *The Innovator's Solution: Creating and Sustaining Successful Growth*. Boston, MA: Harvard Business School Press.

Christensen, D.S., Searle, D.A., & Vickery, C. (1999). The impact of the Packard Commission's recommendations on reducing cost overruns on defense acquisition contracts. *Acquisition Review Quarterly*, Summer, 252–256

Cochrane, C.B., & Hagan, G.J. (2005). *Introduction to Defense Acquisition Management*, Fort Bevoir, VA: Defense Acquisition University Press.

DAU. (2005). *Introduction to Defense Acquisition Management*, (7ᵗʰ Edition). Washington, DC: Department of Defense, Defense Acquisition University Press.

DoD. (1996). *Guide to Integrated Product and Process Development*. Washington, DC: Office of the Undersecretary of Defense for Acquisition and Technology.

DoD. (1995). Economic Analysis for Decision Making. *Defense Instruction Number 7041.3*. Washington, DC: Department of Defense, November 7.

DoD. (2006). The defense acquisition system. *Defense Directive Number 5000.1*. Washington, DC: Department of Defense, May 12.

DoD. (2008). Operation of the Defense Acquisition System. *Defense Directive Number 5000.02*. Washington, DC: Department of Defense, December 2.

Drezner, J.A., Jarvaise, J., Hess, R., Hough, P., & Norton, D. (1993). *An Analysis of Weapon System Cost Growth* (MR-291-AF). Santa Monica, CA: RAND Corporation

DSB. (2006). *Transformation: A Progress Assessment* (Vol. 1). Washington, DC: Defense Science Board.

Gansler, J.S., & Lucyshyn, W. (2005). A *Strategy for Defense Acquisition Research*. College Park, MD: University of Maryland, Center for Public Policy and Private Enterprise, August.

GAO. (2008a). *Defense Acquisitions: Assessments of Selected Weapons Programs* (GAO-08-467SP). Washington, DC: Government Accountability Office.

GAO. (2008b). *Defense Acquisitions: Results of Annual Assessment of DoD Weapons Programs* (GAO-08-674T). Washington, DC: Government Accountability Office.

GAO. (2001). *Best Practices: Better Matching of Needs and Resources Will Lead to Better Weapon System Outcomes*. Washington, DC: General Accounting Office.

Hammer, M., & Champy, J. (1993). *Reengineering the Corporation: A Manifesto for Business Revolution*. New York: Harper Business.

Kaplan, R.S., & Cooper, R. (2008). *Activity-Based Costing*. Boston, MA: Harvard Business School Press.

Kaplan, R.S., & Norton, D.P. (1996). Using the balanced scorecard as a strategic management tool. *Harvard Business Review*, January-February, 75–85.

Kapstein, E.B. (2004). Capturing Fortress Europe: International collaboration and the Joint Strike Fighter. *Survival*, *46*(3), 137–160.

Pennock, M. (2008). *The Economics of Enterprise Transformation: An Analysis of the Defense Acquisition System*, Ph.D. dissertation, School of Industrial & Systems Engineering, Georgia Institute of Technology, Atlanta, GA.

Pennock, M.P., & Rouse, W.B. (2008). The costs and risks of maturing technologies, traditional vs. evolutionary approaches, in *Proceedings of the Fifth Annual Acquisition Research Symposium*. Monterrey, CA: Naval Postgraduate School, pp. 106-126.

Pennock, M.J., Rouse, W.B., & Kollar, D.L. (2007). Transforming the acquistion enterprise: A framework for analysis and a case study of ship acquisition. *Systems Engineering*, *10*(2), 99–117.

Rouse, W.B. (2001). *Essential Challenges of Strategic Management*. New York: Wiley.

Rouse, W.B., & Boff, K.R. (2004). Value-centered R&D organizations: Ten principles for characterizing, assessing & managing value. *Systems Engineering*, *7*(2), 167–185.

Smith, D.V., Searles, D.G., Thompson, B.M., & Cranwell, R.M. (2006). SEM: enterprise modeling of JSF global sustainment, in L.F. Perrone, F.P. Wieland, J. Liu, B.G. Lawson, D.M. Nichol & R.M. Fujimoto, Eds., *Proceedings of the 2006 Winter Simulation Conference*. Piscataway, NJ: IEEE, pp. 1324–1331.

Womack, J.P., & Jones, D.T. (1996). *Lean Thinking: Banish Waste and Create Wealth in Your Corporation*. New York: Simon & Schuster.

Younossi, O., Stern, D.E., Lorell, M.A., & Lussier, F.M. (2005). *Lessons Learned from the F/A-22 and F/A-18E/F Development Programs*. Santa Monica, CA: RAND Corporation.

PART II
Economics Overview

Chapter 4

Human Capital Economics

WILLIAM B. ROUSE

4.1 INTRODUCTION

Historically, economists thought of investments in terms of land, labor, and capital. With the Industrial Revolution, financial capital became increasingly important as it was needed to procure capital equipment. With the more recent Information Revolution, highly skilled labor has become increasingly important. This has led to the emergence and growth of a field of economics termed human capital economics.

"Human capital refers to the stock of skills and knowledge embodied in the ability to perform labor so as to produce economic value. It is the skills and knowledge gained by a worker through education and experience" (Wikipedia, 2009). This chapter addresses the assessment of investments in creating skills and knowledge.

Gary Becker, a pioneer in human capital economics, has asserted that one can invest in human capital (through education, training, health care, etc.) and the returns depend on the rate of return on the human capital one owns (Becker, 1964). Human capital is a means of production and additional investment yields additional output. He argues the following:

- Firms are willing to invest in training workers to develop firm-specific skills that are productive at the current firm but not at other firms.

The Economics of Human Systems Integration: Valuation of Investments in People's Training and Education, Safety and Health, and Work Productivity. Edited By William B. Rouse
Copyright © 2010 John Wiley & Sons, Inc.

- Firms are unwilling to invest in general skills training because workers can simply move to new firms—consequently, workers must bear the costs of general skills training.

Card (1999) provided well-documented estimates of returns, in terms of earnings, of elementary, secondary, and postsecondary education. There is little doubt that one's earnings increase with more education. More recently, Becker (2008) summarized these results:

- Education and training result in higher earnings, even after adjusting for intelligence and netting out direct and indirect costs of schooling.
- Real wages of young high-school dropouts have fallen by more than 25% since the early 1970s.
- The increasing reliance of industry on sophisticated knowledge greatly enhances the value of education, technical schooling, on-the-job training, and other human capital investments.

Another conclusion is that the number of years of schooling is highly correlated between parents and children, but earnings are much less correlated; the earnings of grandparents and grandchildren are hardly related. Thus, there are some subtleties to understanding returns on investments in human capital.

This chapter focuses on assessing returns on investments on human capital. Our main concern is investments by enterprises in the people that they employ, rather than investments by individuals in their own skills and knowledge. Chapter 5 on labor economics addresses competing views of why people invest in schooling and what determines wages, contrasting the constructs of human capital economics with other theories of human and organizational behavior.

4.2 MEASURING RETURNS ON INVESTMENTS

From the perspective of the enterprise, return on investment (ROI) entails much more that the fact that people with increased skills and knowledge earn higher wages. Although assessing the ROI on such investments does present difficulties (Manville, 2003), there have been a range of successful assessments.

4.2.1 Investments in Training

Bartel (2000) focused on measuring employers' ROI for investments in training. Her conclusions include:

- Econometric samples of large databases do not provide much insight, to a great extent because of the difficulties of estimating costs of training, which can be heterogeneous across companies in the sample.

- Studies of individual firms investments in training showed 7% to 50% internal rates of return (using wages as the proxy for productivity), assuming 5% annual depreciation of skills.
- Studies of actual productivity outcomes showed 100% to 200% ROI, although many case studies were seriously methodologically flawed.

Bartel noted that most American companies do not assess the outcomes of their training programs in terms of subsequent increases of productivity.

Bassi et al. (2004) assessed the impacts of firms' investments in human capital on stock prices using firm-level data for more than 400 U.S. companies from 1996 to 1998. The general finding was a relationship between a firm's training investments and its stock performance the next year, more so for training technical skills than business skills. Training in fundamental skills (interpersonal communication, and occupational safety/compliance) had the greatest return. Firms that spent, in a given year, more than $1,000 per worker on training outperformed the S&P 500 Index by almost 50% the next year.

Stroombergen et al. (2002) also discussed the impacts of quantity and quality of education on earnings, test scores, employability, and economic growth.

4.2.2 Investments in Health

Miller and Murphy (2006) discussed the issues and difficulties associated with assessing the ROI associated with investments in employee health. Chapter 13 in this book addresses health and safety investments in the automobile industry. Chapter 14 addresses the issue of medical costs and employee productivity. Both chapters provide some compelling quantitative results.

4.3 HUMAN CAPITAL MANAGEMENT

There is more to managing human capital than simply training, education, health, and safety. This results simply because one cannot own human capital; you can only buy the service of this capital. Thus, people who have both the capacity to perform work and the commitment to perform work are required. In other words, enterprises cannot take for granted that people will be willing to invest their own human capital in the organization.

This recognition has led to the development of the field of human capital management (HCM). Bassi and McMurrer (2007) reviewed more than 20 HCM practices in five categories: leadership, employee engagement, knowledge accessibility, workforce optimization, and learning capacity. They suggest a five-level HCM capability maturity model.

Using this model, they compared sales income growth and accident rates for offices of American Standard with above and below the median maturity in each of the five categories, they found that four of five practices led to greater sales growth and that all five practices led to lower accident rates. In another study, they found that higher maturity scores were associated with higher improvements in math scores in South Carolina schools. Finally, they found that financial services firms with higher maturity scores trend toward greater stock market returns.

In related studies, Pfau and Wyatt (2001) found a positive effect of 21 human capital practices upon stock market performance of 750 large public companies. Ichniowski, Shaw, and Prennushi (1997), in a study of steel finishing lines, found that incentive pay, teams, flexible job assignments, employment security, and training achieved substantially higher levels of productivity than did lines with the more traditional approach, which included narrow job definitions, strict work rules, and hourly pay with close supervision.

More recently, Ichniowski and Shaw (2003) have addressed the overall problem of designing organizations to get the best performance from their workers as a complex managerial issue. They argue that convincing economic analysis of this problem must acknowledge these complexities. It is not simply a matter of finding the optimal level of human capital investment for individual workers. They conclude that an individual worker with a given amount of education and work experience can be a high- or low-quality worker depending on the nature of the work environment.

Davenport (1999) provides extensive guidance in managing human capital. He suggests four key elements of developing the storehouse of human capital required for competitive success:

- Hire the right people.
- Elicit the maximum investment from people.
- Build people's human capital.
- Keep people committed and engaged.

He argues that commitment and engagement pave the way for human capital investment. The key is to get people to invest discretionary human capital that requires intrinsic job fulfillment, opportunity for growth, recognition for accomplishments, and financial rewards. This involves hiring, building, and retaining human capital, a key element of which is employee autonomy plus competence, which leads to greater investment by employees.

Davenport suggests several measures for assessing the extent to which HCM is successful:

- Improvements in strategic capabilities
- Contribution of hiring to strengthening capabilities
- Unit productivity per employee
- Contribution of learning to strengthening key capabilities
- Retention of committed and engaged people in pivotal jobs

Thus, investing in people tends to be multidimensional. The integration of these dimensions constitutes the overall investment for which we would like to project returns.

4.4 INVESTMENT VALUATION

Investments in land, buildings, equipment, and equities result in assets on the enterprise's balance sheet. In contrast, investments in people results in expenses on the income statement and nothing on the balance sheet. What is the "capital" one has acquired when investing in people?

Stroombergen et al. (2002) discuss three broad ways of thinking about the human capital created by investments:

- Capital as a function of earnings: net present value (NPV) of future earnings plus "other non-market benefits" (e.g., enjoyment)
- Capital as the summation of investment: NPV of human capital investments
- Capital as a summation of attributes and capabilities: sum of market and other attributes and capabilities

From an enterprise investment perspective, the first definition is, at best, an indirect measure of value. The second definition reflects the expenditure but not the potential return. The third definition, as it reflects capabilities gained, is more on target. Of course, we have to keep in mind that these are capabilities gained but not owned.

4.4.1 Capital Valuation

What are the likely returns on capabilities gained from human capital investments? Tangible assets and financial assets usually yield returns that are important elements of a company's overall earnings. It is often the case, however, that earnings far exceed what might be expected from these "hard" assets. For example, companies in the software, biotechnology, and pharmaceutical industries typically have much higher earnings than companies with similar hard assets in the aerospace, appliance, and automobile industries, to name just a few. It can be argued that these higher earnings result from greater human capital among software companies, among other things. However, because human capital does not appear on financial statements, it is very difficult to identify and, better yet, project knowledge earnings.

Mintz (1998) summarized a method developed by Baruch Lev for estimating what he termed knowledge capital—what could be argued to be a surrogate for human capital. This article in *CFO* drew sufficient attention to be discussed in *The Economist* (1999) and reviewed by Strassman (1999). In general, both reviews applauded the progress represented by Mintz's article but also noted the shortcomings of his proposed metrics.

The key, Mintz and Lev argued, is to partition earnings into knowledge earnings and hard asset earnings. This is accomplished by first projecting normalized annual earnings from an average of three past years and from estimates for three future years. Earnings from tangible and financial assets were calculated from reported asset values using industry averages of 7% and 4.5% for tangible and financial assets, respectively. Knowledge capital was then estimated by dividing knowledge earnings by a knowledge capital discount rate. Based on an analysis of several knowledge-intensive industries, Mintz and Lev used 10.5% for this discount rate.

Using this approach to calculating knowledge capital, Mintz compared 20 pharmaceutical companies with 27 chemical companies. He determined, for example, a knowledge-capital-to-book-value ratio of 2.45 for pharmaceutical companies and of 1.42 for chemical companies. Similarly the market-value-to-book-value ratio is 8.85 for pharmaceutical companies and is 3.53 for chemical companies.

The key issue within this overall approach is being able to partition earnings. Although earnings from financial assets should be readily identifiable, the distinction between tangible and knowledge assets is problematic. Furthermore, using industry average return rates to attribute earnings to tangible assets does not allow for the significant possibility of tangible assets having little or no earnings potential. Finally, of course, simply attributing all earnings "leftover" to knowledge assets amounts to giving knowledge assets credit for everything that cannot be explained by traditional financial methods. Nevertheless, this approach does provide insights into an important aspect of human capital—human skills and knowledge.

4.4.2 Direct Valuation

Of course, one can simply think in terms of investments and returns, without defining or measuring the capital created. In other words, one does not have to know the value of human capital to assess whether the costs of the investment are warranted relative to the returns on the investment. Given a stream of investments and returns, one can calculate the NPV or internal rate of return (IRR) for investments, whether they are people, equipment, or facilities. See Chapter 7 for these calculations.

Fitz-Enz (2009) has suggested several global measures of returns on human capital investments including the following:

- Human Capital Revenue Factor (HCRF) = Revenue Per FTE
- Human Economic Value Added (HEVA) = (Net Operating Profit After Tax−Cost of Capital) / FTE
- Human Capital Cost Factor (HCCF) = Pay + Benefits + Contingent Labor + Absence + Turnover, where Turnover costs include termination, replacement, vacancy, and learning curve productivity loss
- Human Capital Value Added (HCVA) = (Revenue−(Total Expenses−Pay and Benefits)) / FTE
- Human Capital Return on Investment (HCROI) = (Revenue−(Total Expenses−Pay and Benefits)) / (Pay and Benefits)
- Human Capital Market Value = (Market Value−Book Value) / FTE

Note that most of these measures include revenues minus costs in the numerator and full-time-equivalent people in the denominator. Thus, increasing revenue or decreasing costs or FTEs results in increased human capital value.

This has two limitations. First, investments are seldom so generic that global measures are sufficient. We need to be able to evaluate specific investments. This can make it difficult to translate particular outcomes to global measures such as overall revenue or market value. Second, investment valuation usually needs a more robust set of metrics such as provided by the balanced scorecard (Kaplan & Norton, 2004). For example, extent of strategic fit, leveraging of core competencies, and sustainability of advantage are often key nonfinancial metrics of interest.

4.4.3 Value Mapping

To assess the returns on particular human capital investments, one needs to map how an investment impacts organizational performance and value realized. Fitz-Enz (2009) suggested that this mapping be characterized in terms of processes, changes, impacts, and value. In general, the mapping involves:

- Process performance deficiencies that lead to
- Changes that save time and/or reduce errors that
- Impact processes in terms of less labor, customer satisfaction, and so on, resulting in
- Value in terms of customer spending, faster payment cycles, and so on.

Rucci et al. (1998) provided an excellent example of this type of thinking. They mapped the relationship between investments in improving Sears's retail store employees' attitudes and the market valuation of the company. Their findings can be summarized as follows.

- A compelling place to work leads to a compelling place to shop, which leads to a compelling place to invest.
- Within their model, a 5-unit increase in employee attitude led to a 1.3-unit increase in customer impression and to a 0.5% increase in revenue growth.
- Their empirical data showed that a 4% increase in employee satisfaction prompted a 4% increase in customer satisfaction, yielding a $200 M increase in revenue and a $0.25B increase in market value.

In this case, the investment in improving employee attitude was dwarfed by the returns. The value mapping was critical to fully understanding the relationships among key variables. Chapter 9 provides several illustrations of these types of analyses.

4.5 BEST PRACTICES

Thus far, we have discussed available empirical support for the value of investments in human capital—for specific interventions such as training and health, as well as human capital management practices—and approaches to investment valuation. Beyond metrics, data, and calculations, there is a range of management practices that can be recommended.

An orientation toward human capital suggests several changes in how people are managed:

- Emphasis should shift from efficiency to effectiveness. The goal should be the most effective workforce, not just the most efficient.
- Rather than just assessing costs, the value added by employees should be the central focus.
- Although inputs such as pay and benefits are often scrutinized, the purpose of employing people is their outputs and the value of these outputs.
- Managing people via their activities (i.e., how they spend their time) should be replaced by managing the outcomes people create.

Adoption of these principles requires understanding how the enterprise creates value for customers and other constituencies, how business processes enable providing this value, and how employees' work support these processes. Unfortunately, all too often, this understanding is lacking, and enterprises tend to manage people by how they spend their time rather than by the value of their outcomes.

Fitz-Enz (2009) has suggested 11 principles for human capital management listed as follows:

1. People plus information drive the knowledge economy.
2. Management demands data; data help us manage.
3. Human capital data show the how, the why, and the where.
4. Validity demands consistency; being consistent promotes validity.
5. The value path is often covered; analysis uncovers the pathway.
6. Coincidence may look like correlation but is often just coincidence.
7. Human capital leverages other capital to create value.
8. Success requires commitment; commitment breeds success.
9. Volatility demands leading indicators; leading indicators reduce volatility.
10. The key is to supervise; the supervisor is the key.
11. To know the future, study the past—but don't relive it.

Several of these principles deserve elaboration. The third principle relates to the simple fact that human behaviors are the cause of everything that happens. Another way of saying this, which is popular among those designing complex systems, is as follows: "There are no unmanned systems." At some level, humans have responsibility for everything.

The fourth principle asserts that standard metrics used over a long period of time are accurate. Another way of saying this is as follows: "Measure something and get better at it. If you learn that you should be measuring something else, measure that and get better at it." The key point is that feedback is how an organization—or an individual—learns and gets better.

The fifth principle relates to the value of analysis. Data collection and analysis can ease the difficulty of relating cause to effect. This will help to avoid basing decisions on coincidence, as reflected in the sixth principle.

The seventh principle reflects the notion that you can change human capital, unlike physical and financial capital. People, individually and socially, are the means to change. They can be an affordance, or enabler, or they can be a hindrance. In the latter case, the social network can act like an immune system, rejecting new visions and directions.

The tenth principle recognizes that in supervisor–subordinate relationships, personal relationships are the cornerstone of success. Human capital development does not just concern getting people the right training and annual physicals. The personal care exhibited by a supervisor can be an enormously motivating factor for subordinates.

4.6 CONCLUSIONS

This chapter has provided a brief overview of human capital economics. The central notion is that monies used to train and educate people, as well as to ensure their health and safety, are capital investments rather than just expenditures. This has become increasingly important as our economy has shifted from physical labor to knowledge work. The information economy now values knowledge capital as well as physical and financial capital.

A parallel trend has been the shift from a manufacturing to a service economy. Although the percent of the gross domestic product from manufacturing has held at roughly 25%, increased productivity has led to many fewer manufacturing jobs. Workers have, of course, always participated in a service economy by selling their services to employers. The value of these services increase with general human capital (e.g., literacy or numeracy) and specific human capital (specialized skills).

This chapter discussed measuring returns on investments in human capital using examples from training and health. The broader concept of human capital management was outlined, including the financial benefits of recognized management practices. Clearly, human capital investments pay off in general. However, most enterprises are concerned with whether particular investments will provide attractive returns.

Investment valuation was discussed, in terms of both attaching value to the capital created and direct valuation of investments and returns. Alternative investment metrics were reviewed. The notion of value mapping was introduced as a means for tracing process deficiencies and changes to value created.

Finally, recognized best practices were summarized. These practices rely on financial models, but they extend well beyond purely monetary considerations. To avoid human capital flight, or brain drain, at the enterprise or market level, a variety of human capital management practices are needed to ensure the right human capital is hired, built, and retained.

REFERENCES

Bartel, A. (2000). Measuring the employer's return on investments in training: Evidence from the literature. *Industrial Relations*, *39*(3), 502–524.

Bassi, L., Harrison, P., Ludwig, J., & McMurrer, D. (2004). *The Impact of U.S. Firm's Investments in Human Capital on Stock Prices*. Denver, CO: Bassi Investments.

Bassi, L., & McMurrer, D. (2007). Maximizing your return on your people. *Harvard Business Review*, *85*(3), 115–123.

Becker, G.S. (1964). *Human Capital: A Theoretical and Empirical Analysis, with Special Reference to Education*. Chicago, IL: University of Chicago Press.

Becker, G.S. (2008). Human capital. *The Concise Encyclopedia of Economics*. http://www.econlib.org/librray/Enc/HumanCapital.html.

Card, D. (1999). The causal effects of education on earnings. In O. Ashenfelter & D. Card, Eds., *The Handbook of Labor Economics*. Amsterdam, The Netherlands: Elsevier.

Davenport, T.O. (1999). *Human Capital: What It Is and Why People Invest It*. San Francisco, CA: Jossey-Bass.

Economist. (1999). A price on the priceless: Measuring intangible assets. *The Economist*, 61–62, June 12.

Fitz-enz, J. (2009). *The ROI of Human Capital: Measuring the Economic Value of Employee Performance* (2nd Edition). New York: AMACOM.

Ichniowski, C., Shaw, K., & Prennushi, G. (1997). The effects of human resource management practices on productivity: A study of steel finishing lines. *American Economic Review*, *87*(3) 291–313.

Ichniowski, C., & Shaw, K. (2003). Beyond incentive pay: Insiders' estimates of the value of complementary human resource management practices. *Journal of Economic Perspectives*, *17*(1), 155–180.

Kaplan, R., & Norton, D. (2004). *Strategy Maps: Converting Intangible Assets Into Tangible Outcomes*. Boston, MA: Harvard Business School Press.

Manville, B. (2003). Making sense of human capital economics. *Chief Learning Officer Magazine*, November.

Miller, P., & Murphy, S. (2006). Demonstrating the economic value of investments in health at work: Not just a measurement problem. *Occupational Medicine*, *56*, 3–5.

Mintz, S.L. (1998) A better approach to estimating knowledge capital, *CFO*, 29–37, February.

Pfau, B., & Wyatt, I.K. (2001). Human Capital Edge: 21 People Management Practices Your Company Must Implement (Or Avoid) To Maximize Shareholder Value. New York: McGraw-Hill.

Rucci, A.J., Kirn, S.P., & Quinn, R.T. (1998). The employee-customer-profit chain at Sears. *Harvard Business Review*, 83–97, January-February.

Strassman, P.A. (1999). Does knowledge capital explain market/book valuations? *Knowledge Management*, www.strassman.com/pubs/km, September.

Stroombergen, A., Rose, D., & Ganesh Nana, G. (2002). *Review of the Statistical Measurement of Human Capital*. Auckland, New Zealand: Statistics New Zealand.

Wikipedia. (2009). Human capital. http://en.wikipedia.org/wiki/Human_capital.

Chapter 5

Labor Economics

Nachum Sicherman

5.1 INTRODUCTION

Economists study "markets," and "labor economics" can be viewed as the study of a market with specific, unique characteristics. What makes this market unique is the fact that the "sellers" are the employees who sell their labor services, whereas the "buyers" are the firms that buy or, more correctly, rent the employees' labor services. Indeed, because employers do not "own" their employees and "work" cannot be separated from the "worker," the human aspect becomes a central issue in labor economics. This is what makes the study of labor markets different from the study of other, more standard, markets. Yet, much of the framework used in modern labor economics is the same framework that is employed in other fields of economics, namely microeconomic theory.

Most of the research in labor economics is devoted to "supply-side" considerations (i.e., the provision of labor services by workers).[1] Important questions include how wages and labor conditions affect individuals' decisions concerning occupational choice and the amount of time they want to spend at work; when and why do workers quit their jobs and decide to switch to another job; how

[1]"Labor demand", which can be viewed as a separate subfield of labor economics, is not discussed in this chapter. A good starting point for the interested reader is Hamermesh (1993).

do workers search for a job, and how do they decide when to accept or reject a job offer. Another line of research in labor economics concerns the determinants of wages. Here are examples of two typical questions that have been of central interest in labor economics:

- Why do different individuals have different levels of income?
- Why do individuals' earnings increase over time?

To a noneconomist, such questions may seem trivial. It seems (almost always) so obvious to us that a CEO of a large corporation should earn more than a janitor and that a computer programmer should earn more than a waiter, that we rarely bother to stop and ask "Why?" Most of us are so accustomed to seeing our earnings increase over time that we almost take it for granted. It is possible, however, to arrive at a reasonable and logical explanation for each of these observations. A more careful examination, however, will show that there is more than one possible answer to each of those questions and that different answers have very different policy implications. In the following discussion, I present several alternative answers to these questions, starting with "human capital theory."

5.2 HUMAN CAPITAL THEORY

The most powerful and dominant theory that emerged in the 1960s to explain workers' decisions inside and outside the labor market is "human capital theory."[2] Using standard tools of microeconomic theory, the central assumption of this theory is that human beings invest in their skills and capabilities in a similar way to investors that invest in machines. For example, individuals invest in schooling and on-the-job-training to increase their future earning capacity. Such investments have costs and benefits and can, therefore, be analyzed and studied as any investment decision. The predictions of human capital theory with regard to individuals' earnings, investment in schooling, training, labor mobility, retirement, and numerous other labor market behaviors have been tested, estimated, and confirmed in thousands of studies all over the world.

Although human capital theory is no doubt the dominant theory in modern labor economics, it is not the only one. Not everything that has been observed in the labor market can be explained by human capital theory, and some findings seem to contradict the theory. Also, some alternative theories provide a logical, yet different, explanation for the same observations and questions addressed by labor economists. Let me provide two examples in the following discussion.

5.2.1 Investment in Schooling: Signaling versus Human Capital Theory

A basic assumption of human capital theory is that individuals invest in schooling to acquire skills that will increase their future productivity and, therefore, increase

---

[2]The classic readings on this topic are Becker (1964) and Mincer (1974).

their earnings. They will keep investing in schooling as long as the marginal benefit from this activity is greater than the marginal cost. Indeed, numerous studies in which "wage regressions" are estimated find that the marginal return to an extra year of schooling is similar to the return on other types of investments.

One might take the opposite view, however, by arguing that in some cases, although additional schooling resulted in higher earnings, it does not seem to be the case that what was learned in school has increased the person's productivity on the job. For example, some jobs require a college degree, regardless of major, where it seems that the skills required on the job are very different from those acquired at college. If this is the case, why would an employer pay a higher salary to a person just because they had obtained an academic degree in a subject that does not seem to increase their productivity?

Professor Michael Spence, who won the 2001 Nobel Memorial Prize in Economics, published a seminal paper in 1973, in which he presented his famous "signaling model."

The basic idea of the signaling model is that employers cannot always tell how productive a prospective employee will be. Talented (and productive) employees want to provide a credible signal to the potential employer that they are, indeed, highly productive. To do so they invest in schooling and use the acquired diploma as a signal for their unobserved ability. One question that one might ask is why low-ability employees do not do the same. Why don't they acquire the same signal, present themselves as the high-ability type, and get the higher wages? Of course, if such behavior were possible, the signal would stop being informative and would cease to exist. A necessary condition for the signal to work is that high-ability employees find it beneficial to acquire it, whereas low-ability employees do not. The interested reader can find the conditions necessary for this "separating equilibrium" to exist in the original Spence paper.

Although from the worker's perspective, it might not be important why he or she can increase his or her earnings by obtaining a specific degree, this connection might be relevant for the employer side, especially for those in the organization that determine the schooling requirements for different jobs.

5.2.2 Why Do Wages Increase Over Time?

Human capital theory provides a simple and elegant answer to this question: Productivity increases over time because workers learn new skills and enhance existing skills. They do so either by different forms of formal and informal on-the-job training or simply via "learning by doing." As their productivity is increasing, their wages are also increasing.[3]

The exact relationship between the rate of growth of productivity and the rate of growth of wages is beyond the scope of this chapter. It should be noted, however, that the standard prediction of models based on human capital theory is that

---

[3]For the classic paper that analyzes the optimal division of time between work and on-the-job-training, and the implied wage profiles, see Ben-Porath (1967).

individual wages will rise at a rate that is either similar to or lower than the rate of productivity growth.

Although human capital theory and its predictions have been subject to numerous supporting empirical studies, several notable alternative explanations exist to why individuals' wages increase over time. What has motivated these theories is the fact that, in many cases, the opposite relationship between wages and productivity exists. It is common, for example, to observe wage profiles that are steeper than productivity profiles. It should be noted that, in general, it is not easy to obtain data relevant for the link between wage profiles and productivity profiles because in many cases it is not trivial to measure workers' productivity. There are, however, cases in which workers' productivity does not change over time (at least beyond a certain point), but their wages keep growing. This is clearly an example of wage profiles that are steeper than productivity profiles. Below I present several alternative theories that explain why wage profiles might be steeper than productivity profiles. Understanding these alternative theories can help managers understand the role of wage growth rates as an incentive and selection device.

5.2.2.1 Deferred Payments. Wage profiles that are steeper than productivity profiles could be viewed as a form of deferred payment. Initially workers are paid less than their marginal productivity, but later on in their careers, they get compensated more than they produce. There are several benefits to the employer in offering such a contract:

- **Deter shirking:** Deferred compensation can dissuade workers from shirking on the job by increasing the cost of being fired.
- **Selection:** Workers who do not plan to remain with the employer for an extended period of time might avoid joining a firm that defers payments. Firms that find labor turnover costly and look for ways to reduce it may find deferred compensation to be an effective tool for retention.

Besides wage increases that exceed productivity changes, there are other ways in which a firm can defer payments. An annual bonus is a typical example. If a worker considers quitting, he or she will most likely wait until her annual bonus is received before quitting. Offering workers certain benefits conditional on being with the firm a minimum number of years is another example of such a practice. Incentive stock grants and stock options that become vested only after a period of time, as well as the delayed vesting of retirement benefits, are commonplace forms of deferred compensation.

5.2.2.2 Mandatory Retirement. Although mandatory retirement is illegal in the United States in most civilian jobs, it is still a common practice in many other countries, as well as in the U.S. armed forces.[4] Mandatory retirement in the United

---

[4]Different groups, however, are subject to different rules and different ages of mandatory retirement. In addition, these rules have changed dramatically over the years.

States is banned because it is deemed to constitute a form of age discrimination. For most people outside the United States as well as for members of the armed forces, this institution seems so natural that they never even question its existence. Edward Lazear, in a classic paper (Lazear, 1979) offered the following economic rationale for mandatory retirement: A wage contract where the wage growth rate is steeper than the productivity rate is feasible only when employment ends at a certain point. Otherwise the net cost to the employer of the upward sloping wage profile will increase while the incentive to the worker to retire voluntarily will decrease. Without mandatory retirement, such a contract becomes problematic.

When it comes to HSI, the age of retirement may play an important role. The costs and benefits of the human components of a system are as important as the nonhuman ones. Given the age at which the operators of a system are recruited, the time it takes to train them, the rate at which their proficiency increases (the "learning curve"), all combined with mandatory retirement, determine the length of time left to recoup the returns on investment made by recruiting and training those operators. Therefore, decisions affecting the timing of retirement are as important as the length of time a future fighter jet is expected to be in service.

5.2.2.3 *Empirical Evidence.*

As mentioned, it is not easy to measure workers' productivity. This is especially true in organizations where productivity is not easily translated to financial profits or share prices. As a result, the few published studies on this topic have been limited to a small number of occupations, which used indirect measures of productivity and were often not accurate. To illustrate, a typical, and widely cited, study (Medoff & Abraham, 1981) looked at workers within specific jobs and made the following observations to argue that wages increase while productivity does not: Workers' wages were increasing while managers' performance evaluations of their employees hardly changed. There are two major problems with such a study. First, it does not measure productivity directly. Second, it ignores promotions within the firm. It is possible that the evaluations of workers who remained on the job did not change, although the good workers were promoted.

Another important study (Hutchens & Frank, 1993) found that in occupations such as bus driving, where it is reasonable to assume that productivity does not increase much beyond a relatively short tenure, workers do get substantial wage raises, even after their productivity has leveled off. The importance of these findings is that the standard models of deferred compensation do not seem to provide a sufficient explanation for the findings.

In a study that I conducted with Prof. George Loewenstein (Loewenstein & Sicherman, 1991), we proposed a provocative and alternative explanation: **Wages increase over time because workers like it!** This proposal is provocative because it states that workers will be willing to give up money to experience wage growth over time. For example, workers will prefer a three-year employment contract that pays $90,000 in the first year, $100,000 in the second year, and $110,000 in the third year, to a contract that will pay them $100,000 in each of the three years. Notice that under any positive discount rate, the second contract has a higher net present value.

Why would anyone accept a lower present value of wage income to experience an increasing wage profile? We propose three reasons:

1. **"The Mastery Argument":** Individuals derive satisfaction when they sense they have control of their personal environment. Therefore, if people associate increasing wages with increasing productivity, then by receiving higher wages, they feel an increased sense of control of their environment. Even if there is no link between their wage and productivity, it exists subconsciously. A similar argument can be made with regard to promotions-in-rank that are not necessarily accompanied by a significant increase in responsibilities or assignments (e.g., a fighter pilot promoted from captain to major after certain years of service).

2. **"The Compulsive Spender Argument":** If individuals have, for different reasons, a preference for increasing consumption profiles but lack the self-control required to match their income profile and consumption profile, they might look for such devices that artificially provide this self-control.

3. **Utility from changes in consumption:** This explanation is related to the previous one. Numerous studies have argued that individuals derive utility not only from the level of consumption but also from the rate of change in consumption. This, together with a lack of self-control, might make workers prefer increasing wage profiles.

4. **Utility from anticipation:** In addition to the previous argument, and possibly as an explanation for it, individuals may derive utility from anticipating future consumption (see, for example, Loewenstein, 1987). Again, if it is true, then utility of anticipation, combined with lack of self-control, will make individuals prefer increasing wage profiles.

It is beyond the scope of this chapter to discuss the empirical implications of each of those explanations and how one can test for their validity. The interested reader is referred to the article cited previously. The important message of the study is that individuals find it important to experience their wages increase over time, and this, by itself, should play an important role in job design and wage settings.

5.2.3   The Demand for Skills—General Equilibrium Approach

One of the most important components in the success of any system is the knowledge and skills of the individuals that design, operate, and maintain it. It is, therefore, not surprising that scholars and policy makers have long emphasized the importance of investment in education, skill development, and on-the-job training. With increasing rates of technological change, modern systems are becoming increasingly sophisticated and their success depends more and more on making certain that those who interact with them are capable of fully using their potential.

What I will try to demonstrate in this section is that there is an important aspect that is overlooked when discussing the interaction between human skills and machines. What is absent is the lack of attention to the interplay between the skill requirements of a system and the availability of skills to fulfill such requirements. Economists use the terms "partial equilibrium" and "general equilibrium" to differentiate between the two approaches.

The partial equilibrium approach is to take the existing system as given and then worry about how to find the people with the right skills to operate the system. The designers of a new fighter jet, for example, will design the best machine to fit the specifications set by the entity that ordered the design, without necessarily giving any consideration to the availability of skills that are needed to operate the machine. In the economist's jargon, the "demand" for skills is determined without giving a consideration to the "supply" of skills. In that sense, the demand for skills is exogenous.

The opposite partial equilibrium approach will be to take the existing pool of individuals with their levels of skills as given, and then worry about how to design a system that will best use the available skills in the labor market or, in the case of a defense system, the availability of recruits. An example of such an approach occurs when the designers of a system ignore the possibility that changes on the demand side (i.e., the introduction of new and different systems), with differing skill requirements, is likely to affect individuals' decisions concerning their schooling and training, thus creating changes on the "supply" side.

Such an approach could be especially problematic in cases when the time span between the initial design and the commencement of operations is very long. Think, for example, of a new space plane, where it is very likely that those who will operate it are still in kindergarten when the designers have to make assumptions about the nature of skills that its operators will have. The design implications are very different if one assumes that all recruits will be proficient and comfortable in using a standard computer keyboard as opposed to assuming that new developments in voice recognition software will imply that future recruits will have difficulties in operating devices that use standard keyboards.

But one does not have to go that far to understand the danger of taking the availability of skills, as well as their cost, as given. It is well known that the ability of the armed forces to recruit individuals with certain skills depends on the state of the economy. The cost of recruiting highly skilled individuals is likely to be lower in times of recession and high unemployment rates and higher when the economy is booming with a strong demand for skilled workers. In other words, the relative cost of high-skilled recruits as compared with low-skilled recruits could vary dramatically for various reasons. If we take the broad view of a system, as is emphasized in several chapters in this volume, then the relative prices of skills should play an important role in the design of a system. For example, if one expects an increase in the relative price of highly skilled operators, it might be more cost-effective to invest more in "de-skilling" technologies, technologies that reduce the demand for skilled operators.

The general equilibrium approach will be to acknowledge that both forces are active at the same time and that each is affected and determined by taking into account the other.

5.3 COST–BENEFIT ANALYSIS AND THE VALUATION OF "HUMAN" SYSTEMS

Cost–benefit analysis is one of the economist's key tools of decision making. The basic idea is trivial: A project or a decision should be undertaken if the incremental benefit of the decision is greater than its incremental cost. Indeed, this rule is so obvious that is sounds as good as advising someone to "buy low and sell high."

Applying this rule, however, is not as trivial as it sounds. It requires the decision maker to identify and measure correctly the costs and benefits that are associated with a specific project. This is not an easy task, and there are numerous examples of managers and other decision makers failing to account properly for different costs and benefits. For example, managers frequently include various overhead costs in estimating the cost of a project. When such costs are not affected by whether this project is undertaken (e.g., they are sunk), including them in the cost–benefit analysis will underestimate the net benefit of the project and may result in the incorrect decision to reject it.

Although the previous example is common across sectors, some difficulties are more common in the public sector. When the project analyzed is a defense system, such as a new fighter jet, conducting a cost–benefit analysis is likely to be much more complicated than conducting a similar analysis for a product sold in a competitive private market. Consider, for example, the decision to install a newly improved safety system in a vehicle. It seems very likely that estimating the cost of such a system is not difficult. Estimating its benefits, however, is a very different affair. Possible benefits include lower death and injury rates after an accident. Quantifying such benefits requires, among other things, putting a number or assigning a monetary value to saving life, saving a limb, and so on. And this is only a partial listing of the relevant issues.[5]

Interestingly enough, all these complications become a nonissue in the private sector. When Volvo considers the installation of a new and improved airbag system in its new cars, it does not have to conduct the type of cost–benefit analysis described above. The market will take care of it. What Volvo must determine is how much their customers are willing to pay for such an improvement and thus how the installation of such a new system will affect the sales of new Volvo cars. In the economist's jargon, Volvo has to estimate the effect of such a change on the demand for its product.

It is true that how much more a consumer is willing to pay for a car that has a better airbag system is a function of how much he values safety, which, at least implicitly, must be related to how much he values his own life and the lives of those that will use the car. But unlike the case of a defense system, where a transaction

---

[5]There is a rich literature in economics on estimating the "value of life." The interested reader could start with reading some of Professor Kip Viscusi books and articles. Prof. Viscusi of Harvard University is probably the leading authority in this field.

is conducted outside the market, Volvo does not have to estimate those benefits directly.[6]

A similar problem might develop when trying to calculate the cost of a project. Increasing the safety of a pilot against enemy fire might require increasing the weight of the airplane, thus reducing its range or lowering its speed. Again, although in the private sector such a tradeoff will be reflected in the consumers' willingness to pay, in the public sector, where there is no market to determine the costs and benefits of each of those "characteristics," it is the difficult task of the decision makers to quantify these tradeoffs.

How practical and feasible is it to estimate the costs and benefits associated with each possible decision? For some projects, where there are numerous possible tradeoffs, it is probably unfeasible to identify and quantify each item separately, and compromises must be made. These compromises are usually made by determining or choosing various specifications and requirements of the system without conducting any formal cost–benefit analysis. This does not mean, however, that we should always wave our hands when facing decisions that are associated with hard-to-quantify costs or benefits. If there is an explicit suggestion on the table to modify a system to increase the safety of its operators and an objection is raised, pointing out the associated monetary and nonmonetary costs, via a cost–benefit analysis is the prudent course of action.

Several "tricks" can be employed to avoid some of the difficulties associated with a complete analysis. Returning to the previous example, one could start with calculating only the monetary costs associated with increasing the safety of the operator (e.g., using a more expensive but stronger material for the cockpit hatch). Then, focusing only on the potential benefit, one could conduct a sensitivity analysis assuming different probabilities that the stronger material will save the pilot's life (compared with the alternative material) to figure out the number that one has to put on the value of life to justify the investment. The lower this number is, the more justified the investment is. The health-care literature, for example, could provide benchmark values for what numbers could be considered relatively low.

5.4 SUMMARY

Human systems integration (HSI) is a multidisciplinary field practiced mainly by individuals trained in engineering and management. When integrating economics into engineering, as is done in this volume, the practitioners of HSI commonly adopt tools and concepts from economics and finance, such as cost–benefit analysis, the time value of money, and cash flow analysis, to better assess the economic (as opposed to the technical) viability of projects. Being an economist who is not a practitioner of HSI, I introduced the reader to economic concepts that are not commonly discussed in this literature but are equally relevant.

---

[6]One of the methods by which economists estimate the "value of life" is by using data of automobiles' prices to derive consumers' willing to pay for different safety devices.

Being a labor economist, I find an interesting overlap in the issues and questions studied in both HSI and labor economics. I, therefore, find it important to introduce the reader to these questions and issues and, in doing so, thereby hope to expand the ways in which the practitioner of HSI will address questions and issues in his field. Specifically, I believe that a deeper understanding of the concept of "general equilibrium," and how it differs from a partial equilibrium analysis, will affect the way in which the economic viability of projects and issues such as education, training, and operators' skills are integrated into the study of HSI.

REFERENCES

Becker, G. (1964). *Human Capital*. Chicago, ILThe University of Chicago Press.

Ben-Porath, Y. (1967). The production of human capital over the life cycle. *Journal of Political Economy 75*, 352–365.

Hamermesh, D. (1993). *Labor Demand*. Princeton, NJ: Princeton University Press.

Hutchens, R.M., & Frank, R. H. (1993). Wages, seniority, and demand for rising consumption profiles. *The Journal of Economic Behavior and Organization, 21*.

Lazear, E. (1979). Why is there mandatory retirement? *Journal of Political Economy 87*, 1261–1284.

Loewenstein, G. (1987). Anticipation and the valuation of delayed consumption. *The Economic Journal, 97*, 666–684.

Loewenstein, G. & Sicherman, N. (1991). Do workers prefer increasing wage profiles? *Journal of Labor Economics, 9*, 67–84.

Medoff J. & Abraham, K. (1981). Are those paid more really more productive? The case of experience. *Journal of Human Resources 16*, 186–216.

Mincer, J. (1974). *Schooling, Experience, and Earnings*. New York: Columbia University Press.

Spence, A.M. (1973). Job market signaling. *Quarterly Journal of Economics 87*(3), 355–374.

Chapter 6

Defense Economics

KEITH HARTLEY

6.1 INTRODUCTION: THE SCOPE OF DEFENSE ECONOMICS

Defense economics is a relatively new field within the discipline of economics. It focuses on the economic aspects of defense, conflict, disarmament, and peace (or the economics of war and peace). Even with the end of the Cold War, nations have continued to allocate scarce resources to defense involving the sacrifice of alternatives such as spending on hospitals and schools (the classic guns vs. butter tradeoffs or choices among missiles, schools, and hospitals). Nor have conflicts ended. Since 2000, there have been major conflicts in Afghanistan, Iraq, between Israel and Palestine, and the "War on Terror."

Defense economics analyzes a variety of important policy issues. Examples include the size of defense budgets, their efficiency, and the economics of procuring military equipment and personnel. Other areas include the economics of arms races, military alliances, national defence industries, arms exports, disarmament, peace, and the peace dividend. More recently, defence economists have analyzed conflicts, civil wars, revolutions and terrorism (Sandler & Hartley, 1995, 2007; Tisdell & Hartley, 2008, Chapter 17).

The Economics of Human Systems Integration: Valuation of Investments in People's Training and Education, Safety and Health, and Work Productivity. Edited By William B. Rouse
Copyright © 2010 John Wiley & Sons, Inc.

6.2 WORLD DEFENSE SPENDING

The magnitude of world military spending from 1990 to 2007 is shown in Table 6.1. Since the end of the Cold War in 1990, world military spending in real terms has increased by almost 7%. There were increases in most major regions of the world, especially in North America (i.e., the United States) and Asia-Oceania. Europe differed and showed reduced military spending, particularly in Eastern Europe.

Table 6.2 shows the world's top ten military spending nations in 2007. The United States dominated with a 45% share of the world total, whereas the top ten nations accounted for 76% of world military spending in 2007. These nations also allocated a substantial share of their GDP to defense, with a typical share figure of some 2.5%, within a range of 1%–8.5%. Defense spending also involves the acquisition of labor inputs, comprising military and civilian personnel required for the Armed Forces together with personnel employed in defense industries. Table 6.2 shows the numbers of military personnel in the Armed Forces of the major nations with an aggregate of 7.25 million military personnel in these nations.

Defense ministries have the task of recruiting personnel (labor inputs) and equipment (capital, e.g., weapons) that have to be combined through a military production function to produce a defense output (e.g., nuclear deterrence; capability to fight, say, two regional conflicts). Military personnel can be recruited and retained either as an all-volunteer force (AVF) or through conscription (draft) or some combination of the two. New recruits have to be trained, and the Armed Forces have to face the standard human capital investment choices that develop in the private sector. Human capital investments involve costs and benefits: Training costs are borne by the Armed Forces, but benefits accrue to both the Armed Forces and the individual trainee.

TABLE 6.1 World Military Spending, 1990–2007

Region	US$ billions, 2005 prices and exchange rates	
	1990	2007
Americas:	493	598
including North America	(473)	(562)
Europe:	468	319
including West Europe	(282)	(261)
Eastern Europe	(171)	(41)
Asia and Oceania	110	200
Middle East	53	79
Africa	13	17
World	**1136**	**1214**

Notes:

1. Figures in brackets show nations included in the totals for Americas and Europe.

2. Figures are rounded.

Source: SIPRI military expenditure database: SIPRI (2008).

TABLE 6.2 The Top Ten Military Spending Nations, 2007

		US$ at 2005 prices and exchange rates			
Rank	Nation	Spending ($ billions)	World share (%)	Share of GDP (%)	Numbers of military personnel (000s)
1	USA	547	45	4.0	1498
2	UK	60	5	2.6	181
3	China	58	5	2.1	2105
4	France	54	4	2.4	255
5	Japan	44	4	1.0	240
6	Germany	37	3	1.3	246
7	Russia	35	3	3.6	1027
8	Saudi Arabia	34	3	8.5	224
9	Italy	33	3	1.8	186
10	India	24	2	2.7	1288

Notes:

1. Figures are rounded and based on market exchange rates. Data based on purchasing power parities (PPP) give different rankings. For example, using PPP rates, China is ranked second, Russia is third, India is fourth, and the UK is fifth.

2. Data for China and Russia are estimates. Military personnel data are for 2008.

Sources: SIPRI military expenditure database: SIPRI (2008); IISS (2008).

Typically, private firms will be reluctant to finance transferable skills training because such training has value to large numbers of firms in the economy and in the absence of a contract of slavery, firms are likely to lose their investments to rival firms who have not paid the training costs and can pay a higher wage to the newly trained personnel. However, the Armed Forces have a different and distinctive employment contract. Military personnel are subject to military discipline: They have to go where they are told to any part of the world, and they might never return. Also, they sign a contract committing themselves to the Armed Forces for a specified number of years so that there is a reasonable expectation that the Armed Forces will obtain a return on their costly training investments; this explains why the Armed Forces will finance the training of such transferable skills as drivers, computer operatives, and transport aircraft pilots. Nontransferable skills training will be financed by the Armed Forces because the skills have no value in other uses. Examples of nontransferable military skills include tank gunners, submarine crews, special forces, parachutists, and marines. These issues are part of the defense economics challenge.

6.3 THE DEFENSE ECONOMICS QUESTION

Overall, world defense spending remains substantial and involves significant numbers of personnel. All these resources have alternative uses, and the inevitable question is whether defense spending is a worthwhile investment compared with its alternatives. A key question for defense economists concerns the impact of military spending on an economy: Are the effects beneficial or harmful?

This chapter reviews the available research and literature on the relationship between defense spending and its impact on the wider national economy. Although the focus is on defense spending, the review will also assess the impact of defense research and development (R&D) expenditures on the economy, including the use of scarce scientists in defense work. It starts with an analytical framework and then briefly summarizes the state of knowledge as outlined by Sandler and Hartley (1995), who presented a survey of the major economics literature on economic growth, development, and military expenditure (Sandler & Hartley, 1995, Chapter 8). The 1995 study was then updated with a review of the more recent literature as published in the journal *Defence and Peace Economics*. A conclusion presents the main findings and a research agenda.

6.4   ANALYTICAL FRAMEWORK

This area is dominated by myths, emotion, and special pleading. The approach taken is to review the relevant economic analysis and supporting empirical evidence. A starting point is the belief that defense spending together with defense R&D expenditures "crowd out" valuable civil spending with adverse economic impacts on an economy. Even expressed in this form, the crowding-out hypothesis needs clarification. What are the "adverse economic impacts": Do they refer to growth, physical capital investment, numbers of qualified scientists, engineers and skilled labor (human capital), technical progress, or international competitiveness and exports? Next, questions develop about the causal relationships between defense spending and defense R&D expenditure on the one hand and the various "adverse economic impacts" on the other hand.

6.4.1   UK MoD and Crowding Out

In its 1987 Statement on the Defence Estimates, the U.K. Ministry of Defence (MoD) provided a clear expression of the crowding-out effects of defense R&D. It stated that "... the government shares the underlying concern of those who fear that necessary investment in defense R&D may crowd out valuable investment in the civil sector." It went on to explain that "... Britain's resources of qualified scientists and engineers and the skilled manpower supporting them, are not inexhaustible ... defence and civil work are in competition for the same skills, and it would be regrettable if defence work became such an irresistible magnet for the manpower available that industry's ability to compete in the international market for civil high technology products became seriously impaired" (MoD, 1987, p. 48, para 522). This is the labour market version of the crowding-out hypothesis. Other variants are more general and refer to the crowding-out impacts of defence spending in total and defence R&D spending in particular.

6.4.2 Defense Spending and Crowding Out

This version of the hypothesis focuses on the costs of defense expenditure, especially its adverse impacts on economic growth. On this view, the adverse growth effects of military spending might develop:

1. Defense may divert resources away from public- and private-sector investment that may be more favorable to growth than defense spending.
2. Adverse balance of payments impacts through imports of arms and where resources are diverted away from the civil export sector.
3. Defense, particularly defense R&D, might divert resources from private-sector R&D activities affecting both technology and spinoffs. The resources diverted embrace both physical and human capital.

Such crowding out impacts the need to be assessed critically. First, all civil spending is not allocated to public and private *investment*: A considerable proportion is allocated to consumer spending, some of which takes place overseas. Second, adverse balance of payments impacts can develop from any imports of civil public- and private-sector goods and services; and civil public spending might also have adverse effects on exports. Third, there is the counterfactual: What would happen in the absence of defense spending, including defence R&D spending? For example, it does not follow that without defense R&D, the resources released would be used in private-sector R&D: Some might leave the labor market; some labor might emigrate; some might remain unemployed; and others might obtain jobs outside of R&D (e.g., sales and finance). Against these costs, there are offsetting benefits of military expenditure.

6.4.3 Benefits of Military Spending

The benefits can differ between developed and less developed nations. The possible economic benefits of defense spending include:

1. In periods of high unemployment, both types of countries might experience stimulative effects from defense spending (i.e., defense spending adds to aggregate demand in the economy). However, other types of government expenditure might also stimulate the economy and might offer an even greater stimulus (e.g., construction projects and civil space expenditure).
2. Defense provides direct technology benefits and spinoffs, where spinoffs applied to the civil sector can promote growth.
3. In less developed countries especially, defense spending might promote growth if some expenditure is used to provide social infrastructure (e.g., airports, communication networks, roads, and bridges, all of which contribute to creating national markets).
4. Developing and supporting human capital, especially in less developed nations. A nation's armed forces are provided with nutrition, education,

training, and discipline, with some of these benefits spilling over into the civil-sector labor force.

5. Defense spending provides protection to a nation's citizens where internal and external security promotes beneficial market exchange.

Again, other public and private spending can provide some of these benefits. First, civil public spending programs can stimulate a high unemployment economy (e.g., providing jobs through public works programs such as building roads, bridges, and railways). Second, private R&D provides direct technology benefits and might provide spinoffs (e.g., motor cars and pharmaceuticals). Third, less developed nations can create an infrastructure through civil public spending. Fourth, a nation's population can be educated in state schools and training establishments so that there are alternative nonmilitary methods of creating human capital. Finally, internal and border police might provide the internal and external security needed for the development of markets. Overall, the comparison of costs and benefits of defense spending requires a theory that offers clear predictions about the net effects, or if a fully developed model is complex and ambiguous, then the issue becomes an empirical question (i.e., positive or negative impacts: Do benefits exceed costs and vice versa?). However, before considering empirical issues, there is further scope for developing the model.

6.4.4 Formulating a Model

Some issues involved in the crowding-out hypothesis can be identified through the simple application of the economist's model of a production possibility boundary or frontier. Such a boundary shows that where an economy's resources are fully and efficiently employed, *all* economic activity involves opportunity costs (i.e., alternatives are sacrificed). For an economy positioned on its production possibility frontier (PPF), an increase in defense spending means less civil goods and services for its society (e.g., a sacrifice of social welfare spending for defense). But such effects are not confined to defense. With resources fully and efficiently employed, the economy has many tradeoffs. For example, health spending involves a sacrifice of education, and government expenditure is at the expense of private spending (similarly for other activities such as dining in restaurants and leisure activities). As a result, tests of the crowding-out hypothesis have to recognize that any adverse impacts of defense spending only occur when the economy is on its PPF: If the economy is not on its PPF, then crowding out does not apply. This means that empirical work has to control for the economy's position on its PPF. Similarly, over time with economic growth, an economy's PPF will shift outward and this effect also needs to be recognized in empirical work (e.g., a positive relationship between defense spending and growth might reflect the outward shift of the PPF even though tradeoffs exist on any PPF). Such outward shifts of the PPF also raise the problem of causality, namely, whether defense spending causes growth or whether the relationship is reversed (i.e., growth determining or causing defense spending).

More complex models of the relationship between defense spending and growth have distinguished between supply-side factors (e.g., spinoffs from technology and infrastructure) and demand-side influences [e.g., crowding-out effects on investment and exports (Sandler & Hartley, 1995, Chapter 8)]. Typically, supply-side models show a positive impact of defense spending through externalities, including spinoffs, whereas demand-side models predict a negative impact through crowding-out effects. A more complete model needs to combine both demand- and supply-side influences. One example from Deger and Smith (1983) contained three equations that included growth, savings, and military expenditures. This model assumed that investment share equalled domestic savings in equilibrium (plus net foreign capital flows). However, the military expenditure equation in this model was much more limited and less satisfactory (e.g., it was not based on a demand for a military expenditure function).

6.4.5   Labor Market Crowding Out

The 1987 MoD statement of the crowding-out hypothesis focused on the labor market for qualified scientists and engineers (QSEs). The model effectively assumed that the labor supply for QSEs was perfectly inelastic so that demand increases would fail to induce any supply response. This is an extremely limited view of the labor market for QSEs: It is a static and short-run analysis, allowing for no market adjustment in either the short or the long run. For example, even in the short run with a fixed stock of QSEs, not all are working in R&D: An increase in demand and, hence, higher relative salaries will attract some QSEs from non-R&D work. Similarly, higher relative salaries will induce a long-run supply response through an increased demand for QSE training and skills in universities as well as a possible supply response from foreign-based QSEs (e.g., either through immigration and/or from U.K. firms creating R&D units overseas: Buck et al., 1993).

6.5   LITERATURE REVIEW

This literature review is presented in two parts. The first part is based on Sandler and Hartley (1995), who presented a review of the literature up to 1993. The starting point is the classic study by Benoit (1973, 1978), who found a positive relationship between defense spending and economic growth for 44 less developed countries during the period 1950–1965. This study led to a massive literature on the issue with some studies criticizing Benoit's methodology (e.g., for being *ad hoc* and lacking a theoretical basis) and others developing alternative methodologies (e.g., demand- and supply-side models and combined demand–supply models).

Sandler and Hartley report on 25 economic studies of the Benoit hypothesis from the period 1973–1993 with more than 60% based on less developed countries. There were eight studies of developed nations, including the United States, and these showed all possible relationships between defense spending and growth ranging from positive to negative to no effects! In fact, of the eight studies, only three showed a negative impact of defense spending (three showed a positive impact and two gave no impact), which does not provide convincing support for the crowding-out hypothesis.

Smith (1980) provided a comprehensive study of military spending and investment in 14 large OECD nations over the period 1954–1973. Using time-series, cross-section, and pooled data sets, he estimated a clear negative effect of military spending on investment with a coefficient on military expenditure not significantly different from −1 (i.e., a 1% increase in the share of defense spending in GDP is associated with a 1% reduction in investment shares).

The review by Sandler and Hartley (1995) concluded that:

> Although individual studies on the impact of defense on growth contain *seemingly* contradictory findings, there is greater consistency in the findings than is usually supposed. Models that included demand-side influence found that defense had a negative impact on growth. In contrast, almost every supply-side model either found a small positive defense impact or no impact at all. The findings are amazingly consistent despite differences in the sample of countries, the time-periods and the econometric estimating procedures. Because we suspect that these supply-side models exclude some negative influences of defense on growth, we must conclude that the net impact of defense on growth is negative, but small (Sandler & Hartley, 1995, p. 220).

However, this final sentence suggests scope for further research focusing on some of the negative influences excluded from supply-side models (which are these; how do we know that all positive impacts have been included and only negative impacts excluded?).

The second part of the literature review updates the Sandler/Hartley 1995 survey. It reviews literature published in *Defence and Peace Economics* since 1993 together with other relevant literature. Interestingly, this topic continues to be of major interest among defense economists. Between 1994 and 2009, more than 40 articles on this topic were published in *Defence and Peace Economics*, most focusing on developing nations and many using Granger causality estimating techniques. Some examples of the results are summarized in Table 6.3.

6.5.1   Evaluation

Table 6.3 confirms that since the mid-1990s, there have been a variety of further studies on the defense-growth relationship. Of the studies published in *Defence and Peace Economics* over the period 1995 to January 2009, some 70% were for developing nations and approximately 50% applied Granger causality estimating techniques. The studies provided support for all plausible relationships, namely, positive, negative, or no relationship. A similar range of results was obtained when the sample was restricted to the small number of developed nations. Even more interestingly, the results differed for the same country, such as Turkey and the European Union (EU) where positive, negative, and no causal ordering were found! Another study, not reported in Table 6.3, examined the defense-growth relationship among NATO nations over the period 1951–1988 (Macnair et al., 1995). It

TABLE 6.3 Examples of the Literature, 1995–2009

Authors	Model	Country	Results: Impact of defense spending on growth
Alexander, 1995 (6,1)	Feder	OECD 1966–1988	Negative but small
Madden & Haslehurst, 1995 (6,2)	Granger causality	Australia	No causal relationship in either direction
Kollias, 1997 (8,2)	Granger	Turkey 1954–1993	No causal ordering
Murdoch et al., 1997 (8,2)	Feder-Ram 3 sector	Asia and Latin America	Asia/Latin Am = ME is growth promoting. Latin Am = nondefense government spending gives greater output
Gold, 1997 (8,3)	Granger	USA 1949–1988	No evidence of long-run tradeoff between ME and Investment. Possible short-run tradeoff for 1949–1971. Long-run tradeoff between ME and consumption
Beenstock, 1998 (9,3)	Simulation	Israel 1950–1994	To 1968 = defense was growth-promoting. 1969–1975 = defense harmed growth. 1976–1986 = defense benefited growth
Kollias & Makrydakis, 2000 (11,2)	Granger	Greece 1955–1993	No causal ordering between ME and growth
Dunne et al., 2000 (11,6)	Keynesian simultaneous equation model	S. Africa 1989–1996	Negative impact but coefficients have low significance. No evidence of positive impact
Dunne et al., 2001 (12,1)	Granger	Greece and Turkey 1960–1996	Greece = no evidence of Granger causality.

(*continued overleaf*)

TABLE 6.3   *(Continued)*

Authors	Model	Country	Results: Impact of defense spending on growth
			Turkey = negative impact
Sezgin, 2001 (12, 1)	Deger type demand-supply model	Turkey 1956–1994	Positive impact. Defense spending has no significant impact on savings and trade balance
Scott, 2001 (12,4)	OLS regressions	UK 1974–1996	Crowding out of private investment
Atesoglu, 2002 (13,1)	New macro model; Granger	USA 1947–2000	Positive impact of defense spending. But effects of nondefense government spending are even larger
Al-Yousif, 2002 (13,3)	Granger	Arab Gulf 1975–1998	ME/growth relationship cannot be generalized across countries
Morales-Ramos, 2002 (13,5)	Demand-supply; Granger	Defence R& D: UK, France, Germany, Japan, USA	R&D = no impact in UK, Japan, USA. Positive impact in France and on savings (investment) in Germany
Murdoch & Sandler, 2002 (13,6)	Solow growth model	Civil wars: Africa; Asia; Latin America	Civil wars have strong negative impact on growth of income per capita
Galvin, 2003 (14,1)	Demand-supply model	64 developing nations	Negative impact on growth and savings ratio; but effects greater for middle-income nations

TABLE 6.3   (*Continued*)

Authors	Model	Country	Results: Impact of defense spending on growth
Cuaresma & Reitschuler, 2004 (15,1)	Solow growth model; nonlinearity	USA 1929–1999	Nonlinear relationships estimated = positive externalities for moderate levels of ME; but positive impact disappears at high levels of ME
Klein, 2004 (15,3)	Demand-supply model	Peru 1970–1996	Negative impact of ME on growth
Yildirim et al., 2005 (16,4)	Feder	Turkey & Middle East 1989–1999	ME enhances growth; Defense sector more productive than civilian sector
Aslam, 2007 (18,1)	Ram	59 developing countries, 1972–2000	Different results for different regions
Lee & Chen, 2007 (18,3)	Production function; Granger	27 OECD and 62 non-OECD nations, 1988–2003	Positive impact for OECD; negative impact for non-OECD and for whole panel
Kollias et al., 2007 (18,1)	Fixed-effects model	EU, 1961–2000	Positive impact
Mylondis, 2008 (19,4)	Barro growth model	EU, 1960–2000	Negative impact

Notes:

1. All references to *Defence and Peace Economics*: Figures in brackets are for volume and issue number. Examples are from the period 1995 to January, 2009. Other relevant references from *Defence and Peace Economics* are shown in the Reference list for this chapter.

2. ME is military expenditure.

found evidence of a positive relationship between defense and growth (although the explanatory value of the equations was low). Overall, the studies since the mid-1990s show no clear and convincing support for a negative impact of defense spending on growth; nor has there been a clear refutation of the Benoit hypothesis. One assessment of the literature concluded that the relationship "cannot be generalized across countries and over time. Among other things, it depends on the econometric methodology and specification employed in its empirical investigation as well as the time periods covered by the different studies" (Kollias et al., 2007, p. 75).

After all this academic research effort, the results are disappointing but perhaps not surprising in view of the diversity and complexity of all economies. A focus on new econometric estimating techniques (Granger causality) has diverted effort from the more demanding task of resolving the proper economic theoretical foundation for studying the defense-growth relationship. Nor has much attention been given to assessing the accuracy, reliability, and consistency of the data embracing either long time-series and/or a variety of vastly different nations at varying stages of development. For example, defense-spending data varies in its definition and coverage; it includes nations with conscript forces as well as volunteer forces (raising problems of estimating the market value of armed forces personnel under conscription); each nation's mix of labor, capital, and technology inputs and their efficiency will differ; and over time, nations will face different threats, which affects their defense spending. Even the simple notion of defence spending is never addressed by the research into defense-growth relationships. Different types of defense spending are likely to have differential economic impacts (e.g., spending on defense R&D, spending on equipment, and spending on foreign or domestic equipment) as well as differential lags before the spending impacts the economy (e.g., some defense spending such as R&D has long lead times). Membership of a military alliance is a further complication. These factors are ignored in most defense-growth studies.

Similar problems develop when efforts are made to construct macroeconomic models and growth models that are then applied to a variety of diverse economies at varying stages of economic development with different degrees of capacity utilization and a range of variation in their microeconomic features (e.g., state vs. private ownership, subsidies and protection, policy on competition and monopoly, as well as entrepreneurship). One authority has commented: "It is ... a very pervasive problem of the literature on the determinants of economic growth that there are so many potential explanatory variables, and so many missing data problems, that systematic analysis comparing alternative explanations is difficult to achieve" (Hughes, 2003, p. 8). Finally, where multiequation models are used, the defense equation is often limited and simplistic, bearing little relation to the standard demand for military expenditure models that characterize defense economics.

The next section reviews the literature on crowding out from defense R&D. Defense R&D is one component of defense spending and as such is likely to differ between developed and less developed nations. Developed nations usually have a national defense industrial base that receives defense R&D funding, hence, such funding is a source of difference between developed and less developed nations, confirming that defense spending is not a homogeneous entity. Defense spending is an aggregate that contains a variety of elements each with different potential impacts on the defense-growth relationship.

6.6   DEFENSE R&D AND CROWDING OUT

Defense R&D provides beneficial externalities in the form of spinoffs or spillovers. There is no shortage of examples, including the jet engine, radar, composite materials, and the Internet, with applications to such industries as health, motor cars, and Formula 1 racing cars. But such examples provide no indication of the market value of spinoffs nor whether these market values differ between defense and civil R&D (e.g., what is the market value of the Internet?).

There are some studies that have measured spillovers from all R&D. Their conclusions are that the social returns to R&D exceed the private returns. One U.S. study found that the social returns to R&D within six industries exceeded private returns by 20–200% (Bernstein and Nadiri, 1991); another U.S. study concluded that "The magnitudes of the social rates of return on R&D capital (net of depreciation) in industries which have relatively larger R&D spending propensities are about 25% to 115% greater than the net private rate of return (which is approximately 11.5%: Nadiri, 1993)." The U.K. DTI published a summary of the research in the field showing social returns to R&D considerably greater than private returns (i.e., social returns of 65% against private returns of 25%: DTI, 2003, p. 28). However, most of these studies were dated and were published over the period 1974–1993. A more recent study using U.S. data confirmed that social returns to R&D are approximately 3.5 times larger than the private returns (Bloom et al., 2005). However, all these studies were for R&D in general. It might be argued that defense R&D is no different and so is likely to offer substantial social returns, but a contrary view is that defense R&D will offer lower social returns because of its requirements for secrecy and associated restrictions on dissemination and knowledge transfer.

There have been relatively few detailed studies of defense R&D and crowding out. One study by Morales-Ramos (2002) reported results for five nations, namely, the United Kingdom, France, Germany, Japan, and the United States for 1971–1996 (i.e., 1966–1996 for the United Kingdom); it also distinguished between defense R&D and other components of defense spending. For the United Kingdom, there was no evidence of either defense R&D or non-R&D defense having any significant impact on growth. For the remaining nations, there was some evidence that defense R&D had significant and positive impacts on growth for France and on savings (investment) for Germany. However, the overall conclusion of the study was that the results were limited by statistical estimating problems (Morales-Ramos, 2002).

A similar U.K. study examined the 1987 MoD R&D crowding-out hypothesis. This study tested for R&D crowding out affecting either or both expenditure and manpower. For expenditure, there was no evidence to suggest a simple linear long-run relationship between defense and civil R&D spending and, therefore, no simple crowding out between government-funded U.K. defense R&D and civil R&D expenditure. For manpower, it was found that in 1989, defense R&D staff were paid less than their civilian equivalents in absolute and relative terms, suggesting that the defence sector was not attracting QSEs through higher salaries. But, civil industry might have to offer higher pay rates to compensate for the possible nonmonetary benefits of defense work, such as job security, status, and the opportunity for high-technology work (Buck et al., 1993). Overall, these tests of R&D crowding out were limited by data availability and by the lack of data on the life-cycle job patterns of QSEs (e.g., where do they go when leaving defense R&D: are their skills transferable?)

Another study estimated the contribution of technical change to multifactor productivity growth in 16 OECD countries over the period 1980–1998, using annual data (Guellec and Potterie, 2001). The model was based on a Cobb–Douglas production function with various measures of R&D capital stock as independent variables (i.e., business R&D, foreign R&D, and public R&D capital stocks). One set of equations introduced defense R&D variables, and these were estimated to have small but negative effects on multifactor productivity growth. However, the overall model of multifactor productivity growth was not obviously related to any standard economic model of growth and the defense variables were introduced in an *ad hoc* fashion into a broadly *ad hoc* model.

6.7 CONCLUSION

Plausible explanations can be given for a positive or negative impact of defense spending on growth, but such explanations are no substitute for a properly specified economic model showing the causal relationships. Models of economic growth do not include a defense variable so that economists have often simply added a defense term to such equations without analysis of the causal mechanisms. Growth models also need to include all other relevant influences on growth. Nor has attention been given to the composition of defense spending (i.e., the allocation among equipment, personnel, R&D, and other budget items: the distinction between current and capital spending) and to possible lag effects (e.g., for defense equipment and R&D spending). The divergent results in this field reflect the use of different econometric methods, different combinations of variables, different time periods, and an heterogeneous set of countries. In recent years, the fashion has been to use Granger causality estimating methods to the neglect of the underlying economic model and its causal relationships (Granger causality is about statistical relationships and not economic causal relationships).

A 2005 critical review of models of military expenditure and growth argued that the results in the defence economics literature have been dominated by the Feder–Ram model (finding a significant effect of military spending on growth). In contrast, the mainstream economic growth literature has not found military spending to be a significant determinant of growth. This review concluded that the Feder–Ram model ". . . is prone to theoretical misinterpretation, and the usual interpretations are mistaken; it suffers severe econometric problems, particularly simultaneity bias and lack of dynamics; and it provides too narrow a list of possible influences on growth" (Dunne et al., 2005, p. 459).[1]

Various proposals can be made for the next stage of research in this field. There remains considerable scope for focusing on the defense-growth relationship in developed nations such as France, Germany, Italy, Sweden, the United Kingdom,

---

[1] The authors concluded that the Augmented Solow and Barro models are the more promising areas for future research. A review of the literature is presented in Sandler and Hartley (1995, Chapter 8). This describes the supply-side models (Feder–Ram) and demand-side models (Keynesian) as well as a combined demand- and supply-side model.

and the United States. In-depth studies are required of the time-series experience of each nation with a careful disaggregation of the components of defense spending with specific focus on defense R&D. Possible research questions are as follows: What are the impacts of defense R&D, equipment spending, and personnel spending on growth; and what are the impacts on growth of a nation's defense industrial base?

For example, the author examined U.K. economic growth statistics over the period 1970–2000 with special interest in the 1990s as a period of a Peace Dividend after the end of the Cold War. For the period 1970–1979, U.K. GDP grew by 24%; for 1980–1989, it grew by 29%; and for 1990–1999, it grew by 22%. In other words, these simple descriptive statistics provide no support for higher growth associated with lower defense spending, but much more analysis and further research is needed before a definitive conclusion can be reached.

Empirical work in this field has been dominated by macroeconomic studies. There is scope for microeconomic studies comparing the growth and performance of defence and civilian firms. Here, there are major data problems, and further work in this area might require interview case studies (which would also identify transmission mechanisms for technology spillovers; see, for example, Watkins, 2005). Nor should any analysis neglect the main aims of defense spending, namely, peace, protection, and security for a nation's population.

REFERENCES

Alexander, R.J. (1995). Defence spending: Burden or growth-promoting, *Defence and Peace Economics*, 6(1), 13–25.

Al-Yousif, Y.K. (2002). Defense spending and economic growth: some empirical evidence from the Arab Gulf Region, *Defence and Peace Economics*, 13(3), 187–197.

Aslam, R. (2007). Measuring the peace dividend: Evidence from developing economies, *Defence and Peace Economics*, 18(1), 39–52.

Atesoglu, H.S. (2002). Defense spending promotes aggregate output in the United States—Evidence from cointegration analysis, *Defence and Peace Economics*, 13(1), 55–60.

Beenstock, M. (1998). Country Survey XI: Defence and the Israeli economy, *Defence and Peace Economics*, 9(3), 171–222.

Benoit, E. (1973). Defense and Economic Growth in Developing Countries. Boston, MA: Lexington Books.

Benoit, E. (1978). Growth and defense in developing countries. *Economic Development and Cultural Change*, 26(2), 271–80.

Bernstein, J. & Nadiri, M.I. (1991). Product demand, cost of production, spillovers and the social rate of return to R&D. *NBER Working Paper No 3625*. Cambridge, MA: National Bureau of Economic Research.

Bloom, N., Schankerman, M., & Reenen, J.V. (2005). Identifying technology spillovers and product market rivalry. *CEPR, Paper 4912*. London, UK: Center for Economic and Policy Research.

Buck, D., Hartley, K., & Hooper, N. (1993). Defence research and development, crowding-out and the peace dividend, *Defence Economics*, 4(2), 161–178.

Cuaresma, J.C. & Reitschuler, G. (2004). A non-linear defense-growth nexus? Evidence from the US economy, *Defence and Peace Economics*, *15*(1), 71–82.

Deger, S. & Smith, R. (1983). Military expenditures and growth in less developed countries. *Journal of Conflict Resolution*, *27*(2), 335–353.

DTI. (2003). *Prosperity for All: The Strategy: Analysis*. London, UK: DTI.

Dunne, P., Nikolaidou, E., & Roux, A. (2000). Defence spending and economic growth in South Africa: A supply and demand model. *Defence and Peace Economics*, *11*(6), 573–585.

Dunne, P., Nikolaidou, E. & Vougas, D. (2001). Defence spending and economic growth: A causal analysis for Greece and Turkey. *Defence and Peace Economics*, *12*(1), 5–26.

Dunne, P., Smith, R. & Wilenbockel, D. (2005). Models of military expenditure and growth: A critical review. *Defence and Peace Economics*, *16*(6), 449–462.

Galvin, H. (2003). The impact of defence spending on the growth of developing countries: A cross-section study. *Defence and Peace Economics*, *14*(1), 51–59.

Gold, D. (1997). Evaluating the trade-off between military spending and investment in the United States. *Defence and Peace Economics*, *8*(3), 251–266.

Guellic, D. & Potterie, B.P. (2001). R&D and productivity growth: Panel data analysis of 16 OECD countries. *OECD Economic Studies*, *33*(II), 103–126.

Hughes, A. (2003). Knowledge transfer, entrepreneurship and economic growth: Some reflections and implications for policy in The Netherlands, *Working Paper No. 273*. Cambridge, U.K.: ESRC Centre for Business Research, University of Cambridge.

IISS. (2008). *The Military Balance 2008*. London, U.K.: International Institute for Strategic Studies.

Klein, T. (2004). Military expenditure and economic growth: Peru 1970–1996. *Defence and Peace Economics*, *15*(3), 275–288.

Kollias, C. (1997). Defence spending and growth in Turkey 1954–1993: A causal analysis. *Defence and Peace Economics*, *8*(2), 189–204.

Kollias, C. & Makrydakis, S. (2000). A note on the causal relationship between defence spending and growth in Greece 1955–1993. *Defence and Peace Economics*, *11*(2), 173–184.

Kollias, C. Mylonidis, N. & Paleologou, S.-M. (2007). A panel data analysis of the nexus between defence spending and growth in the European Union. *Defence and Peace Economics*, *18*(1), 75–85.

Lee, C.-C., & Chen, S.-T. (2007). Do defence expenditures spur GDP? A panel analysis from OECD and non-OECD countries. *Defence and Peace Economics*, *18*(3), 265–280.

Macnair, E., Murdoch, J., Pi, C., & Sandler, T. (1995). Growth and defense: Pooled estimates for the NATO Alliance, 1951–1988. *Southern Economic Journal*, *61*(3), 846–860.

Madden, G.G. & Haslehurst, P.I. (1995). Causal analysis of Australian economic growth and military expenditure: A note. *Defence and Peace Economics*, *6*(2), 115–121.

MoD. (1987). *Statement on the Defence Estimates*, 1987. London, U.K.: Ministry of Defence, HMSO.

Morales-Ramos, E. (2002). Defence R&D expenditure: The crowding-out hypothesis. *Defence and Peace Economics*, *13*(5), 365–383.

Murdoch, J.C., Pi, C.R., & Sandler, T. (1997). The impact of defense and non-defense public spending on growth in Asia and Latin America. *Defence and Peace Economics*, *8*(2), 205–224.

Murdoch, J.C. & Sandler, T. (2002). Civil wars and economic growth: A regional comparison. *Defence and Peace Economics*, *13*(6), 451–464.

Mylonidis, M. (2008). Revisiting the nexus between military spending and growth in the European Union. *Defence and Peace Economics*, *19*(4), 265–272.

Nadiri, M.I. (1993). Innovations and Technological Spillovers, *NBER Working Paper No 4423*, Cambridge, MA: National Bureau of Economic Research.

Sandler, T., & Hartley, K. (1995). *The Economics of Defense: Cambridge Surveys of Economic Literature*. Cambridge, UK: Cambridge University Press.

Sandler, T. & Hartley, K., eds. (2007). *Handbook of Defense Economics: Volume 2, Defense in a Globalised World*. Amsterdam, The Netherlands: North Holland, Amsterdam.

Scott, J.P. (2001). Does UK defence spending crowd-out UK private sector investment? *Defence and Peace Economics*, *12*(4), 325–336.

Sezgin, S. (2001). An empirical analysis of Turkey's defence-growth relationship with a multi-equation model (1956–1994). *Defence and Peace Economics*, *12*(1), 69–86.

Smith, R. (1980). Military expenditure and investment in OECD countries, 1954–1973. *Journal of Comparative Economics*, *4*, 19–32.

Tisdell, C. & Hartley, K. (2008). *Microeconomic Policy: A New Perspective*. Cheltenham, UK: Elgar.

Watkins, T. (2005). Do Workforce and Organizational Practices Explain the Manufacturing Technology Implementation Advantages of Small Defense Contractors Over Non-Defense Establishments? Lehigh University, Bethlehem, PA.

Yildrim, J., Sezgin, S., & Ocal, N. (2005). Military expenditure and economic growth in Middle Eastern countries: A dynamic panel data analysis. *Defence and Peace Economics*, *16*(4), 283–295.

Chapter 7

Engineering Economics

WILLIAM B. ROUSE

7.1 INTRODUCTION

The objective of this chapter is to summarize and illustrate the fundamentals of engineering economics. The focus is on the concepts, principles, models, methods, and tools that are necessary to understanding subsequent chapters, as well as address the economics of human systems integration (HSI) in general. Engineering economics builds on theories and principles of economics as presented in Chapters 4–6.

However, engineering economics is much more pragmatic and focused in that the primary concern is making specific decisions about allocations of resources to creation and operation of capabilities, processes, facilities, and so on. In other words, engineering economics is less concerned with decision making in general than with framing, analyzing, and making specific decisions. For example, rather than asking whether research and development (R&D) is a good investment in general, the question typically of interest is whether to invest in a particular project.

A central theme in this chapter is the difference between monies invested to create future returns versus monies expended for operating costs. Monies invested in upstream HSI can yield substantial returns in terms of downstream savings. Savings from decreased future operating costs, decreased operational mishaps, and decreased long-term health costs should be viewed as returns on HSI investments. Hence, the economic valuation of HSI should employ investment models.

The Economics of Human Systems Integration: Valuation of Investments in People's Training and Education, Safety and Health, and Work Productivity. Edited By William B. Rouse
Copyright © 2010 John Wiley & Sons, Inc.

However, costs associated with people—recruiting, selection, training, safety, health, and so on—are often viewed as operating costs. Furthermore, R&D associated with these aspects of a system is often perceived as an operating expense or, more often, simply not pursued because new ideas and innovations are expected to occur elsewhere and then be adopted once successful. In this chapter, it is argued that improved HSI can be viewed as an investment as indicated above.

It is important to note, however, that many costs related to humans are operating costs. This includes wages and benefits, which are often the majority of an enterprise's operating costs. In contrast, those monies invested to increase productivity, decrease operating costs, and decrease longer term human-related liabilities are best viewed as investments. The case studies in later chapters of this book provide illustrations of this approach.

This chapter covers costs, cost estimation, costs of money, effects of uncertainty, and investment analysis, as well as how these topics are approached differently in the public sector. This chapter does not address a range of topics within engineering economics that are not central to subsequent chapters, for example, depreciation, taxes, and so on. Although these topics are certainly important for accounting, preparation of financial statements, and paying taxes, they are not central to the discussions in this book and are well covered elsewhere (Newman et al., 2008; White et al., 2008).

7.2   HSI INVESTMENTS

The many chapters in this book amply illustrate the operating costs and investments associated with HSI. In this chapter, the range of possibilities is only considered briefly. Thinking broadly (Rouse, 2007, 2009), these costs and investments relate to recruiting, selecting, training, and aiding humans who design, produce, operate, maintain, and manage systems. Also of concern is the safety and health of these people, as well as the habitability, sustainability, and survivability of their work environments.

Operating costs are associated with these activities and, of course, wages and benefits for the particular people involved. Investments are often associated with the efficiency and/or effectiveness of these activities. Sources of labor efficiencies include changes in the personnel mix, standardization, specialization, methods improvements, better use of equipment, changes in resource mixes, product and service redesign, and shared best practices.

We can think of these possibilities in terms of less labor hours per transaction or less expensive labor hours per transaction, and possibly no labor per transaction. Fewer hours are achievable via individual learning. Less expensive hours are achievable by, for example, substituting less skilled personnel for highly skilled personnel. In this case, the experts might be used as orchestrators of cadres of much less expensive personnel.

Labor elimination is often technology enabled. For example, Web-based scheduling and account management can enable people to substitute their labor for that of service providers, as has been experienced in the airline, banking, and retail industries. People often find this much more satisfactory than dealing with multilevel phone systems typical for many service providers.

Human-related improvements of system effectiveness tend to be more context specific as performance measures typically vary with contexts, (e.g., bombs on targets versus lives extended via health-care interventions). However, in virtually all contexts, one would like to avoid the negative consequences of human errors. These consequences almost always undermine system effectiveness.

Humans are usually included in systems because they are flexible information processors that can adapt to circumstances perhaps unforeseen in design. Humans also have good manipulative skills and abilities to feel responsible for system performance, the latter seldom seen with automation. Humans' adaptations are almost always supportive of system effectiveness. However, occasionally these adaptations have undesirable consequences. In these circumstances, we attribute these consequences to human error.

Human error is a ubiquitous explanation of airplane crashes, process plant shutdowns, and a wide variety of accidents and mishaps. One approach to this problem is to automate operations if possible or, if not possible, to proceduralize operations so that humans cannot deviate from correct task sequences. However, this approach assumes away the very reason for using flexible human information processing in the first place. There is a wide variety of systems and domains for which autonomous operations are not feasible and/or acceptable.

Another approach is needed to reduce the frequencies of consequential errors and/or develop systems that are error tolerant in the sense that the undesirable consequences of errors do not propagate. This design philosophy focuses not on deviations from "correct" task sequences but instead on the occasionally undesirable consequences of deviations. The goal is error-tolerant systems.

Reduction and/or tolerance can be accomplished with a variety of mechanisms, including selection, training, equipment design, job design, and aiding. Because no single mechanism is sufficient, a mixture of mechanisms may be needed. An important question concerns how one should allocate resources among these mechanisms to achieve acceptable frequencies of consequential errors. In other words, where should one invest monies to minimize the undermining of system effectiveness by errors?

We approached this problem by developing a mathematical model of the effects of resources on error reduction/tolerance mechanisms (Rouse, 1985, 2007). This model included numerous parametric relationships among resource investments in training, aiding, and so on, and human behaviors and errors. Extensive sensitivity analyses were performed, in part because of the lack of definitive data on many aspects of these mechanisms. One result was particularly noteworthy. Across 80 sets of parameter variations, aiding received from one sixth to one half of the total resources allocated.

This result is not really surprising. The other mechanisms focus on reducing the likelihood of all the errors that might possibly occur. Aiding, for the most part, focuses on errors that have occurred, with support that helps recovery and avoidance of consequences. These results provide clear and strong evidence for the benefits of error tolerance. An approach to designing an error-tolerant system is discussed at length elsewhere (Rouse, 1990, 2007; Rouse & Morris, 1987).

Thus, we have two examples of HSI investments that are carried forth in this chapter:

- Efficiency: Reduction of the amount and/or cost of labor required by a system
- Effectiveness: Reduction and/or mitigation of the consequences of human errors

The investments for both examples concern the R&D to determine how best to enhance efficiency and/or effectiveness, development of the interventions to accomplish these ends, and the costs of deploying these interventions. The operating costs include the ongoing costs of employing these interventions and, of course, the wages and benefits of personnel of interest. These HSI investments would yield returns in terms of reduced operating costs as well as in terms of reduced costs associated with the consequences of human errors.

7.3   COSTS AND COST ESTIMATION

Cost estimates are central to economic analysis. One needs to know what labor and materials cost, how these expenditures vary in time, and how returns are associated with investments in humans, equipment, facilities, and so on. The availability of data on which to base cost estimates is often a significant issue. Ideally, one would just retrieve such data from enterprise databases. However, this ideal is seldom realized.

Cost estimation tends to be complicated when considering new systems that provide new capabilities in new ways. Nevertheless, there is often baseline data from similar systems. For example, labor costs for maintaining a new aircraft are likely to relate to the costs of current aircraft in terms of dollars per hour for wages and benefits. In some cases, there is relevant industry and/or engineering standards, for example, labor times for automobile maintenance activities.

Another approach is to employ parametric models that relate activities to resources to costs and performance. One can then employ sensitivity analysis to identify key cost-performance tradeoffs. In this way, one determines where a lack of data is most problematic. Typically, only a small subset of relationships is sufficiently sensitive to uncertainty—relative to the decisions of interest—to warrant data collection efforts.

The costs of interest often relate to future operations, maintenance, and sustainment. Uncertainties can be magnified as one ponders the future of systems that will remain in use for many years or decades. One needs to consider inflation in the costs of labor and materials. Learning curves—for production, operations, maintenance, and sustainment—can result in decreased unit costs of products and services. Depending on the source of the learning (e.g., enhancements of productivity via proprietary technology vs. productivity enhancements that are broadly adapted), decreased unit costs may result in increased profits (for proprietary technology) or in decreased costs to consumers (nonproprietary technology).

When costs, as well as prices and profits, change over time, these estimates should be discounted to present values, as explained in the following discussion. This enables comparison of alternatives in terms of present values rather than in terms of time series of projections. Of course, as we will discuss, such discounting can present both conceptual and practical difficulties.

7.3.1 Activity-Based Costing

A common practice in most enterprises very much complicates cost estimation. This practice involves the use of an "overhead rate" that attributes costs to labor and materials beyond their direct costs. Overhead typically includes the costs of management, finance and accounting, legal, and human resources, as well as the costs of facilities and equipment. In large organizations in particular, it is not unusual for overhead rates to be 100% to 300% or more. Not surprisingly, those managing production refers to these costs as "burden."

Beginning in the late 1980s, Robin Cooper and Robert Kaplan advocated the notion of activity-based costing as an approach to addressing cost-management issues (Kaplan & Cooper, 2008). This approach involves characterizing an enterprise's processes, determining the activities associated with these processes, and assessing how these activities consume resources. The goal is to attribute all resource consumption to process-related activities.

Ideally, the overhead rate would then be zero. This ideal is seldom achieved because of difficulties attributing costs such as the chief executive's salary and benefits. However, as long as the resulting overhead rate is relatively small, the activity-based cost estimates usually provide significant insights into where profitable value is created and where it is not. It s not unusual to identify large consumers of (former) overhead that cannot be justified now that the real costs are known.

The lack of activity-based costing can present difficulties when justifying nascent initiatives such as new products and services that do not need the full spectrum of enterprise services but are nevertheless burdened with these costs via the bloated overhead rate typical of large enterprises. Consequently, such initiatives are perceived as money losers when, if costs were attributed appropriately, they are actually profitable. This is an important aspect of the "innovator's dilemma" (Christensen, 1997).

7.3.2 Life-Cycle Costing

As indicated in the introduction to this chapter, expenditures for human systems integration, in particular, and systems engineering, in general, are often made with the intent of reducing later costs of operations, maintenance, and sustainment. The goal is to optimize life-cycle costs—or total costs of ownership—rather than trying to minimize R&D and production costs.

The motivation for this goal is obvious. As an example, if we want to maximize the economic contributions of people to society, we do not want to minimize the costs of their education and health care. The latter is all too easy, but the economic and societal consequences of large numbers of illiterate and diseased people are unacceptable.

Blanchard (2008) has developed a methodology for life-cycle costing. A central construct in his methodology is a cost breakdown structure, which is akin to a work breakdown structure, which links costs with all the activities associated with a system from R&D to production and then operations, maintenance, sustainment, retirement, and disposal.

A wide range of studies of life-cycle costs have concluded that the costs subsequent to R&D and production account for roughly 75% of the total costs of ownership, with half of these costs relating to personnel costs. Thus, up-front increases in R&D related to human performance and productivity in their jobs and tasks subsequent to system deployment may have the potential to yield substantial life-cycle savings. The cost estimation challenge is to be able to project human-related costs many years into the future.

A fundamental issue associated with a life-cycle cost perspective concerns who "owns" the future. The program or product manager responsible for R&D and production of a complex system may be aware of the long-term cost implications of near-term decisions, but he or she will not be around when long-term consequences occur. In contrast, we invest in our children, despite near-term hardships, because we expect to experience the benefits of the long-term consequences of these investments. Instead, we see decision making associated with complex systems often, in effect, significantly discounting long-term consequences because there is no one "at the table" that owns these consequences.

7.4   COSTS OF MONEY

Economics is the science that focuses on how people allocate resources to produce, distribute, and consume goods and services. In this chapter, the primary focus is on allocating money. In this section, we address the valuation of time series of monetary outflows (expenditures) and inflows (incomes).

When considering an allocation of money to an investment or some activity, deciding about such an allocation usually involves comparing the choice at hand with other possible choices. One simple choice is to put the money in the bank or in low-risk bonds. These investments earn interest—often low but relatively riskless. The valuation of this alternative provides a baseline against which to compare the range of alternatives available. Other alternatives will usually provide greater returns, typically with greater risks.

The notion of interest is central to economics. When you lend money (e.g., put it in a bank or buy a bond), you expect to earn interest on this loan. Similarly, when you borrow money, you expect to pay interest on the loan. In situations where you borrow money in order to invest it, you hope that the return you earn on this investment exceeds the interest you are paying on the loan. Another way of saying this is that you want your return to exceed your cost of capital.

This seems reasonable but can get complicated in practice. One complicating factor is the need to pay interest on a loan now while the return on the investment does not occur until later. Paying an amount of money now is not equivalent to receiving the same amount later. This is because the amount paid now could have been earning interest had one not paid it, whereas the amount not received until later could not earn interest until it was received. Thus, the value of money is time dependent.

The time value of money is central to engineering economics. In general, resources invested now are worth more than the same amounts gained later. This results from the costs of the investment capital that must be paid, or foregone, while waiting for subsequent returns on the investment. The time value of money is represented by discounting the cash flows produced by the investment to reflect the interest that would, in effect at least, have to be paid on the capital borrowed to finance the investment.

Equations 7.1 and 7.2 summarize the basic calculations of the discounted cash flow model. Given projections of costs, $c_i, i = 0, 1, \ldots N$, and returns, $r_i, i = 0, 1, \ldots N$, the calculations of net present value (NPV) and internal rate of return (IRR) are straightforward elements of financial management (Brigham & Gapenski, 1988). The only subtlety is choosing a discount rate, (DR,) to reflect the current value of future returns decreasing as the time until those returns will be realized increases.

$$\text{NPV} = \sum_{i=0}^{N} (r_i - c_i)/(1 + \text{DR})^i \tag{7.1}$$

$$\text{IRR} = \text{DR such that} \sum_{i=0}^{N} (r_i - c_i)/(1 + \text{DR})^i = 0 \tag{7.2}$$

It is possible for DR to change with time, possibly reflecting expected increases in interest rates in the future. Equations 7.1 and 7.2 must be modified appropriately for time-varying discount rates.

The metrics in equations 7.1 and 7.2 are interpreted as follows:

- NPV reflects the amount one should be willing to pay now for benefits received in the future. These future benefits are discounted by the interest paid now to receive these later benefits.
- IRR, in contrast, is the value of DR if NPV is zero. This metric enables comparing alternative investments by forcing the NPV of each investment to zero. Note that this assumes a fixed interest rate and reinvestment of intermediate returns at the internal rate of return.

One should proceed with an investment if NPV is greater than zero or the IRR is greater than the cost of capital, assuming of course that there are not alternative investments with higher NPV or IRR. However, most organizations have much higher "hurdle rates" than zero for NPV or the cost of capital for IRR. This results from perceived uncertainties associated with investments.

Organizations "hedge" these uncertainties by using significantly higher hurdle rates and/or increasing DR beyond their cost of capital. The cost of capital for a company is typically 8% to 12% and for government 4% to 6%. However, it is not unusual for companies to employ discount rates of 20% to 50% and government agencies to use 8% to 10%. In this way, they heavily discount the longer term returns in equations 7.1 and 7.2. As is explained in the following discussion, this is not the best approach to address uncertainties. However, the discount rate is the only free parameter in equations 7.1 and 7.2 and, therefore, decision makers adjust it to hedge uncertainties.

7.5 EFFECTS OF UNCERTAINTY

There are no sure investments. Government bonds may be as close as it gets. For everything else, there are many uncertainties that affect what things really cost, what prices actually are paid, and what returns one's investments earn. Beyond the particular uncertainties associated with specific investments, there are broader economic uncertainties such as inflation and economic cycles.

Figure 7.1 summarizes a range of uncertainties. Markets may or may not have intentions to buy, have the necessary money, or actually make purchase decisions. Technologies may or may not work, be affordable, or perform in practice as demonstrated in R&D. Suppliers' prices are also uncertain, as is the timing of everything that must come together to result in business success. In light of all these uncertainties, the projections of revenues and profits should be expressed in terms of probability distributions rather than in terms of point estimates.

In the aerospace and defense context, contractors may develop technologies that are not mature or affordable, or they may not provide the expected performance benefits or cost savings. The government (or airline) may not proceed to contract for system production, or it may procure many fewer units than originally projected. Supplier prices may have risen, perhaps because of delays in procurement decisions. If such delays amount to multiple years, the mission for which the system was intended may change significantly or disappear.

Beyond the design, development, and deployment of a product or system depicted in Figure 7.1, there are total life-cycle costs of operations, maintenance, sustainment, and retirement of systems. As indicated, these costs often amount to 75% of the total costs of ownership of a system. The uncertainties associated with these costs relate to economic uncertainties surrounding future wages, benefits, and other personnel-related costs. There are also uncertainties associated with consumables, such as fuel, spare parts, and rent.

These uncertainties can be modeled as point probabilities or as probability distributions, depending on the nature of the phenomenon. In many situations, it can be difficult to obtain sufficient data to enable estimating probabilities or fitting distribution functions. For such situations, parametric models can be used and sensitivity

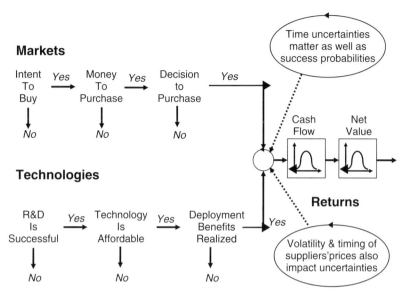

FIGURE 7.1 Types of uncertainties.

analyses employed to assess the impacts of parameter variations. As indicated, organizations can benefit significantly from ongoing collection of data to support such modeling efforts.

One reason for representing the uncertainties associated with an investment is to enable hedging against the risks posed by these uncertainties. One way to hedge risks is to buy insurance. Another way is to buy energy futures, for example, that hedge against sharp rises in fuel costs. Both of these approaches represent ways of securing a benefit if the random outcomes of all the probability distributions align in ways to result in failure.

An alternative approach is to partition the investment decision into stages such that one can terminate the investment at the end of any stage that is unsuccessful and in the event that the rosy predictions that prompted the investment in the first place are no longer attractive because of the passage of time resulting in changed circumstances. The management flexibility to terminate projects, thereby limiting downside risks, has economic value in itself, as is explicated in the next section.

7.6 INVESTMENT ANALYSIS

When one puts money in banks, bonds, or equities, the concern is not just with the costs of the deposit or purchase. One is also concerned about the likely returns in terms of interest, dividends, and appreciation. In contrast, one does not expect a return on payments of rent, utility bills, and other operating costs.

The distinction between investments and costs is central to most enterprises, as exemplified by investments appearing on balance sheets and operating costs appearing on income statements. An investment results in an asset on the balance sheet, whereas an operating cost appears as an expense on the income statement. In general, one would like to increase assets but decrease operating costs.

Enterprises often invest in new equipment that will enable employees to be more productive. Such HSI investments to increase productivity enable, in turn, higher profit margins and/or lower prices to gain market share. If, on the other hand, such investments are viewed as operating costs (which, incidentally, the Internal Revenue Service does not allow), then enterprises might not be willing to procure such productivity enhancements because this would result in large costs on their income statements that, in turn, would undermine the firm's profits and earnings per share, probably resulting in lower share prices.

Interestingly, if the enterprise invests in equipment to increase productivity, this equipment becomes an asset on the balance sheet. However, if the enterprise invests in training people to enhance productivity, this investment does not appear as an asset. It is typically viewed as an operating cost.

As discussed in Chapter 4, Mintz (1998), citing Baruch Lev, addressed this discrepancy by defining knowledge capital in contrast with tangible and financial capital. He showed how the earnings and market valuation of pharmaceutical and software companies cannot be explained by the tangible and financial assets on their balance sheets. The imputed returns on invested capital would be far too high if only tangible and financial capital is in the denominator. This denominator, Mintz argued, should also include the knowledge capital resident in the heads of employees. Despite the merits of this argument, there is not, as yet, a generally agreed upon way to determine knowledge capital, and hence, it does not appear on the balance sheet.

7.6.1 Economic Models

Economic theories and models provide a basis for valuation of investments. An enterprise's production function, f, is a specific mapping from or between the M input variables X to the production process and the output quantity produced, denoted by q.

$$q = f(X) = f(x_1, x_2, \ldots, x_M) \tag{7.3}$$

If the firm prices these products or services at p per unit, then revenue is given by pq. If each input x_i has "wage" w_i, then profit $\prod$ is given by

$$\prod(X) = pf(X) - W^T X - \text{fixed costs} \tag{7.4}$$

In principle, the firm can maximize profit by determining the levels of X that maximize $\prod$ using optimizations methods. In addition, an enterprise's production function can provide insights into how to leverage increasing returns to scale—the more units produced, the less expensive they become, which enables acquiring more units, which makes them even less expensive.

The construct of production functions can be broadly applied to design and development, manufacturing, operations, maintenance, and sustainment. For the case of manufacturing, production can be defined in the traditional sense. For design and development, production concerns the provision of engineering services to create a product or system. For operations, maintenance, and sustainment, production involves services associated with the use of the product or system.

Investments involve enhancing the relationship between X and q, [i.e., changing f(x)]. This might involve, for example, reducing labor requirements or perhaps enabling the use of lower cost labor, which would be reflected in decreased elements of W. The result would be increased $\prod$. A central issue concerns the extent of the increase in $\prod$ relative to the investment required to change f(X) and/or W. In the context of HSI, the concern is whether the cost of better HSI (i.e., improving f(X) and/or W) provides enough value in terms of $\prod$ to justify this cost.

It can be seen from equation 7.4 that $\prod$ also depends on prices, denoted by p. In some cases, prices are fixed, or when multiple enterprises are offering competing products and services, it may be possible to determine prices via a market equilibrium model. However, it is much more common for prices to be uncertain, especially multiple years into the future. There are several ways we can address such uncertainties.

7.6.2 Capital Asset Pricing Model

One approach is to employ the Capital Asset Pricing Model (CAPM; Mullins, 2008). CAPM relates the return on an asset (RA) to the risk-free rate of return (RF) and the correlation (β) of asset returns to market returns (RM) via the following equation:

$$RA = RF + \beta(RM - RF) \tag{7.5}$$

Once RA is determined, one can project future cash flows, calculate the NPV of these cash flows (equation 7.1), and if this NPV exceeds the required investment, again expressed in present value, then the investment makes economic sense.

To apply this model to HSI where returns are realized in terms of reduced downstream costs, we need to think in terms of the "market" for cost-saving initiatives and the typical returns from this market. One might find that there are different markets for savings in operations, maintenance, and sustainment, or different betas for each of these relative to the whole market.

Conceptually at least, it would seem that such models might work. The difficulty would be in finding data on which to base estimates of RM and β. It can be difficult to get data appropriate to one cost-saving initiative. Obtaining such data for the "market" of all such initiatives might be daunting. Thus, we need an approach that can focus on one particular investment opportunity.

7.6.3 Real Option Models

Consider the investment opportunities in Figure 7.2. In Figure 7.2(a), one can invest in R&D and, if the technology is successful and the market is still attractive,

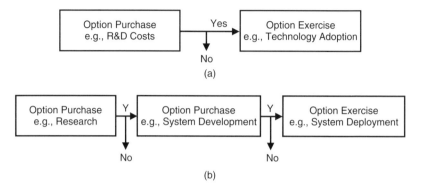

FIGURE 7.2   (a) Two-stage option. (b) Three-stage option.

proceed to adopt the technology in a product line; otherwise, the investment can be terminated. In Figure 7.2(b), there is a three-stage investment, with two possibilities for terminating the investment.

The investments, operating costs, and profits from each of these investments occur over several years, perhaps even many years. Thus, we need to employ equation 7.1 to calculate the present value (NPV) for each of these investments. However, this model assumes that we proceed with all stages of the investment, regardless of what happens after each stage. The NPV does not reflect the value of being able to terminate an investment if it "heads south." For this reason, the NPV is a conservative estimate of the value of the investment.

We need a model that attaches value to the ability to terminate investments. Thus, the "purchase" of the first stage of the investment in Figure 7.2(a) results in gaining an "option" on the second stage. This option gives one the right, but not the requirement, to exercise the option and purchase the second stage. In Figure 7.2(b), investment in the first stage provides the right to the second and third stages. Investment in the second stage provides the right to the third stage.

The value of an option equals the discounted expected value of the asset (EVA) at maturity, conditional on this value at maturity exceeding the option exercise price (OEP), minus the discounted option exercise price, all times the probability that, at maturity, the asset value is greater than the option exercise price (Smithson, 1998). The net option value (NOV) equals the option value calculated in this manner minus the discounted option purchase price (OPP). In equation form,

$$NOV = [(EAV \text{ at Maturity} \mid Value > OEP) - OEP] \text{ Prob.} \qquad (7.6)$$

$$(Value > OEP) - OPP$$

NOV is, in general, much less conservative than NPV because the downside risk is hedged by the right to terminate the investment. Simply, one does not exercise an option unless it still makes sense at the point that one can make this decision. All one loses when "walking away" is OPP. Option pricing models were originally developed for valuation of financial assets. When these models are applied to tangible assets, the options are referred to as "real options."

Theoretical treatments of real options, as well as computational methods, can be found in Dixit and Pindyck (1994) and Luenberger (1997), whereas more conceptual expositions can be found in Amram and Kulatilaka (1999) and Guerrero (2007). Real options are pursued in much greater depth in Chapter 10 of this book.

To illustrate the nature of real options, consider the two case studies summarized in Table 7.1 and drawn from Rouse and Boff (2004). For case study A, $420M was invested in R&D to "purchase" an option on a technology that, when deployed 10 years later for $72M, would yield roughly $750M of operating savings when compared with the current way of operating. The NOV of $137M represents the value of this option in excess of what they needed to invest.

For case study B, $109M was invested in R&D to "purchase" an option to deploy this technology in the marketplace four years later for an expected investment of approximately $1.7B. The expected profit was roughly $3.5B. The NOV of more than $0.5B reflects the fact that this option was purchased for much less than it was worth.

It is instructive to compare these two examples intuitively. For case study A, the option value of roughly $560M (i.e., the R&D investment plus the NOV) represents more than two thirds of the net present difference between the expected cost savings from exercising the option and the investment required to exercise it. In contrast, for case study B, the option value of more than $600M represents roughly one third of the net present difference between the expected profit from exercising the option and the investment required to exercise it. This difference (two thirds for A and one third for B) results from greater uncertainty for case study B in the 10+-year time period when most profits would accrue for either investment.

The source of this greater uncertainty is important to understand. The quotient of expected profit (or cost savings) divided by the investment required to exercise the option is different for these two examples. This quotient is roughly 10.0 for the case study A investment and 2.0 for the case study B investment. Thus, the likelihood of the option being "in the money" is significantly higher for the A than the B. There is much greater uncertainty about B yielding returns. This is why the option value is two thirds for A and one third for B.

7.6.4 Multiattribute Utility Models

The models discussed thus far only consider the financial requirements and consequences of investments, beginning with a theory of the firm (i.e., production

TABLE 7.1 Two Case Studies Using Real Options

	Option Purchase			Option Exercise			
	Investment	NPV ($M)	Duration (Years)	Investment	Exercise NPV ($M)	NPV Profit ($M)	NOV
A	R&D	420	10	Deploy System	72	749	137
B	R&D	109	4	Initiate Offering	1688	3425	546

functions) and then discussing ways to value investments. Economics also has theories of the consumer that focus on how consumers make purchase decisions. The previous discussion only considered price. However, in general, consumers consider multiple attributes, which we can denote by $y_1, y_2, \ldots, y_M$. Multiattribute utility theory considers the relationship between the set of attributes of an alternative, Y, and the consumer's relative preferences for this alternative. Equation 7.7 expresses this model.

$$U(Y) = U[u(y_1), u(y_2), \ldots u(y_L)] \tag{7.7}$$

This equation is premised on the utility of an alternative being composable from the utility functions for each attribute, $u(y_i)$. Keeney and Raiffa (1993) explore a range of possible forms of U[.] including linear (weighted sums) and multilinear (weighted sums with interaction terms), together with the assumptions necessary to justify alternative forms.

The attributes of an alternative can be deterministic, such as price, power, weight, and so on, or they may be uncertain, such as operating costs, mean time between repairs, among others. For the latter, the expected value of equation 7.7 is used, which requires knowledge of the probability density functions for the uncertain attributes. One also needs the utility functions, $u(y_i)$, to perform this calculation.

In many situations, there is no single decision maker whose preferences should be reflected by the utility functions. Instead, there may be K stakeholders whose preferences are of interest. This leads to the multiattribute, multistakeholder model as shown in equation 7.8.

$$U = U[U_1(Y), U_2(Y), \ldots U_K(Y)] \tag{7.8}$$

Keeney and Raiffa (1993) also discuss linear and multilinear forms for this equation. Determining the parameters in this equation requires knowledge of preferences across stakeholders (e.g., the relative importance of each stakeholder). This implies some higher level stakeholder whose preferences only involve the preferences of others. This could be the group of stakeholders as a whole or perhaps a "benevolent dictator."

7.6.5   Game Theory

Theories of the firm and consumer have been discussed. Also of interest is a theory of the market where firms compete and consumers seek to maximize their utility. Economists tend to look for conditions under which "market equilibrium" will occur, that is, where prices, profits, market shares, and so on are such that it is not in any stakeholder's interest to change the situation.

Elaboration of game theory is beyond the scope of this chapter. However, it is useful to illustrate the value of applying this theory to HSI-related decisions. Pennock et al. (2007b) studied the process whereby technologies are adopted by military acquisition programs. In recent years, such programs have tended to have substantial cost overruns and significant schedule slippages. The result is that older, less capable technologies remain in the field longer. Thus, the average fielded capability gets older.

As this phenomenon plays out, advocates for particular aspects of the capability of interest (e.g., radar or sonar) tend to push for adoption of newer, less mature technologies, perhaps feeling that this will be the only chance to add this technology given everything takes so long. Pennock formulated the decisions each stakeholder makes using game theory. His analysis shows that the collective decisions of each of these completely rational stakeholders results in undermining the acquisition program in terms of cost and schedule, and especially fielded capabilities.

This situation represents a "tragedy of the commons" in that individuals, all trying to do their best with regard to the interests for which they are responsible, make decisions that yield a collective compromising of the objectives they are each seeking. Game theory shows that this result is natural, not the result of incompetency or sloth. The acquisition "game" as it stands does not incentivize the behaviors needed to accomplish the overarching objectives.

7.7 PUBLIC-SECTOR ECONOMICS

Public budgets are allocated to defense—or education, libraries, and so on—with justifications that these allocations will yield "public good." It is often concluded that "returns" on these expenditures are too unpredictable to enable viewing these expenditures as investments. Consequently, budget expenditures are viewed as the operating costs of the agencies making the expenditures.

It is also often argued that returns are primarily noneconomic benefits such as safety, health, accessibility, usability, and so on. The emphasis then becomes one of securing the best cost/benefit tradeoff or the most benefits within fixed costs. An alternative approach is to focus on investments that yield fixed benefits with reduced costs—the savings then become the cash flows for the economic models discussed earlier.

The application of engineering economics concepts, principles, models, methods, and tools to public-sector issues and tradeoffs is discussed at length by White et al. (2008) and, to less an extent, by Newman et al. (2008). In this section, discussion is limited to cost/benefit analysis, cost-effectiveness analysis, and real options models applied to public-sector investments.

7.7.1 Cost/Benefit Analysis

Cost/benefit analysis concerns comparing the apparently disparate aspects of an investment (Mayhew & Bias, 1994; Rouse & Boff, 2003, 2006). This requires establishing scales of benefits and costs that enable comparisons across attributes. Utility theory offers an approach to such comparisons. This approach is discussed in this section. Chapter 9 provides another view of cost/benefit analysis.

Cost/benefit analysis should always be pursued in the context of particular decisions to be addressed. A valuable construct for facilitating an understanding of

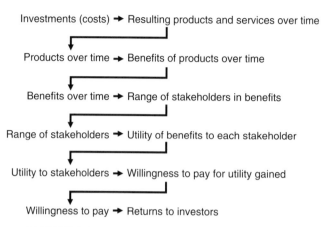

FIGURE 7.3   Value chain from investments to returns.

the context of an analysis is the value chain from investments to returns. More specifically, it is helpful to consider the value chain from investments (or costs), to products, to benefits, to stakeholders, to utility of benefits, to willingness to pay, and finally to returns on investments. Figure 7.3 depicts this value chain.

The process starts with investments that result—or will result—in particular products and services over time. Products need not be tangible end products; they might be knowledge, skills, or technologies. These products and services yield benefits, also over time. A variety of people—or stakeholders—have a stake in these benefits. These benefits provide some level of utility to each stakeholder. The utility perceived—or anticipated—by each stakeholder affects their willingness to pay for these benefits. Their willingness to pay affects their "purchase" behaviors that result in returns for investors.

The central methodological question concerns how one can predict the inputs and outputs of each element of this value chain. A variety of models has been developed for addressing this need for prediction. These models are very interesting and offer much potential. However, they suffer from a central shortcoming. With few exceptions, there is an almost overwhelming lack of data for estimating model parameters, as well as a frequent lack of adequate input data. As indicated earlier in this chapter, use of data from baselines can help, but the validity of these baselines depends on new systems and products being very much like their predecessors. Overall, the paucity of data dictates development of a more qualitative methodology whose usefulness is not totally determined by availability of hard data.

The remainder of this section outlines a seven-step methodology for cost/benefit analysis (Rouse & Boff, 2003, 2006):

1. Identify stakeholders in alternative investments.
2. Define benefits and costs of alternatives in terms of attributes.
3. Determine utility functions for attributes (benefits and costs).
4. Decide how utility functions should be combined across stakeholders.

5. Assess parameters within utility models.
6. Forecast levels of attributes (benefits and costs).
7. Calculate expected utility of alternative investments.

7.7.2   Step 1: Identify Stakeholders

The first step involves identifying the stakeholders who are of concern relative to the investments being entertained. Usually this includes all of the people in the value chain summarized earlier. This might include, for example, those who will provide the resources that will enable a solution, those who will create the solution, those who will implement the solution, and those who will benefit from the solution.

7.7.3   Step 2: Define Benefit and Cost Attributes

The next step involves defining the benefits and costs involved from the perspective of each stakeholder. These benefits and costs define the attributes of interest to the stakeholders. Usually, a hierarchy of benefits and costs emerges, with more abstract concepts at the top [e.g., viability, acceptability, and validity (Rouse, 1991, 2007)] and concrete measurable attributes at the bottom.

7.7.4   Step 3: Determine Stakeholders' Utility Functions

The value that stakeholders attach to each of these attributes is defined by stakeholders' utility functions—see equation 7.7. The utility functions enable mapping disparate benefits and costs to a common scale. As indicated, a variety of techniques are available for assessing utility functions (Keeney & Raiffa, 1993).

7.7.5   Step 4: Determine Utility Functions Across Stakeholders

Next, one determines how utility functions should be combined across stakeholders—see equation 7.8. At the very least, this involves assigning relative weights to different stakeholders' utilities. Other considerations such as desires for parity can make the ways in which utilities are combined more complicated. For example, equation 7.8 may require interaction terms to ensure all stakeholders gain some utility.

7.7.6   Step 5: Assess Parameters of Utility Functions

The next step focuses on assessing parameters within the utility models. For example, utility functions that include diminishing or accelerating increments of utility for each increment of benefit or cost involve rate parameters that must be estimated. As another instance, estimates of the weights for multistakeholder utility functions have to be estimated. Fortunately, there are a variety of standard methods for making such estimates.

7.7.7   Step 6: Forecast Levels of Attributes

With the cost/benefit model fully defined, one next must forecast levels of attributes or, in other words, benefits and costs. Thus, for each alternative investment, one must forecast the stream of benefits and costs that will result if this investment is made. Quite often, these forecasts involve probability density functions rather than point forecasts. Utility theory models can easily incorporate the impact of such uncertainties on stakeholders' risk aversions. On the other hand, information on probability density functions may not be available or may be prohibitively expensive. In these situations, the beliefs of stakeholders and subject matter experts can be employed, perhaps coupled with sensitivity analysis (see Step 7) to determine where additional data collection may be warranted.

7.7.8   Step 7: Calculate Expected Utilities

The final step involves calculating the expected utility of each alternative investment. These calculations are performed using specific forms of equations 7.7 and 7.8. The financial attributes of these models may involve using one or more of equations 7.1 through 7.6. This step also involves using sensitivity analysis to assess, for example, the extent to which the rank ordering of alternatives, by overall utility, changes as parameters and attribute levels of the model are varied.

7.7.9   Use of the Methodology

Some elements of the cost/benefit methodology just outlined are more difficult than others. The overall calculations are straightforward. The validity of the resulting numbers depends, of course, on stakeholders and attributes having been identified appropriately. It further depends on the quality of the inputs to the calculations.

These inputs include estimates of model parameters and forecasts of attribute levels. As indicated, the quality of these estimates is often compromised by lack of available data. Perhaps the most difficult data collection problems relate to situations where the impacts of investments are both uncertain and very much delayed. In such situations, it may not be clear which data should be collected and when they should be collected.

A recurring question concerns the importance that should be assigned to differences in expected utility results. If alternative A yields $U(A) = 0.648$ and alternative B yields $U(B) = 0.553$, is A really that much better than B? In fact, is either utility sufficiently great to justify an investment?

These questions are best addressed by considering past investments. For successful past investments, what would their expected utilities have been at the time of the investment decisions? Similarly, for unsuccessful past investments, what were their expected utilities at the time? Such comparisons often yield substantial insights.

Of course, the issue is not always A versus B. Quite often the primary question concerns which alternatives belong in the portfolio of investments, and which do not. Portfolio management is a fairly well-developed aspect of new product development, e.g., (Cooper et al., 1998) and (Gill et al., 1988). Well-known and recent books on R&D/technology strategy pay significant attention to portfolio selection and management [e.g., (Roussel et al., 1991), (Matheson & Matheson, 1998), (Boer, 1999), and (Allen, 2000)]. In fact, the conceptual underpinnings of option pricing theory are based on notions of market portfolios (Amram & Kulatilaka, 1999).

Most portfolio management methods rely on some scoring or ranking mechanism to decide which investments will be included in the portfolio. Expected utility is a reasonable approach to creating such scores or ranks. This is particularly useful if sensitivity analysis has been used to explore interactively the basis and validity of differences among alternatives.

A more sophisticated view of portfolio management considers interactions among alternatives in the sense that synergies between two alternatives may make both of them more attractive (Allen, 2000; Boer, 1999). Also correlated risks between two alternatives may make both of them less attractive. A good portfolio has an appropriate balance of synergies and risks.

In principle at least, the notions of portfolio synergy and risk can be handled within multiattribute utility models. This can be addressed by adding attributes that are characteristics of multiple rather than individual alternatives. In fact, such additional attributes might be used to characterize the whole portfolio. An important limitation of this approach is the likely significant increase in the complexity of the overall problem formulation. Indeed, this is an issue in general when multiattribute utility models are elaborated to better represent problem complexities.

Beyond these technical issues, it is useful to consider how this cost/benefit methodology should affect decision making. To a very great extent, the purpose of this methodology is to get the right people to have the right types of discussions and debates on the right issues at the right time. If this happens, the value of people's insights from exploring the multiattribute model usually far outweighs the importance of any particular numbers.

The practical implications of this conclusion are simple. Very often, decision making happens within working groups who view computer-generated, large-screen displays of the investment problem formulation and results as they emerge. Such groups perform sensitivity analyses to determine the critical assumptions or attribute values that are causing some alternatives to be more highly rated or ranked than others. They use "What if.. ?" analyses to explore new alternatives, especially hybrid alternatives.

This approach to investment decision making helps to decrease substantially the impact of limited data being available. Groups quickly determine which elements of the myriad of unknowns really matter—where more data are needed, and where more data, regardless of results, would not affect decisions. A robust problem formulation that can be manipulated, redesigned, and tested for sanity provides a good way for decision-making groups to reach defensible conclusions with some level of confidence and comfort. Chapter 15 presents a case study where this approach was employed.

7.7.10   Cost-Effectiveness Analysis

For some types of public-sector investments, the benefits are fixed, perhaps contractually in requirements for particular deliverables or levels of service. The question then becomes one of identifying the most cost-effective means of providing this fixed level of benefits. Cost-effectiveness analysis provides an approach to answering this question (Blanchard, 2008; Sage & Rouse, 2009).

Cost-effectiveness analysis is usually concerned with life-cycle costs or total costs of ownership. This raises many of the issues indicated in the earlier discussion of costs and cost estimation. A life-cycle or cost of ownership perspective is important for HSI investments. As noted, upstream HSI investments can yield downstream returns in terms of reduced costs of operations, maintenance, and sustainment.

Cost-effectiveness analysis often focuses on tradeoffs between the near term and long term. Often, in practice at least, near-term savings are sought at the price of long-term costs. This is particularly problematic in the public sector where there are limited market forces to correct for poor investment decisions. The public usually has to pay the taxes to live with past decisions.

One possible explanation of decisions to increase substantially long-term costs to achieve modest near-term savings is use of a very high discount rate—far beyond the risk-free rate typically used for public-sector investments. Another explanation is the fact that those responsible for the long term are not privy to near-term decisions. As indicated in earlier discussions, this reflects the lack of a balance sheet for public-sector investments.

7.7.11   Real Options in the Public Sector

This section discusses how real option models can be adapted to public-sector investments where returns on investments differ from traditional returns in private-sector markets. Chapter 10 addresses real option models in some depth. Chapter 15 presents a full case study of a public-sector application of real options.

Perhaps the most significant issue in analysis of public-sector investments is the concept of return on investment. As argued, future cost savings—relative to what costs would have been without the investment—can readily be viewed as a return on the investment. There certainly can be difficulties attributing particular savings to specific investments as well as concerns about abilities to "capture" these savings. These factors can increase uncertainties substantially. Such increases of uncertainty make option-based approaches to investment more attractive than traditional approaches.

In analyzing public-sector investments, it is important to understand what influences the magnitude, timing, and uncertainty associated with returns on investments. Consider the enterprise of military shipbuilding (Pennock et al., 2007a). This enterprise is facing serious cost challenges. Shipbuilding costs have increased enormously in the past three decades, far beyond inflation during this period. It certainly can be argued that these more expensive ships are much more capable than earlier ships. Thus, you may need fewer ships. It is possible, however, that increased costs will cause the number of ships you can buy to decrease faster than new capabilities reduce the number of ships needed. Thus, these cost challenges cannot be dismissed. This situation raises the question of where investments in shipbuilding should be focused.

As shown in Figure 7.4, the enterprise of interest includes a set of stakeholders and issues much broader than those directly associated with the ships of interest. Congress, the armed services, defense contractors, and workforce organizations have a significant impact on the magnitude and timing of returns associated with alternative investments.

As noted in Chapter 1, these stakeholders affect the shipbuilding enterprise in a variety of ways:

- Congressional interests and mandates (e.g., jobs and other economic interests)
- Service interests and oversights (e.g., procedures, documentation, and reviews)
- Incentives and rewards for contractors (e.g., cost-plus vs. firm fixed price)
- Lack of market-based competition (e.g., hiring and retention problems)
- Aging workforce and lack of attraction of jobs (e.g., outsourcing limitations and underutilization of capacity)

The example stakeholders in Figure 7.4 and their varied interests tend to introduce significant uncertainties into the enterprise in terms of both magnitudes and timing of returns. Such uncertainties strongly impact the value of potential investments in ships themselves, as well as the investments in improved processes for acquiring ships. For example, an acquisition that requires many decades until the system is

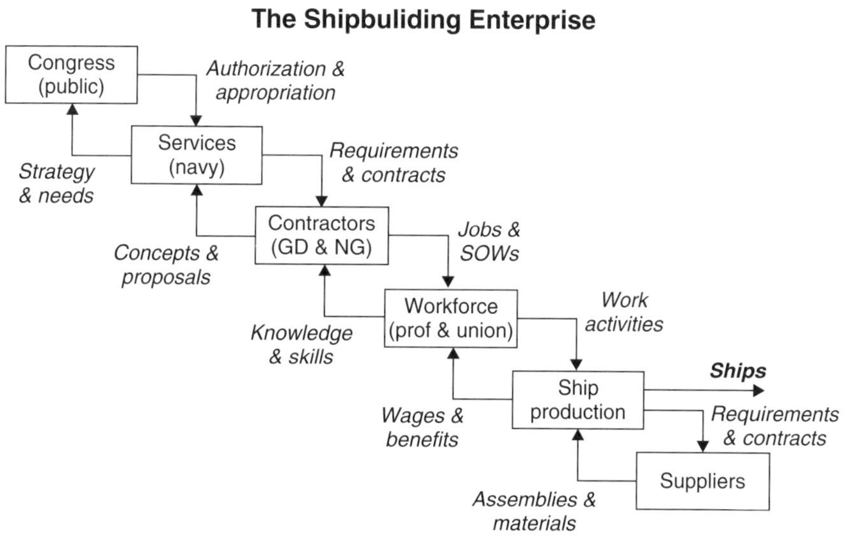

FIGURE 7.4 The overall enterprise of military shipbuilding.

deployed, results in costs and returns that are highly discounted relative to the time when an acquisition decision is made.

A variety of studies has shown that the adoption of commercial practices is not the "silver bullet" that will provide ships faster and cheaper—much more of the costs of commercial ships is in the hull than for military ships. However, there are many opportunities for fundamental change beyond the ship itself. The overall ship building enterprise could be transformed by changes in organizational processes for policy, authorization, appropriation, acquisition, development, and deployment, or of technical processes for design, production, operations, maintenance, and repair.

Thus, for example, one might accelerate the processes associated with authorization, appropriation, requirements, and contracts, while decreasing the uncertainties surrounding these processes (e.g., number of ships acquired). These changes will impact the magnitude and timing of expected cash flows. In particular, costs savings from such streamlining will be larger and realized more quickly. Notice that the examples of change just discussed do not necessarily result in the acquisition of a different ship than what would have been obtained in the slower, more uncertain way.

It is important to note that decreasing time and uncertainty also tends to affect more than just the magnitude and timing of cash flows. Accelerating processes usually decreases "requirements creep" because there is a smaller time window within which technologies can change and key stakeholders can change and/or change their minds. Furthermore, decreased time often results in decreased costs because of their being fewer calendar days over which labor costs can be charged. These impacts will reduce opportunities for rescoping of needed capabilities (e.g., changing missions and requirements), intensity of oversight (e.g., number of development reviews), and workforce sustainment (e.g., number of person-hours per ship).

7.7.12 An Example

Assume that the U.S. Navy would like to transform the way it acquires ships and, therefore, proposes several changes that will streamline the development and design process and reduce rework. Thus, the Navy has the option to transform its ship acquisition enterprise. In order to determine whether the Navy should initiate transformation, an option model was developed (Pennock et al., 2007a).

To mitigate technical risks of unsuccessful transformation, it was assumed that there would be a three-stage process:

- Stage 1: Concept development and feasibility analysis. This stage is relatively short and inexpensive. If the transformation idea proves to be infeasible in this stage, the Navy can terminate the project at no additional cost.
- Stage 2: Pilot testing the changes on the acquisition of a single ship. If the project fails in this stage, rework costs will be required to rectify the situation and complete the acquisition of the ship.

TABLE 7.2   Stage Parameter Values

Stage	Stage Cost ($ billions)	P (Success)	Rework Cost ($ billions)	Duration (years)
1	0.001	0.4	0	0.5
2	0.01	0.6	1	3
3	0.1	0.8	10	N/A

- Stage 3: Implementing the transformation across the whole shipbuilding enterprise. If the transformation fails in this stage, a substantial cost in rework is incurred.

Table 7.2 summarizes the staging parameter values for this example.

Using the real options model developed, we found that the NOV of this transformation option is approximately $0.61 billion. If we were to calculate the traditional NPV when considering this technical risk, we would find that the value of the transformation project is approximately −$6.43 billion. That means that we would expect to incur a substantial loss by initiating this project. Here we can see the discrepancy between the NOV and the NPV. The NPV is too conservative because it fails to account for the risk mitigation inherent in staging. So, in this example, a decision maker using NPV as the decision criterion would reject a potentially beneficial program.

The example can be expanded by introducing increased market risk (i.e., allowing for uncertainty in cash flows). Option values will inherently increase because options will only be exercised if the upside occurs. If the downside occurs, options will simply not be exercised. The resulting NOV is $5.94 billion, a value that is almost ten times greater than without the market risk. Hence, risk can be valuable if you can take advantage of the upside while avoiding the downside.

7.8   CONCLUSIONS

Complex systems can be very expensive to research, design, develop, and deploy. They are often even more expensive to operate, maintain, sustain, and retire. Overall, 30% to 40% of the life-cycle costs can be associated with the human and organizational aspects of these systems. Upstream investments in human systems integration can yield substantial downstream savings in life-cycle costs. From this perspective, HSI can often be viewed as an investment whose returns accrue in terms of operational savings.

This chapter has focused on engineering economics and the concepts, principles, models, methods, and tools that can support analysis of HSI investments and operating costs. Engineering economics enables a much more rigorous approach to articulating the investment value of HSI than has traditionally been employed. In general, the relatively small cost of HSI is, in fact, an investment that will yield substantial subsequent returns. These returns, in terms of cost savings, can be employed to acquire larger numbers of units of a system of interest, or possibly to design and develop the next-generation system.

A key success factor in being able to articulate the value of HSI in terms of modest upstream investments yielding substantial downstream returns is the capability to attach value to long-term consequences that, in turn, depends on someone, or some organization, having responsibility for the future. This is central to the success of investment analysis, whether the investment is defense, health care, or education.

REFERENCES

Allen, M.S. (2000). *Business Portfolio Management: Valuation, Risk Assessment, and EVA Strategies*. New York: Wiley.

Amram, M., & Kulatilaka, N. (1999). *Real Options: Managing Strategic Investment in an Uncertain World*. Boston, MA: Harvard Business School Press.

Blanchard, B.S. (2008). Cost management. In A.P. Sage & W.B. Rouse, Eds., *Handbook of Systems Engineering and Management* (2nd Edition). New York: Wiley.

Boer, F.P. (1999). *The Valuation of Technology: Business and Financial Issues in R&D*. New York: Wiley.

Brigham, E.F., & Gapenski, L.C. (1988). *Financial Management: Theory and Practice*. Chicago, IL: Dryden.

Christensen, C.M. (1997). *The Innovator's Dilemma: When New Technologies Cause Great Firms to Fail*. Boston, MA: Harvard Business School Press.

Cooper, R.G., Edgett, S.J., & Kleinschmidt, E.J. (1998). Best practices for managing R&D portfolios. *Research Technology Management*, *41*(4), 20–33.

Dixit, A., & Pindyck, R. (1994). *Investment Under Uncertainty*. Princeton, NJ: Princeton University Press.

Gill, B., Nelson, B., & Spring, S. (1996). Seven steps to new product development. In M.D. Rosenau, Jr. Ed., *The PDMA Handbook of New Product Development*. New York: Wiley.

Guerrero, R. (2007). The case for real options made simple. *Journal of Applied Corporate Finance*, *19*(2), 39–49.

Kaplan, R.S., & Cooper, R. (2008). *Activity-Based Costing*. Boston, MA: Harvard Business School Press.

Keeney, R.L., & Raiffa, H. (1976). *Decisions with Multiple Objectives: Preferences and Value Tradeoffs*. New York: Wiley.

Luenberger, D.G. (1997). *Investment Science*. Oxford, UK: Oxford University Press.

Matheson, D., & Matheson, J. (1998). *The Smart Organization: Creating Value Through Strategic R&D*. Boston, MA: Harvard Business School Press.

Mayhew, D.J., & Bias, R.G., eds. (1994). *Cost Justifying Usability*. San Francisco, CA: Morgan Kaufmann.

Mintz, S.L. (1998) A better approach to estimating knowledge capital, *CFO*, February, 29–37.

Mullins, D.W., Jr. (2008). *Does the Capital Asset Pricing Model Work?* Boston, MA: Harvard Business School Press.

Newman, D.G., Eschenbach, T.G., & Lavelle, J.P. (2008). *Engineering Economic Analysis* (10th Edition). New York: Oxford University Press.

Pennock, M.J., Rouse, W.B., & Kollar, D.L. (2007a). Transforming the acquisition enterprise: A framework for analysis and a case study of ship acquisition. *Systems Engineering*, *10*(2), 99–117.

Pennock, M.J., Rouse, W.B., & Kollar, D.L. (2007b). Development vs. deployment: How mature should a technology be before it is considered for inclusion in an acquisition program? *Proceedings of the Fourth Annual Conference on Acquisition Research*. Monterrey, CA.

Rouse, W.B. (1985). Optimal allocation of system development resources to reduce and/or tolerate human error. *IEEE Transactions on Systems, Man, and Cybernetics*, *SMC-15*(5), 620–630.

Rouse, W.B. (1990). Designing for human error: Concepts for error tolerant systems. In H.R. Booher, Ed., *MANPRINT: An approach to systems integration*. New York: Van Nostrand Reinhold.

Rouse, W.B. (1991). *Design for Success: A Human-Centered Approach to Designing Successful Products and Systems*. New York: Wiley.

Rouse, W.B. (2007). *People and Organizations: Explorations of Human-Centered Design*. New York: Wiley.

Rouse, W.B. (2009). Engineering perspectives on healthcare delivery: How can we afford technological innovation in healthcare? *Systems Research and Behavioral Science*, *26*, 1–10.

Rouse, W.B., & Boff, K.R. (2003). Cost/benefit analysis for human systems integration: Assessing and trading off economic and non-economic impacts of HSI. In H.R. Booher, Ed., *Handbook of human systems integration*. New York: Wiley.

Rouse, W.B., & Boff, K.R. (2004). Value-centered R&D organizations: Ten principles for characterizing, assessing & managing value. *Systems Engineering*, *7*(2), 167–185.

Rouse, W.B., & Boff, K.R. (2006). Cost/benefit analysis for human systems investments: Assessing and trading off economic and non-economic impacts of human factors and ergonomics. In G. Salvendy, Ed., *Handbook of Human Factors and Ergonomics*. New York: Wiley.

Rouse, W.B., & Morris, N.M. (1987). Conceptual design of a human error tolerant interface for complex engineering systems. *Automatica*, *23*(2), 231–235.

Roussel, P.A., Saad, K.N., & Erickson, T.J. (1991). *Third Generation R&D: Managing the Link to Corporate Strategy*. Cambridge, MA: Harvard Business School Press.

Sage, A.P., & Rouse, W.B. (2009). *Economic Systems Analysis and Assessment*. New York: Wiley.

Smithson, C.W. (1998). *Managing Financial Risk: A Guide to Derivative Products, Financial Engineering, and Value Maximization*. New York: McGraw-Hill.

White, J.A., Case, K.E., & Pratt, D.B. (2008). *Principles of Engineering Economic Analysis* (5th Edition). New York: Wiley.

## PART III

Models, Methods, and Tools

Chapter **8**

Parametric Cost Estimation for Human Systems Integration

RICARDO VALERDI AND KEVIN LIU

8.1 INTRODUCTION

Humans are critical to the success at every stage of the life cycle of complex systems. The International Council of Systems Engineering (INCOSE) defines human systems integration (HSI) as the interdisciplinary technical and management processes for integrating human considerations within and across all system elements; it is an essential enabler to systems engineering practice (INCOSE, 2007). In the defense industry, HSI is a comprehensive management and technical approach for addressing the human element in weapon system development and acquisition (U.S. Air Force, 2008a). By taking into account the interests of designers, operators, maintainers, and other human stakeholders, HSI can improve system performance and minimize ownership costs. Published case studies and best practices have highlighted the technical and economic benefits of successful HSI, particularly when HSI is incorporated with other systems engineering activities early in the acquisition process (Booher, 1997; Landsburg et al., 2008).

When considering the economics of human systems integration, it is useful to think about it in terms of three levels of costs:

1. The cost of doing HSI within systems engineering
2. The cost of satisfying HSI requirements and performing HSI support activities
3. The total ownership cost (and savings) impact of HSI investment

The Economics of Human Systems Integration: Valuation of Investments in People's Training and Education, Safety and Health, and Work Productivity. Edited By William B. Rouse
Copyright © 2010 John Wiley & Sons, Inc.

This chapter focuses on the first level of cost because it plays a critical role in the appropriate allocation of resources that lead to HSI success. The greatest impacts on system total ownership cost result from decisions made early in the acquisition cycle. Depending on the system, HSI considerations can dictate whether a system will stay within budget and accomplish its mission. Most organizations do not have reliable approaches for estimating the cost of doing HSI; some simply allocate between 2% and 4.2% of total acquisition cost to HSI (U.S. Air Force, 2008a). Although this approach provides minimum and maximum values for HSI investments on large programs, it does not provide insight into why certain programs need more or less HSI effort.

Although the first level of cost deals with the cost of engineering a system to take HSI into consideration, the second level addresses the other factors that contribute to acquisition cost. For example, the Department of Defense expects program managers to consider Doctrine, Organization, Training, Materiel, Leadership, Personnel, and Facilities (DOTMLPF) prior to beginning any acquisition project (CJCS, 2007).

The third level of cost considers both the cost and cost savings of doing HSI. This life-cycle view is essential in evaluating which HSI activities should be performed on a system. If organizations focus on the third level, they are explicitly taking an economic perspective that enables them to determine the return on investment of HIS efforts. This third level is the most difficult to quantify because HSI spans multiple interdependent domains that each impact cost.

To reach an understanding of HSI total ownership costs, one important question must first be answered: *What is the right amount of HSI for a given system?* Consider the notional example of a cockpit redesign. Making a cockpit more intuitive and less prone to user error could reduce mishaps and improve performance. To achieve these level 3 cost savings, level 2 costs might include cockpit electronics, training, and simulation facilities. However, level 2 and 3 costs could not be calculated without first considering level 1 efforts such as human factors analyses, tradeoff studies, and requirements engineering.

The question of "how much HSI?" is not based on economics alone. It is also driven by the technical requirements and overall complexity of the system. To understand the impact of requirements and system complexity, HSI must be treated as a subset of systems engineering. In this light, the objective of this chapter is to provide an approach that can help answer the question of "how much" through the adaptation of a systems engineering cost model (COSYSMO) to HSI. Without a reliable approach for level 1 costs, any discussion about the subsequent levels will be inherently limited.

Industry best practices and government policies claim that HSI is most effective when it is integrated as part of systems engineering activities early in the life cycle (Mack et al., 2007; Wallace et al., 2007). However, it can be difficult to generate an accurate estimate of HSI costs and return on investment without taking into account total systems engineering effort. This chapter shows that the level 1 costs of HSI can be estimated as a function of the total cost of systems engineering. Additionally, we demonstrate that HSI's impact on systems engineering is furthered by a better appreciation for HSI's impact on the number and complexity of system requirements. A variety of sources, including related literature, a case study, and an industry-validated cost model help validate our approach.

This chapter is organized into seven sections. The first provides a brief introduction of HSI, its origins, and why it is critical to understand its role in systems engineering. The second section provides an overview of generally accepted cost estimation methods. The third section provides a detailed explanation of parametric cost estimation, particularly why it is well suited for estimating HSI. The fourth section introduces the Constructive Systems Engineering Cost Model (COSYSMO), a parametric model used to estimate a systems engineering effort. The fifth section describes a case study of a complex system that illustrates a best practice of HSI. The sixth section provides an example cost estimate using COSYSMO, highlighting key elements of HSI. The final section establishes a set of recommendations for improved parametric cost estimation of HSI.

8.1.1 Origins of HSI

HSI has its origins in the field of human factors engineering (HFE), with which it is commonly confused. Human factors is the science of understanding the properties of human capability and the application of this understanding to the design and development of systems and services. Although human factors has arguably been studied since the very beginning of scientific inquiry, the technological advances of the Industrial Revolution drove modern research on how humans could best interact with machines. During this time period, innovations were also made in work and schedule management. At the time, these efforts were known as industrial engineering (Nemeth, 2004).

The challenges and requirements of industry leading up to the beginning of the 20th century grew significantly during the first and second World Wars. In response, the United States and United Kingdom both funded efforts to understand human impacts on performance (Nemeth, 2004). It is difficult to pinpoint the exact "beginning" of the field of human factors engineering, as HFE activity has been documented throughout the 20th century, sometimes under different names (Meister, 2000). However, the Human Factors and Ergonomics Society, with which HFE is commonly associated, was incorporated in 1957 (HFES, 2009).

HFE is the field that human systems integration grew from and continues to be one of its central elements. However, human systems integration as it is practiced expands on human factors engineering by incorporating a broader range of human considerations such as occupational health, training, and survivability over the system life cycle.

In 1981 and later in 1985, the U.S. General Accounting Office (since renamed the Government Accountability Office) released reports calling on the U.S. Army to improve integration of manpower, personnel, and training (MPT) into its systems acquisitions processes (GAO 1981, 1985). In response, the U.S. Army developed the Manpower and Personnel Integration (MANPRINT) program, its first directorate for HSI issues. The term itself was coined in 1984 and became an official Army Directorate in 1987 (U.S. Army, 2007).

Human factors work in the United Kingdom, Canada, Australia, and New Zealand happened in parallel with work done in the United States (U.S. Air Force, 2008a). However, as with human factors before it, current efforts to define and apply HSI to systems engineering are led by the U.S. Department of Defense. Likewise, the early adopters of HSI have been the U.S. military services and major defense contractors. Other significant programs are underway at the U.S. Federal Aviation Administration (FAA) and National Aeronautics and Space Administration (NASA), at the U.K. Ministry of Defense (MoD), and in the commercial sector.

8.1.2 The Need for Better Cost Estimation

HSI practitioners often vary in their definitions of, perspectives on, and approaches to HSI. The similarity of the designation "human systems integration" to other fields, such as human factors engineering, human factors integration, human performance enhancement, human–computer interaction, and so on can also cause confusion. As discussed, HSI evolved from the study of human factors. Human factors tools are typically used to evaluate a design later in the acquisition process. Unfortunately, this means that many engineers tend to view HSI as a means of identifying problems with a design, rather than as an enabler of good design (Booher, 2003). Although HSI analysis in the later phases of acquisition is an important part of HSI success, HSI considerations early in the life cycle can lead to lower costs (Wallace et al., 2007) and shorter acquisition cycles (Mack et al., 2007). In this chapter we provide an approach that is consistent with this philosophy because it can be used by program managers (PMs) early in the acquisition process to estimate the costs of HSI. Understanding these costs will allow PMs to better incorporate HSI considerations into their systems and to formulate effective plans for HSI activities throughout the acquisition cycle.

In the last 20 years, the U.S. military has been a strong advocate of HSI and has made it a key component of its acquisition life cycle. For this reason, we will focus our discussion on defense systems, although it should be noted that similar concepts can be applied to other types of systems.

Figure 8.1 depicts the Defense Acquisition Management System and summarizes the different phases of military acquisition. Publications from the U.S. Armed Services emphasize that the HSI effort should begin prior to Milestone A, during the Pre-Systems Acquisition phase. To comply, program managers and systems engineers both within and outside of the defense community need tools that can help them develop cost estimates with limited information. Aside from early planning and risk management, the benefit of defining HSI early in the life cycle is that less engineering effort is needed to implement changes. Such a relationship follows the S-curve of commitment of system-specific knowledge and cost (Blanchard & Fabrycky, 2005) where the cost of making changes sharply increases as a function of time.

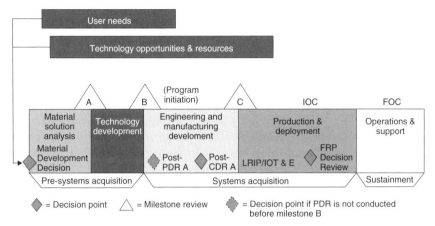

FIGURE 8.1 The Defense Acquisition Management System, adapted from U.S. Department of Defense, 2008.

To put the three cost levels described in the Introduction section into the context of the Defense Acquisition Management System, level 1 costs (doing HSI within systems engineering) and level 2 costs (satisfying HSI requirements and performing HSI support activities) are spread throughout the life cycle, while level 3 costs (total ownership cost (and savings) impact of HSI investment) are realized during the latter phases of the life cycle during Operations and Support.

Many different cost estimation techniques exist that are applicable throughout the system life cycle. The most common approaches are reviewed in the next section.

8.2 OVERVIEW OF ESTIMATION APPROACHES

Cost estimation helps program managers and systems engineers to plan their work, predict costs, and better understand the scope of the systems they develop. Cost estimation is especially important when developing systems of high complexity, cost, and duration. The best guidance on cost estimation techniques comes from organizations that have expertise in developing and acquiring these classes of systems. Industry and government guidebooks provide a rich source for best practices, lessons learned, tools, and cost estimation processes (GAU, 2009; ISPA, 2004; NASA, 2008; U.S. Army, 2002; U.S. Air Force, 2008b; U.S. Department of Defense, 1992).

Numerous cost estimation methods exist, most of which can be classified into one of the eight described here. These methods vary in both maturity and sophistication, but their application along different phases of a system's life cycle provides useful tools for sense-making in organizations. It has been shown that the best cost estimates are developed when several or all of the methods are used in combination (Jørgensen, 2004). A hybrid approach that considers each method is the best way to capture HSI impacts that a single method may overlook. As parametric cost estimation is the focus of this chapter, it is described in greater detail in a subsequent section.

8.2.1 Analogy

The estimation by analogy method capitalizes on the institutional memory of an organization to develop its estimates. This type of estimate is typically used when only one or very few historical systems similar to the new system exist. The method works best when many similarities between old and new systems exist, as in when a new system is developed using components of previous systems.

Case studies are an instrument of estimation by analogy; they represent an inductive process, whereby estimators and planners try to learn useful general lessons by extrapolation from specific examples. They examine in detail elaborate studies describing the environmental conditions and constraints that were present during the development of previous projects, the technical and managerial decisions that were made, and the final successes or failures that resulted. They then determine the underlying links between cause and effect that can be applied in other contexts. Ideally, they look for cases describing projects similar to the project for which they will be attempting to develop estimates and apply the rule of analogy that assumes previous performance is an indicator of future performance. Well-documented cases studies from other organizations doing similar kinds of work can also prove very useful so long as their differences are identified.

Later in this chapter, we provide a case study that highlights HSI domains. If an organization was developing a system similar in scope and complexity, then our case study would be a valuable comparison.

8.2.2 Bottom-up/Activity-Based Costing

The bottom-up cost estimation approach begins with the lowest level cost component and rolls it up to the highest level for its estimate. This method produces the most accurate estimates of cost but also requires the most data and is the most labor-intensive to create. A bottom-up estimate of a system's cost is created using costs reported from lower level components.

Lower level estimates are typically provided by the people who will be responsible for doing the work. This work is usually represented in the form of a work breakdown structure (WBS), which makes this estimate easily justifiable because of its close relationship to the activities required by the project elements. This can translate to a fairly accurate estimate at the lower level. The disadvantages are that this process can place additional burden on workers and is typically not uniform across entities. In addition, every level may be victim to a layer of conservative management reserve which can result in an over estimate. The approach also requires detailed cost and effort data from throughout the system, so the method cannot be used early in the development cycle.

Later in this chapter we provide an example systems engineering WBS and discuss its connection to HSI activities.

8.2.3 Expert Opinion

The expert opinion method simply involves querying experts in a specific domain and taking their subjective opinion as an input. The obvious drawback to this technique is that the estimate is only as good as the experts' opinions, which can vary greatly from person to person. Expert opinion is not always included as a scientifically valid estimation method because estimates generated using only expert opinion are the most difficult to justify and are typically only used when no other methods are available.

The benefits of this method are that experts can provide a quick estimate with minimal investment in the absence of empirical data. They can also account for other variables, such as customer demands or technology availability that other approaches may overlook. Unfortunately, having many years of experience does not always translate into the right expertise. Moreover, because this technique relies on human judgment, it has low reliability because even the most highly competent experts can be wrong.

Expert opinion is most useful for confirming and informing other cost estimation methods. For example, parametric models are often calibrated using a combination of expert opinion and historical data. The analogy method is most effective when an expert determines how best to map one system to another. The bottom-up approach depends on experts to conduct low-level analyses of cost. A common technique for capturing expert opinion is the Delphi method, which was improved and renamed Wideband Delphi (Boehm, 1981; Dalkey, 1969). These methods reduce natural human bias, improving the usefulness of data collected from experts.

8.2.4 Heuristics

Heuristic reasoning has been commonly used by engineers to arrive at quick answers to technical problems. Practicing engineers, through education, experience, and examples, accumulate a considerable body of contextual information. These experiences evolve into instinct or common sense that is seldom recorded. These can be considered insights, lessons learned, common sense, or rules of thumb, which are brought to bear in certain situations. In more precise terms, heuristics are strategies using readily accessible, although loosely applicable, information to control problem-solving in human beings and machines. Heuristics are common in psychology, philosophy, law, and engineering. Systems engineering cost estimation heuristics and rules of thumb have been developed by researchers and practitioners (Boehm et al., 2000; Honour, 2002; Rechtin, 1991; Valerdi, 2008a) as shortcuts for decision making.

Ultimately, heuristics are based on experience and often provides valuable results. However, they face the same shortfalls as expert opinion: heuristics based on past experiences may not accurately describe changing environments and heuristics are only as good as the experiences upon which they are built. As with expert opinion, heuristics are best used in combination with other cost estimation techniques.

8.2.5   Top Down and Design to Cost

The top down or design to cost (DTC) technique is most typically used when budget restrictions on a system are predefined and non-negotiable. It can be useful when a certain cost target must be reached regardless of the technical features. However, the approach can often miss the low-level nuances that can emerge in large systems. It also lacks detailed breakdown of the subcomponents that make up the system. It is up to managers and executives to ensure that standards or targets for cost set early during development are not exceeded.

In the defense acquisition community, the DTC philosophy is used to set cost targets and to make program managers more cost-conscious early in the acquisition life cycle. The method can also encompasses the use of incentives and/or awards to encourage achievement of specific production or operation and support (O&S) cost goals (Gille, 1988).

8.3   PARAMETRIC COST ESTIMATION

Parametric cost estimating dates back to World War II (NASA, 2002). The war caused a demand for military aircraft in numbers and models that far exceeded anything the aircraft industry had manufactured before. Although there had been some rudimentary work to develop parametric techniques for predicting cost, there was no widespread use of any cost estimating technique beyond a bottom-up buildup of labor hours and materials. A type of statistical estimating for the cost of airplanes was suggested in the *Journal of Aeronautical Science* (Wright, 1936). Wright provided equations that could be used to predict the cost of airplanes over long production runs, a theory that came to be called the learning curve. By the time the demand for airplanes had exploded in the early years of World War II, industrial engineers were using Wright's learning curve to predict the unit cost of airplanes. Today, parametric cost models are used for estimating a much broader spectrum of systems (Jones, 2007; NASA, 2008; USCM, 2002), including unmanned satellites, launch vehicles, solid rockets, digital signal processors, ground operations, nuclear space power, and IT systems.

Parametric models are widely used today because they provide quantifiable measures of a system's likely success and allow users to quickly see the impacts of their choices on the overall system.

8.3.1   Cost Estimating Relationships

The parametric cost estimation approach is the most sophisticated and most difficult to develop. Parametric models generate cost estimates based on mathematical relationships between independent variables (i.e., requirements) and dependent variables (i.e., effort). The inputs characterize the nature of the work to be done, plus the environmental conditions under which the work will be performed and delivered. The definition of the mathematical relationships between the independent and dependent variables is at the heart of parametric modeling. These relationships are known as cost estimating relationships (CERs) and are usually based on statistical analyses of large amounts of data. Regression models are used to validate the CERs and operationalize them in linear or nonlinear equations. Developing CERs requires a detailed understanding of the factors that affect the phenomenon being modeled, the assumptions of the model in use, and the units of measure provided by the model.

The main advantage of using parametric models is that, once validated, they are fast and easy to use. Parametric models do not require as much information as other methods, such as activity-based costing and estimation by analogy, and can provide fairly accurate estimates. Parametric models can also be tailored to a specific organization's CERs. However, some disadvantages of parametric models are that they are difficult and time consuming to develop and require a significant amount of clean, complete, and uncorrelated data to be validated properly.

The basic unit of measure for most cost models is a *person-month*. Although many parametric models are referred to as *cost* models, they are actually *effort* models because they are designed to provide an estimate of the human effort required to successfully deliver a system. In the United States, the person-month unit is equivalent to 152 *person-hours* as shown by the following logic. In one year there are 52 available workweeks. Subtract two weeks for vacation, two weeks for holidays, one week for sick leave, and one week for training. This leaves 46 weeks of available work. Assuming 40 hours per week, this results in:

$$\frac{(46 \text{ weeks / year}) \times (40 \text{ hours / week})}{(12 \text{ months / year})} = 153 \text{ hours / month} \qquad (8.1)$$

Rounded down to the nearest even number to make calculations easier and to capture the fact there are other reasons—such as travel—that a person may not be able to work, the number that is typically used is 152 hours. For some countries in Europe that follow a shorter workweek, the number of hours per person-month is 138, which means they assume that there are 36 hours of available work time each week.

The next section describes a validated parametric cost estimation model for systems engineering.

8.4 THE CONSTRUCTIVE SYSTEMS ENGINEERING COST MODEL

The COSYSMO is a parametric model used to estimate a systems engineering effort. As COSYSMO was designed to estimate systems engineering effort early in the acquisition process, it is ideally suited to include HSI considerations. The coupling between systems engineering and HSI is based on the premise that both:

- Involve intellectual engineering work

- Often lead to intangible outcomes or work products
- Have a similar relationship between system complexity metrics and effort

The systems engineering work done specifically to support HSI is often known as human systems engineering (HSE) (Beaton, 2008). The intellectual output of systems engineering, like HSI, can be difficult to quantify. Historically, common work artifacts such as system specifications, architectures, interface control documents, risk management, and test procedures have not been correlated with required effort. For this reason, systems engineering is better suited for a parametric approach where its effort can be estimated as a function of system complexity.

The critical role of systems engineering on large complex systems is widely recognized but not well understood (GAO, 2003a; Young, 2003). Additionally, the discipline of systems engineering does not have well-established metrics (Valerdi & Davidz, 2009) or methods to estimate return on investment (Boehm et al., 2008). To help address these shortfalls, COSYSMO is meant to estimate the cost of systems engineering activities throughout the life cycle with an acceptable degree of accuracy.

Before users begin to work with COSYSMO, there should be an awareness of the inherent assumptions embedded in the model. The first is that the function of systems engineering explicitly exists in an organization. In some organizations, systems engineering exists as a formal role, whereas in others, it is combined with the activities done by hardware or software engineers and even program management. In either case, the clear identification of the systems engineering function is necessary in order to take advantage of COSYSMO. The second assumption is that the organization develops large-complex systems similar to the ones developed by the organizations that participated in the definition of the model. COSYSMO was validated with input from BAE Systems, Boeing, General Dynamics, L-3 Communications, Lockheed Martin, Northrop Grumman, Raytheon, and SAIC. If the systems under consideration are similar in complexity, scope, and cost to the ones developed by these organizations, then COSYSMO will be directly applicable.

COSYSMO is also useful to government organizations that acquire systems from these types of contractors; the model can be used to (1) evaluate estimates provided in proposals, (2) manage existing systems engineering efforts, or (3) benchmark systems engineering performance across organizations. Academic researchers and industrial analysts can also use COSYSMO to model phenomena such as systems engineering reuse and productivity or perform independent cost assessments. The parameters in the model also serve as a fundamental set of metrics that help quantify systems engineering performance.

The operational concept for COSYSMO is illustrated in Figure 8.2. To use the model, estimators need to understand the expected technical capabilities of the system to be developed and make basic assumptions about the organization performing the technical work. COSYSMO requires no complex calculations on the part of the user. System characteristics are simply assigned complexity ratings such as "easy" or "difficult," and the appropriate effect on effort is calculated based on the CER. However, COSYSMO does allow more advanced users to calibrate

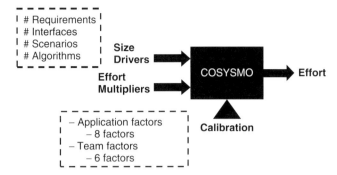

FIGURE 8.2 COSYSMO operational concept.

the model to their specific organizations in order to increase the model's accuracy. The specific parameters to COSYSMO are described in the next section.

8.4.1 Model Form

In the universe of systems engineering models, a cost model like COSYSMO belongs to a very specific class. It is considered to be a *property model* as classified in the Model-Based System Architecting and Software Engineering (MBASE) framework (Boehm & Port 1999). This is because COSYSMO focuses on the cost properties of systems and the tradeoffs between elements that affect systems engineering. The basic CER embedded in COSYSMO includes additive, multiplicative, and exponential parameters as shown in Equation 8.2.

$$PM = A \times Size^E \times \prod_{i=1}^{n} EM_i \tag{8.2}$$

where

PM = effort in person-months

A = calibration constant derived from historical project data

Size = determined by computing the weighted sum of the four size drivers

E = economy/diseconomy of scale; default is 1.0

n = number of cost drivers (14)

EM$_i$ = effort multiplier for the ith cost driver; nominal is 1.0

The general rationale for whether a factor is additive, exponential, or multiplicative comes from the following criteria (Boehm et al., 2005):

A factor is additive if it has a local effect on the included entity. For example, adding another source instruction, function point entity, requirement, module, interface, operational scenario, or algorithm to a system has mostly local additive effects.

From the additive standpoint, the impact of adding a new item would be inversely proportional to its current size. For example, adding one requirement to a system with ten requirements corresponds to a 10% increase in size, whereas adding the same single requirement to a system with 100 requirements corresponds to a 1% increase in size.

A factor is multiplicative if it has a global effect across the overall system. For example, adding another level of service requirement, development site, or incompatible customer has mostly global multiplicative effects. Consider the effect of the factor on the effort associated with the product being developed. If the size of the product is doubled and the proportional effect of that factor is also doubled, then it is a multiplicative factor. For example, introducing a high security requirement to a system with ten requirements would translate to a 40% increase in effort. Similarly, a high security requirement for a system with 100 requirements would also increase by 40%.

A factor that is exponential has both a global effect and an emergent effect for larger systems. If the effect of the factor is more influential as a function of size because of the amount of rework from architecture, risk resolution, team compatibility, or readiness for SoS integration, then it is treated as an exponential factor.

The size drivers and cost drivers of COSYSMO were determined via a Delphi exercise by a group of experts in the fields of systems engineering, software engineering, and cost estimation. The definitions for each of the drivers, while not final, attempt to cover those activities that have the greatest impact on estimated systems engineering effort and duration. These drivers are discussed in more detail the next two sections.

8.4.2   Size Drivers

It can be empirically shown that developing complex systems like a satellite ground station represents a larger systems engineering effort than developing simple systems, such as a toaster. To differentiate the two, four size drivers were developed to help quantify their relative complexities. The role of size drivers is to capture the functional size of the system from the systems engineering perspective. They represent a quantifiable characteristic that can be arrived at by objective measures.

As the focus of COSYSMO is systems engineering effort, its size drivers need to apply to software, hardware, and systems containing both. They are as follows: (1) *Number of System Requirements*, (2) *Number of System Interfaces*, (3) *Number of System-Specific Algorithms*, and (4) *Number of Operational Scenarios*. Another categorization of complexity levels is used for each, as shown in the *Number of Requirements* example in Table 8.1. The assumption is that these drivers are also reliable predictors of HSI effort. A more detailed discussion on the use of the *Number of Requirements* driver to estimate HSI effort is provided later in this chapter.

The assignment of complexity levels to size drivers is based on past experience with similar systems. To facilitate this assessment, a corresponding definition and

TABLE 8.1 Number of System Requirements Rating Scale

Easy	Medium	Difficult
Simple to implement	Familiar	Complex to implement or engineer
Traceable to source	Can be traced to source with some effort	Hard to trace to source
Little requirements overlap	Some overlap	High degree of requirements overlap

rating scale was developed for each size driver. The rating scale is divided into three sections: easy, medium, and difficult, corresponding to a complexity weight for each of the three levels. Cost drivers also are characterized in terms of a rating scale, but the focus is to describe the range of multiplicative effects they can have across the entire system.

8.4.3 Cost Drivers

A group of 14 effort multipliers have been identified as significant drivers of systems engineering effort. These are used to adjust the nominal person-month effort of the system under development. Each driver is defined by a set of rating levels and corresponding multiplier factors. The nominal level always has an effort multiplier of 1.0, which has no effect on the CER. Off-nominal ratings change the overall estimated effort based on predefined values.

Assigning ratings for these drivers is not as straightforward as the size drivers mentioned previously. The difference is that most cost drivers are qualitative in nature and require subjective assessment. A list of the 14 cost drivers is provided in Table 8.2 with the corresponding data items or information needed to assess each driver. The assumption is that these cost drivers are reliable predictors of HSI effort. A case study demonstrating how these cost drivers can influence HSI is provided later in this chapter.

The size and cost drivers described previously provide a mechanism to model the complexity of a project and to estimate systems engineering effort. As HSI is treated as a subset of systems engineering, it is adequate to adapt COSYSMO to estimate HSI effort. However, it is also important to consider the similarities between systems engineering and HSI beyond the size and cost drivers. The scope of the technical effort is highly dependent on the relevant domains of HSI for each system.

8.4.4 HSI Activities in the Context of Systems Engineering

Performing cost estimates also requires an understanding of the scope of work to be done, described in terms of a WBS. Several WBS lists for systems engineering exist, but a widely accepted one is the ANSI/EIA *Processes for Engineering a System* (1999). This set of systems engineering activities, shown in Table 8.3, is a

TABLE 8.2   Fourteen Cost Drivers and Corresponding Data Items

Driver Name	Data Item
Requirements understanding	Subjective assessment of the understanding of system requirements
Architecture understanding	Subjective assessment of the understanding of the system architecture
Level of service requirements	Subjective difficulty of satisfying the key performance parameters (i.e., reliability, maintainability, manufacturability, etc.)
Migration complexity	Influence of legacy system (if applicable)
Technology risk	Maturity, readiness, and obsolescence of technology
Documentation to match life-cycle needs	Breadth and depth of required documentation
# and diversity of installations/platforms	Sites, installations, operating environment, and diverse platforms
# of recursive levels in the design	Number of applicable levels of the work breakdown structure
Stakeholder team cohesion	Subjective assessment of all stakeholders and their ability to work together effectively
Personnel/team capability	Subjective assessment of the team's intellectual capability
Personnel experience/continuity	Subjective assessment of staff experience in the domain and consistency on the project
Process capability	CMMI level or equivalent rating
Multisite coordination	Location of stakeholders and coordination barriers
Tool support	Subjective assessment of SE tools

CMMI = capability maturity model integration; SE = systems engineering.

useful framework for conceptualizing systems engineering in a project. The standard is reproduced here to emphasize the relevance of HSI to systems engineering throughout the system life cycle.

The primary objective of HSI in system acquisition is to influence design with requirements and constraints associated with human performance and accommodation. The way in which this is accomplished is through several initiatives (Malone & Carson, 2003):

- Identify human performance issues and concerns early in system acquisition.
- Define the roles of humans in system operations and maintenance early in system development.
- Identify deficiencies and lessons learned in baseline comparison systems.
- Apply simulation and prototyping early in system design to develop and assess HSI concepts.

TABLE 8.3   Systems Engineering Activities (ANSI/EIA 632, 1999)

Fundamental Processes	Process Categories	Activities
Acquisition and Supply	Supply Process Acquisition Process	(1) Product Supply (2) Product Acquisition (3) Supplier Performance
Technical Management	Planning Process	(4) Process Implementation Strategy (5) Technical Effort Definition (6) Schedule and Organization (7) Technical Plans (8) Work Directives
	Assessment Process	(9) Progress Against Plans and Schedules (10) Progress Against Requirements (11) Technical Reviews
	Control Process	(12) Outcomes Management (13) Information Dissemination
System Design	Requirements Definition Process	(14) Acquirer Requirements (15) Other Stakeholder Requirements (16) System Technical Requirements
	Solution Definition Process	(17) Logical Solution Representations (18) Physical Solution Representations (19) Specified Requirements
Product Realization	Implementation Process	(20) Implementation
	Transition-to-Use Process	(21) Transition to Use
Technical Evaluation	Systems Analysis Process	(22) Effectiveness Analysis (23) Tradeoff Analysis (24) Risk Analysis
	Requirements Validation Process	(25) Requirement Statements Validation (26) Acquirer Requirements (27) Other Stakeholder Requirements (28) System Technical Requirements (29) Logical Solution Representations
	System Verification Process	(30) Design Solution Verification (31) End-Product Verification (32) Enabling Product Readiness
	End-Product Validation Process	(33) End-product Validation

- Optimize system manning, training, safety, survivability, and quality of life.
- Apply human-centered design.
- Apply human-centered test and evaluation.

These initiatives are high-level descriptions of what needs to be done. The WBS in Table 8.3 provides a more detailed view of how this work can be accomplished. For instance, the first initiative involving human performance issues and concerns early in the life cycle can be carried out by several detailed activities listed in Table 8.3: technical plans, system technical requirements, implementation, transition to use, and so on.

HSI considerations in systems engineering are brought to life through a case study that illustrates how an organization carried out HSI successfully. The next section describes a case of an actual large military acquisition program where HSI made a significant impact to systems processes, life-cycle cost, and system performance.

8.5 HUMAN SYSTEMS INTEGRATION CASE STUDY: F119-PW-100 ENGINE

As discussed in previous sections of this chapter, a primary user of parametric cost estimation is the U.S. Department of Defense and its major contractors. Large defense projects require a significant systems engineering effort that can quickly drive up costs. At the same time, defense projects typically have high requirements for survivability, safety, and other human considerations. The DoD is interested in human systems integration as a means of reducing cost (Wallace et al., 2007), shortening acquisition cycles (Mack et al., 2007), and improving system performance (DoD, 5000.02). We chose to do an analysis of a major defense program from both the HSI and the cost estimation perspectives in order to improve our understanding of how parametric cost estimation tools like COSYSMO can account for HSI activities in systems engineering. A longer version of this case study is published in Liu et al. (2009) and also discussed in Liu et al. (2010).

8.5.1 Methodology

This case study documents HSI activities performes during the development of Pratt & Whitney's F119 engine, which powers the $143M Lockheed Martin F-22 Raptor fighter aircraft (Drew, 2008). The F-22 raptor fulfills the air superiority role in the Air Force by using a package of technologies to deliver "first look, first shot, first kill capability in all environments" (U.S. Air Force, 2008c). Although the Air Force HSI Office was not formalized until 2007, much of the work performed on the F-22 (see Figure 8.3) and the F119 in the 1980s and 1990s spans the domains of HSI, making the F119 a best source of practices in HSI for the Air Force.

In designing the study, we followed Yin's (2003) approach for identifying five important components to case study design: (1) a study's questions, (2) its

FIGURE 8.3   The F-22 Raptor (Dunaway, 2008).

propositions, (3) its units of analysis, (4) the logic linking the data to the propositions, and (5) the criteria for interpreting the findings.

The case study was designed around three central research questions:

(1) How did Pratt & Whitney determine how much HSI effort would be needed?

(2) How much did HSI effort eventually cost?

(3) How did HSI fit into the larger systems engineering picture?

Because we sought to describe *how* the F119 became a best practice of HSI, we designed our study as a single-case descriptive study. Our proposition was that HSI effort could be isolated from the larger systems engineering effort spent. We hoped to establish a quantitative relationship between HSI cost and systems engineering cost. We sought to analyze the early development of the F119, from concept development until major engineering and manufacturing development (EMD). Although HSI activities would continue to be important after EMD, the HSI activities that affected the design of the F119 largely occurred prior to EMD. The engineering organization responsible for HSI on the F119 at Pratt & Whitney was our unit of analysis.

As historical data on specific costs associated with HSI activities was not available either because data were not kept or the records could not be found, we depended on Pratt & Whitney employees familiar with the F119 to build an understanding of its development. We conducted a series of interviews with Pratt & Whitney engineers who were active in the development of the F119, in both technical and management roles. Because our central proposition was that HSI cost could be isolated from systems engineering cost, in our interviews, we focused both on life-cycle cost measurement as well as on systems engineering and HSI methodology. With this information, we could establish the general cost estimation approaches (see previous sections of this chapter) that Pratt & Whitney used and reach conclusions about HSI's role in systems engineering. We concluded the case study by validating our results using existing literature on the F119 and the F-22 and by comparing the results of our interviews from multiple engineers.

8.5.2   Early Air Force Emphasis on Reliability and Maintainability

The Defense Resources Board approved the creation of the Advanced Tactical Fighter (ATF) program in November 1981 to create a military jet that would be able to guarantee air superiority against the Soviet Union. This fighter was meant to replace the F-15 Eagle, which had previously filled this role. A team composed of Lockheed, Boeing, and General Dynamics competed against Northrop Grumman to develop the fighter. In 1991, the ATF contract was awarded to the Lockheed team's F-22, powered by Pratt & Whitney's F119 engine (shown in Figure 8.4). Then Secretary of the Air Force Donald Rice noted that an important consideration in the awarding of the contract was the fact that the F-22's engines offered superior reliability and maintainability (Bolkcom, 2007).

The Air Force placed an emphasis on reliability and maintainability from the beginning of the ATF program as well as of the Joint Advanced Fighter Engine (JAFE) program—the program to develop the engine for the ATF. In June 1983, four general officers representing the Army, Navy, and Air Force signed a joint agreement to "emphasize to the DoD and defense contractor communities the critical importance of improving operational system availability by making weapon system readiness and support enhancement high priority areas for all our research and development activities" (Keith et al., 1983, pg 1). Later that year, the director of the JAFE program sent a memorandum to participants in the program, including Pratt & Whitney, asking them to consider that more than 50% of the Air Force budget was then devoted to logistics, and that the problem would only worsen (Reynolds, 1983).

To address this increase in logistics cost and determine ways to develop creative solutions, the Air Force created the Reliability, Maintainability & Sustainability

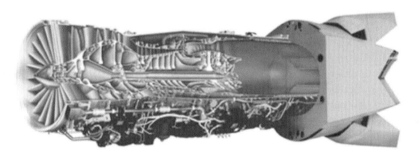

FIGURE 8.4   Cutaway of the F119 engine (Pratt & Whitney, 2003).

(RM&S) program in 1984 (Gillette, 1994). Besides reducing life-cycle cost, the RM&S program also sought to address the reliability and durability problems that had plagued Pratt & Whitney's previous F100 engine, which powered the Air Force's F-15 Eagle. Developed in the 1970s, the F-15 was developed specifically to counter the Russian MiG-25. Therefore, emphasis was placed on performance during the development of both the F-15 and the F100. Unfortunately, the high performance of the F100 meant that the engine was more prone to failure and downtime. By the 1980s, the Russian air superiority threat was no longer as pressing as when the F-15 was developed and supportability was emphasized over performance. As a result, the Air Force wanted improved RM&S not only on the F119 engine but also on development of the F-22 as a whole. Specific supportability goals for the F-22 were announced as early as 1983 (Aronstein et al., 1998).

8.5.3 Understanding Customer Needs

The F-22 engine competition was not the only instance in which Pratt & Whitney had competed with General Electric. Both companies had developed engines to power the Air Force's F-16 Fighting Falcon. In the end, GE provided the majority of engines for that platform. Pratt & Whitney saw success in the JAFE program as critical to the company's ability to continue to compete in the military engine market. For the F119 engine, Pratt & Whitney decided to not only to meet the Air Force's RM&S requirements but also to emphasize designing for the maintainer throughout all aspects of the program. The company's approach exemplified the best practices of what is now known as human systems integration.

Pratt & Whitney conducted approximately 200 trade studies as contracted deliverables for the Air Force. Pratt & Whitney engineers also estimated they had conducted thousands of informal trade studies for internal use. These trade studies used evaluation criteria, including safety, supportability, reliability, maintainability, operability and stability, and manpower, personnel, and training (Deskin & Yankel, 2002).

Figures of merit were developed for the trade studies to define a consistent set of criteria against which to assess the trade studies. Pratt & Whitney engineers used these figures of merit to determine which engineering groups would participate in each trade study.

As is often the case in the development of complex defense systems, responsibilities for the various domains of HSI are distributed among many different organizations at Pratt & Whitney. Of the nine domains of HSI[1], seven were represented in Pratt & Whitney's engineering groups. Maintainability, Survivability, Safety, Training, and Materials were all engineering groups at Pratt & Whitney. manpower, personnel, and human factors engineering were taken into account by the Maintainability group. Human factors engineering also impacted the Safety group. Occupational health was considered by both the Safety group and the Materials group, which dealt with hazardous materials as one of its responsibilities.

---

[1]Environment, safety, manpower, personnel, training, human factors, habitability, survivability, and occupational health.

Although there was an Environmental Health and Safety (EH&S) group at Pratt & Whitney, it dealt with EH&S within the organization itself and did not impact engine design. Habitability was not an important consideration in the engine design.

8.5.4 Integrated Product Development

The major requirements for RM&S came from the Air Force. These requirements went to both GE and Pratt & Whitney and were decomposed to lower level requirements internally. The JAFE program in particular was intended to improve RM&S by "reducing the parts count, eliminating maintenance nuisances such as safety wire, reducing special-use tools, using common fasteners, improving durability, improving diagnostics, etc." (Aronstein et al., 1998, pg 247). Although General Electric made significant RM&S improvements to its F120 engine during this time period, Pratt & Whitney centered its competitive strategy on RM&S superiority.

During the JAFE competition, Pratt & Whitney participated in the Air Force's "Blue Two" program. The name refers to the involvement of maintenance workers in the Air Force—"blue-suiters." The program brought Pratt & Whitney engineers to Air Force maintenance facilities so that the engine designers could experience first hand the challenges created for maintainers by their designs. Maintainers showed how tools were poorly designed, manuals had unclear instructions, and jobs supposedly meant for one person took two or more to complete safely.

One of the most important requirements for the F119 was that only five hand tools should be used to service the entire engine. All line replaceable units (LRUs) would have to be "one-deep," meaning that the engine would have to be serviceable without removal of any other LRUs and each LRU would have to be removable using a single tool within a 20-minute window (Gillette, 1994). Maintenance would have to be possible while wearing hazardous environment protection clothing. Maintenance tasks would have to accommodate maintainers from the 5th percentile female and 95th percentile male as shown in Figure 8.5 (Aronstein et al., 1998, pg 226). In addition:

> Built-in test and diagnostics were integrated with the aircraft support system, eliminating the need for a special engine support system. Lockwire was eliminated, and torque wrenches were no longer required for "B" nut installations. The engine was designed with built-in threadless borescope ports, axially split cases, oil sight gauges, and integrated diagnostics. Other improvements were a modular design..., color-coded harnesses, interchangeable components, quick disconnects, automated integrated maintenance system, no component rigging, no trim required, computer-based training, electronic technical orders, and foreign object damage and corrosion resistant. These advances were intended to reduce operational level and intermediate level maintenance items by 75% and depot level tools by 60%, with a 40% reduction in average tool weight.

These innovations were only possible by the use of the integrated product development (IPD) concept. Whereas on previous projects, engineering groups at Pratt & Whitney each worked in their own respective disciplines, under IPD,

FIGURE 8.5 Tool design to accommodate maintainers (Gillette, 1994).

teams of engineers from varying disciplines were able to provide design engineers with the perspectives they needed to see the full impacts of their design decisions.

8.5.5 Continuing Accountability and Enforcement of HSI

Adoption of the IPD concept brought various stakeholders together early in the design process and ensured multidisciplinary input through design and development. As a matter of policy, whenever a design change needed to be made, the originating group would submit the change to be reviewed by a configuration control board (CCB). CCBs were composed of senior engineers from multiple engineering groups. At CCB meetings, each group with a stake in a particular design change would explain the impacts of that change to the chair of the CCB, typically a design engineer. The chair would then weigh the different considerations of the design change and either approve/disapprove the change or recommend additional analysis be performed.

In instances when Air Force requirements needed to be changed, the originating group would submit a Component Integration Change Request (CICR), which would then be internally debated much like with design changes. CICRs were typically initiated when it was determined that a particular requirement might not be in the best interests of the customer or when one requirement conflicted with another. Once a CICR was finalized internally by all of Pratt & Whitney's engineering groups, it was presented to the Air Force, which would then make the final decision on whether a requirement could be eliminated, modified, or waived. The processes for design and requirement change ensured that the work of one group did not create unforeseen problems for another. However, change requests were typically made in response to problems that originated during development.

Other processes were introduced at Pratt & Whitney to ensure continued commitment to HSI. All part design drawings were required to be annotated with the tools needed to service that part. This helped to achieve the goal of being able to service the entire engine with only five hand tools (in the end, the F119 required five two-sided hand tools and one other tool, which was still a significant improvement over previous engines).

Several full-scale mock-ups of the F119 were commissioned. These mock-ups came at a considerable cost (more than $2M a piece, whereas the cost of an engine was then approximately $7M) but allowed engineers to test whether their designs had actually achieved maintainability goals. Engineers were asked to service LRUs on the mock-ups by hand to ensure that they were each indeed only "one-deep." When an LRU was shown to not meet that requirement, the teams responsible for those LRUs were asked to redesign them.

8.5.6   HSI Efforts Lead to Competition Success

Leading up to the major EMD contracts awarded in 1991, Pratt & Whitney conducted 400 distinct demonstrations of the F119's RM&S features. The F119 also accrued more than 110,000 hours of component tests and 3,000 hours of full-up engine tests, representing a 30 times increase in total test hours over its predecessor, the F100 (Aronstein et al., 1998). Pratt & Whitney was willing to spend significant effort on demonstrating the F119's RM&S features because the company had recently been beat out by GE in their competition to provide engines for the Air Force's F-16 Fighting Falcon and therefore saw the JAFE competition as its last chance to stay in the military engine market.

Both Pratt & Whitney and General Electric were awarded contracts worth $290 million to complete the EMD phase of competition. The companies were given independence as to the number and type of tests that would be run on their engines, while the Air Force provided safety oversight. As a result, Pratt & Whitney chose to log approximately 50% more test hours than General Electric (Aronstein et al., 1998).

GE chose to emphasize the performance of its F120 engine over RM&S, although the F120 did meet the Air Force's RM&S requirements. The F120 was the world's first flyable variable cycle engine (Hasselrot & Montgomerie, 2005). This meant that the F120 could change from turbofan to turbojet configuration to achieve maximum performance in multiple flight situations. The F120 was tested in both Lockheed's YF-22 and Northrop Grumman's YF-23 prototypes, demonstrating better maximum speed and supercruise than Pratt & Whitney's F119 in both cases (Aronstein et al., 1998). The dry weight of the F119 is classified, making it impossible to calculate its exact thrust-to-weight ratio. However, Pratt & Whitney advertises the F119 as a 35,000-lb thrust class engine, putting it into the same thrust class as the F120 (Gunston, 2007).

Despite the F120's superior performance in the air and higher thrust-to-weight ratio, on April 23, 1991, the Air Force chose the combination of Pratt & Whitney's

F119 and Lockheed's YF-22 to be developed into the F-22. Pratt & Whitney had repeatedly demonstrated a better understanding of the Air Force's RM&S needs, investing more time and money into demonstrations and internal efforts than its competitor. It also avoided the increased risk of developing a variable cycle engine, at the time considered a relatively new and untested technology. By 1991, the Air Force's RM&S program was less focused on reducing downtime and more concerned with reducing life-cycle costs. Pratt & Whitney had presented a management plan and development schedule that the Air Force considered sensitive to their needs (Aronstein et al., 1998). On August 2, 1991, contracts worth $11 billion were awarded to Lockheed and Pratt & Whitney (Bolkcom, 2007) demonstrating the Air Force's commitment to HSI. Pratt & Whitney's portion was worth $1.375 billion alone (Aronstein et al., 1998).

8.5.7 Case Study Observations

In this case study, we document an example of successful human systems integration. It is clear that HSI strongly influenced the development of Pratt & Whitney's F119 turbofan engine from early in the acquisition life cycle through engineering manufacturing and development. It is also clear that many traditional systems engineering activities were impacted.

The Air Force's early and continuing emphasis on RM&S was captured via requirements. In 2003, the GAO advocated for more equal consideration of reliability and maintainability in requirements definition (GAO, 2003b). Our case study showed that the Air Force already understood this principle a decade prior. The Air Force's initial guidance to emphasize RM&S shaped the design approach of all of its contractors. The specific activities Pratt & Whitney did to ensure HSI was represented design—such as trade studies, integrated product development, and engine mock-ups—were described in the case study.

Conversations with Pratt & Whitney engineers indicated that by the time HSI requirements were integrated into the engine, the cost of specific HSI activities could no longer be distinguished from other systems engineering costs. Pratt & Whitney estimated the cost of the F119 using their records of costs from a previous engine, an example of estimation by analogy. When new requirements needed to be met in response to RM&S concerns, the projected cost of those requirements was simply added to historical costs. The projected costs were estimated using a combination of expert opinion and modeling.

This case study represents a first step toward a quantitative understanding of the costs of HSI in the context of systems engineering. We could not isolate HSI costs from systems engineering. Instead we showed that the two disciplines were tightly coupled, reinforcing the fact that COSYSMO can be adapted to address HSI considerations. Because we observed that HSI activities were mostly driven by specific requirements, in the next section, we discuss in more detail how requirements are incorporated into COSYSMO and provide an example of how COSYSMO can be used to estimate the costs of HSI.

8.6   EXAMPLE COST ESTIMATE USING COSYSMO

The Pratt & Whitney F119 engine case study provides a good example of the role of multiple HSI domains in successful system design. To demonstrate how HSI effort can be estimated, we provide an example calculation based notionally on the information presented in the case study. This is done in two parts. The first part is a description of how the *Number of Requirements* parameter in COSYSMO should be used to ensure consistent estimates. The second part provides an illustration of how COSYSMO helps arrive at a calculation of effort and cost of systems engineering and HSI.

8.6.1   Number of Requirements Size Driver

The *size drivers* subsection of this chapter discusses the general role of size drivers in COSYSMO. A system like the F-22 fighter jet has thousands of requirements that are decomposed into requirements pertaining to subsystems. In this case, we treat the F119 engine as a subsystem of the F-22 fighter. Naturally, not all requirements for the F119 have the same level of complexity. Some may be more complex than others based on how well they are specified, how easily they are traceable to their source, and how much they overlap with other requirements. A simple sum of the total number of requirements would not be a reliable indicator of functional size of the F119. Instead, the sum of the requirements requires a complexity weight to reflect the corresponding complexity of each requirement. The meaning and implications of complexity play an important role in estimating systems engineering and HSI effort.

Of the four COSYSMO size drivers, *Number of Requirements* is the best measure of HSI effort within systems engineering. However, stakeholders often disagree on how to designate a requirement's complexity. This is due in part to the different types of requirements (i.e., functional, operational, and environmental) that are used to define systems and their functions, the different levels of requirements decomposition used by organizations, and the varying degree of quality of requirements definition (how well they are written).

HSI adds to the challenges in defining the *Number of Requirements* size driver. Many problems faced when trying to count requirements consistently for the

TABLE 8.4   Number of System Requirements Rating Scale

Easy	Medium	Difficult
Simple to implement	Familiar	Complex to implement or engineer
Traceable to source	Can be traced to source with some effort	Hard to trace to source
Little requirements overlap	Some overlap	High degree of requirements overlap

purposes of cost estimation are more complicated when the requirements span multiple HSI domains. In addition, the definition of an HSI requirement can vary between stakeholders (Liu et al., 2009). The COSYSMO definition of the *Number of Requirements* size driver is provided in the box. Its corresponding rating scale was shown previously as Table 8.1 and reproduced here as Table 8.4 for convenience.

Number of System Requirements
This driver represents the number of requirements for the system-of-interest at a specific level of design. The quantity of requirements includes those related to the effort involved in system engineering the system interfaces, system specific algorithms, and operational scenarios. Requirements may be functional, performance, feature, or service-oriented in nature depending on the methodology used for specification. They may also be defined by the customer or contractor. Each requirement may have effort associated with it, such as verification and validation, functional decomposition, functional allocation, and so on. System requirements can typically be quantified by counting the number of applicable shalls/wills/shoulds/mays in the system or marketing specification. Note: Some work is involved in decomposing requirements so that they may be counted at the appropriate system-of-interest.

The definition of a system requirement does not explicitly state which HSI domains are relevant, but this should be noted during the process of identifying requirements. In some cases, requirements are readily available and do not require much effort to allocate to systems engineering and HSI. In other cases, determining the number of requirements is not as straightforward. It is helpful to understand the requirements decomposition process for the benefit of systems engineering and HSI.

8.6.2 Decomposition of Requirements

A system specification may contain many different types of technical requirements varying in nature and complexity. Their decomposition[2] from high-level objectives to low-level technical detail is a critical front-end process often referred to as top-down requirements analysis (Malone & Carson, 2003). In this process, it is convenient to think of requirements as belonging to either functional or nonfunctional types.

Functional requirements are the fundamental or essential subject matter of the system. They describe what the product has to do or what processing actions it is to take. An example of a functional requirement is as follows: "The engine shall provide a thrust-to-weight ratio of T." Each functional requirement should have a

---

[2]It is important to distinguish between *decomposed* and *derived* requirements. For the purposes of COSYSMO, we focus on *decomposed* requirements because they are a better proxy for HSI effort.

criterion or use case. These serve as benchmarks to allow the systems engineer to determine whether the implemented product has met the requirement.

Nonfunctional requirements are the properties that the system must have, such as performance and usability. These requirements are as important as functional requirements to a product's success but are not always weighted accordingly (GAO, 2003b). HSI requirements are more likely to be perceived as nonfunctional requirements because they describe usability as well as operational and maintainability characteristics of the system. Other nontechnical requirements that may have significant influences on the complexity of a project include cost/schedule constraints, business-related forces, and political issues. They are not directly treated in the COSYSMO model but are important nonetheless.

The classification of requirements into functional and nonfunctional types does not capture all of the nuances in describing the complexity of a system. Additional work is involved in decomposing requirements so that they may be counted at the appropriate system-of-interest. We provide rules to help clarify the definition and adjustment factors while providing consistent interpretations of the size drivers for use in cost estimation. Other data items or sources may be available on certain projects depending on the processes used in the organization. For example, system requirements may be counted from the requirements verification matrix or from a requirements management tool.

The challenge with requirements is that they can be specified by either the customer or the contractor. In addition, requirements can be specified at different levels of decomposition and with different levels of sophistication. Customers may provide high-level requirements in the form of system capabilities, objectives, or measures of effectiveness; these need to be translated into requirements by the contractor and decomposed into numerous levels as illustrated by the framework in Figure 8.6.

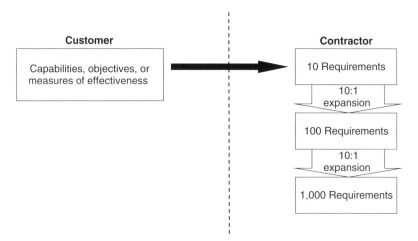

FIGURE 8.6   Notional example of requirements decomposition.

For the purposes of this example, the expansion ratio from one level of require-ment decomposition to the other is assumed to be 10:1. Different systems will exhibit different levels of requirements decomposition depending on the appli-cation domain, the customer's ability to write good system requirements, and the complexity of the system. The requirements flow framework in Figure 8.6 provides a baseline for the process of counting requirements.

Additional rules were designed during the development of COSYSMO to increase the reliability of requirements counting by different organizations on different systems regardless of their application domain. The five rules are as follows:

1. **Determine the system-of-interest.** For an airplane, the system-of-interest may be the avionics subsystem, the engine, or the entire airplane depending on the perspective of the organization interested in estimating HSI. This key decision needs to be made early on to determine the scope of the COSYSMO estimate and to identify the requirements that are applicable for the chosen system.

2. **Decompose system objectives, capabilities, or measures of effectiveness into requirements that can be tested, verified, or designed.** The decomposition of requirements must be performed by the organization using COSYSMO. The level of decomposition of interest for COSYSMO is the level in which the system will be designed and tested.

3. **Provide a graphical or narrative representation of the system-of-interest and how it relates to the rest of the system.** This step focuses on the hierarchical relationship between the system elements. This information can help describe the size of the system and its levels of design. It serves as a sanity check for the previous two steps. In some cases, DODAF diagrams are an adequate approach (U.S. Department of Defense, 2007).

4. **Count the number of requirements in the system/marketing specification or the verification test matrix for the level of design in which systems engineering is taking place, given the desired system-of-interest.** The focus of the counted requirements needs to be for HSI within systems engi-neering. Lower level requirements may not be applicable if they have no effect on HSI domains. Requirements may be counted from the Requirements Verification Trace Matrix (RVTM) that is used for testing system require-ments. The same rules apply as before: all counted requirements must be at the same design level, whereas lower level requirements should not be counted if they do not influence HSI effort.

5. **Determine the complexity of requirements.** Once the quantity of require-ments has been determined, a complexity rating of *easy, medium*, or *difficult* must be determined. The numerical weights for these factors were determined using expert opinion through the use of a Delphi survey (Valerdi, 2008b).

The objective of the five steps is to lead users down a consistent path of similar logic when determining the number of system requirements for the purposes of HSI effort in COSYSMO. It has been found that the level of decomposition described in rule number 2 may be the most volatile because of the diversity of roles played by HSI. To alleviate this, an example for software use case decomposition was adopted (Cockburn, 2001).

The basic premise behind the Cockburn hierarchy is that different levels exist for specific system functions. Choosing the appropriate level can provide a focused basis for describing the customer and developer needs. A metaphor is used to describe four levels: *sky level, kite level, sea level*, and *underwater level*. The summary level, or *sky level*, represents the highest level that describes the system scope.

We can apply the hierarchy to the example of the F119, as shown in Figure 8.7. A *sky level* goal for the system would have been to "deliver air-to-air combat capability." The stakeholders of the system, the U.S. Air Force, articulated this as their fundamental need that in turn drives a collection of user level goals. A *kite level* goal provides more detailed information as to "how" the *sky level* goal will be satisfied. For example, the Air Force emphasized reduced life-cycle cost and better reliability, maintainability, and supportability. The *sea level* goals represent a user level task that is the target level for counting HSI requirements in COSYSMO. It may describe maintenance, training, and human factors requirements that will enable the accurate estimation of HSI effort, also providing more information on how the higher goals at the *kite level* will be satisfied. The *sea level* is also important because it describes the environment in which the HSI community interacts with end users and stakeholders. The last step below *seal level* is the *underwater level*, which is of most concern to the developer. Most of the actions performed by Pratt & Whitney to achieve Air Force HSI requirements fall into this level.

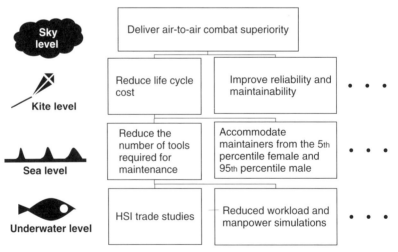

FIGURE 8.7 Cockburn's hierarchy as related to F119 use case levels (adapted from Cockburn, 2001).

Going down the hierarchy from *sky* to *underwater* provides information on "how" a particular requirement will be satisfied by the system, whereas going up the hierarchy provides information on "why" a particular lower level requirement exists.

It is worthwhile to point out that the different levels shown in this hierarchy depend greatly on the system-of-interest, the first rule of requirements decomposition, as discussed earlier in this section. For example, if the system-of-interest had been a single component on the F119 rather than the engine itself, a *kite level* goal may have been to reduce maintenance tools, whereas a *sea level* goal might have been to reduce the number of types of bolts used on the component. COSYSMO can be applied to different systems of interest, but it should be kept in mind that the model was calibrated to large defense and aerospace projects and would need a new calibration to work for different types of systems.

8.6.3   Cost Estimation Calculation

The assumptions in the F119 case study mirror a typical scenario in which a customer provides a system specification and requests an estimate of HSI effort from a contractor. At this stage, it is assumed that the information needed to populate the COSYSMO inputs is readily available, although this may be the exception rather than the rule.

For the purposes of this exercise we have extracted some details from the F119 cases study to demonstrate how an HSI cost estimate could have been performed using COSYSMO. To begin with, the systems engineering effort must be estimated using the information provided in the system specification. At this stage, we exclude any HSI considerations so that we can later demonstrate the marginal cost of doing HSI. For the purposes of systems engineering, we reference the system specification, which contains several hundred requirements. The first step is to decompose these requirements provided by the customer down to the appropriate level for systems engineering. After decomposition, let us assume that the requirements provided by the customer yield 500 systems engineering requirements at the *sea level*.

The next step is to allocate the decomposed requirements into one of the three available complexity levels in COSYSMO. Through additional dialog with the customer, a review of the system specification, and discussion with experts in who have worked on similar systems, it is determined that the 500 decomposed requirements can be allocated as 200 *easy*, 200 *medium*, and 100 *difficult* requirements.

These quantities are entered into COSYSMO as size drivers. At this stage, the model provides an initial systems engineering person-month estimate of 300 person-months based solely on the size parameters, as shown in Figure 8.8. To keep this example simple, we assume that all of the cost drivers are unaffected, which means they remain at their default rating of "nominal" and have no effect on the effort calculation.

For the purposes of this example, it is assumed that the project being estimated includes the standard systems engineering life-cycle phases and a standard systems

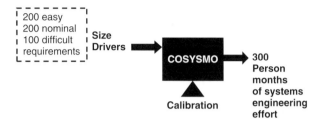

FIGURE 8.8   Systems engineering estimate with COSYSMO.

engineering work breakdown structure, as shown in Table 8.3. Tailoring this predetermined scope is necessary to ensure the relevance of the cost estimate. Moreover, the estimate for systems engineering effort can be translated into a need for people and time. For instance, 30 person-months might mean 10 systems engineers would need to be assigned for a period of 30 months to complete all of the necessary tasks.

The introduction of HSI considerations would change the initial estimate of 300 person-months in several ways. First, the introduction of HSI requirements would increase the requirements count. Second, HSI considerations would affect some of the cost drivers. Thinking back to the F119 engine, suppose that the introduction of HSI requirements increased the total requirements count by 20, half of which were considered medium and half of which were considered difficult. This would bring the total systems engineering and HSI requirements to 200 *easy*, 210 *medium*, and 110 *difficult* requirements.

Additional information about the F119 could be used to adjust the estimate using some of the cost drivers listed in Table 8.2. First, a high maintainability requirement was stated by the primary stakeholder (Aronstein, et al., 1998; Reynolds, 1983). This key performance parameter results in a *high* degree of *level of service requirements*. Second, there would likely have been a need for a sophisticated suite of tools to perform HSI safety and occupational health analyses. Tools are critical for performing modeling and simulation in HSI and can lead to significant payoffs (Malone et al., 1998) and reduction in life-cycle cost (Stanco & Malesich, 1999). Assuming tools with a high maturity level could be acquired or developed, a *high* rating could be assigned to the *tools support* cost driver.

Both *level of service requirements* and *tools support* are cost drivers of COSYSMO that have a multiplicative impact on cost, as described in this chapter. Setting the first driver to *high* would increase systems engineering effort, whereas setting the second driver similarly would decrease effort. The amount of impact is defined within COSYSMO based on calibration from large defense and aerospace projects. This additional project information would also provide deeper insight into the project's potential performance as well as into possible risk factors that could introduce schedule or cost variation. In summary, the information obtained from the system specification—supplemented by additional dialog with the customer and discussion with experts familiar with similar efforts—would provide the necessary information to populate HSI-specific information in COSYSMO.

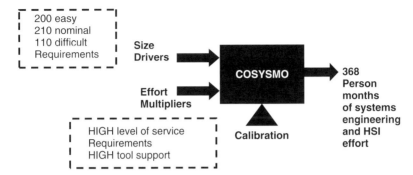

FIGURE 8.9   Systems engineering estimate with COSYSMO (includes HSI).

The additional HSI requirements and the two cost drivers are combined to provide an estimate in COSYSMO of 368 person-months, as shown in Figure 8.9. This represents an increased systems engineering effort of 22% in order to satisfy HSI requirements. The absolute number is not as important as the relative increase from HSI requirements because it helps illustrate the impact of HSI considerations on systems engineering.

The power of cost estimation often comes not from the number generated by a model but from the process through which the estimate is reached. This section highlighted some of the key ways in which HSI can affect systems engineering cost and shows how a parametric cost estimation tool like COSYSMO can translate those impacts into relative differences in cost. It also draws attention to the importance of understanding systems engineering requirements and their decomposition. The next section offers some concluding thoughts and provides guidelines for future development.

8.7   CONCLUSION AND GUIDELINES FOR PROFESSIONAL PRACTICE

Developing and calibrating cost models not only improves the validity of the model itself but also contributes to an iterative learning cycle. Lessons learned have been captured during both the model development (Valerdi et al., 2004) and the industrial validation (Valerdi et al., 2007) of COSYSMO. Those lessons learned have led to significant improvements to the model. The lessons most pertinent to HSI and our case study of the F119 are discussed as follows:

1. A standardized WBS and dictionary provides the foundation for decisions on what is within the scope of the model for both data collection and estimating.
   Most HSI and human factors modeling tools address factors such as fatigue, reliability, and safety. The ability of COSYSMO to model the effort required of HSI practitioners and systems engineers to complete HSI tasks within systems engineering is dependent on the ability to define the scope of

HSI activities accurately. A better understanding of what is meant by HSI and "HSI requirement" will help ensure consistent interpretation and estimation.

2. The collection of the size driver parameters requires access to project technical documentation as well as project systems engineering staff that can help interpret the content.

   The guidelines for counting requirements and example cost estimate provided in this chapter can assist in producing consistent evaluations of system requirements. As awareness and standardization in the field of requirements engineering improve, the cost estimates produced by COSYSMO will become more accurate. When defining and elaborating HSI requirements, representatives from all relevant domains of HSI should provide input.

3. Understanding COSYSMO's usability will lead to more reliable inputs to the model especially at the early phases of the life cycle where there is little project information available.

   The discussion thus far has been under the assumption that organizations have HSI requirements readily available and have the ability to interpret their relative complexity. The diverse range of HSI domains makes this a difficult task because of the separation of interests that divide engineering specialties. As these interests converge and begin to operate under a single HSI umbrella, the estimation of HSI effort will become more accurate.

4. Detailed counting rules can ensure that size drivers, specifically requirements, are counted consistently across the diverse set of systems engineering projects, hence improving the model's application across organizations.

   This chapter has repeatedly emphasized the importance of consistently counting requirements. Counting of HSI requirements can improve as a result of better understanding of the impacts across domains of an HSI requirement by both the organization doing the counting and the HSI community at large.

5. Guidance on rating complexity easy, medium, or difficult is necessary to ensure consistent cost estimation across organizations.

   Although the observations in this chapter tend to focus on consistent counting of requirements, it may be the case that the current *Number of Requirements* complexity rating cannot sufficiently capture the impact of HSI requirements. Input from subject matter experts has suggested a number of possible directions for future research. These include counting a requirement multiple times in relation to the number of HSI domains it affects, assigning different weights to functional and nonfunctional requirements, and internalizing the effect of HSI in the *Number of Major Interfaces* size driver.

In this chapter, we introduced several cost estimation approaches, highlighted one implementation of parametric cost estimation, and showed how that tool could be used to increase understanding of the cost of HSI within systems engineering. We mostly focused on the concept of *Number of System Requirements* because our case study of the F119 engine identified requirements development and decomposition as having been a driving factor behind successful HSI practice. We also focused on two cost drivers that were clearly linked to HSI requirements. Our example showed how changing the ratings of those cost drivers affected systems engineering cost.

Future research will undoubtedly show that HSI impacts size and cost drivers beyond the ones specifically addressed in this chapter. However, the work presented here demonstrates both the importance and the feasibility of applying cost estimation approaches to HSI. This information should not be considered a comprehensive guide but one way of approaching the level 1 costs of HSI within systems engineering, and a stepping stone to understanding the cost of satisfying HSI requirements and performing HSI support activities (level 2 costs) and the total ownership cost and savings impact of HSI investment (level 3 costs).

Perhaps the most important lesson from this chapter is the fact that HSI will continue to be an important factor in system acquisition in terms both of system performance and life-cycle cost in the foreseeable future. Historically, HSI requirements have been afforded less importance because they are represented as nonfunctional requirements. The COSYSMO model provides a framework for measuring the impact of nonfunctional requirements on HSI and on the system.

COSYSMO counts HSI requirements as equal to any other requirement. HSI requirements could even possibly be weighted more heavily than other requirements because they have the potential to span multiple domains, which would be represented with a *high* complexity rating. The COSYSMO model establishes a means for quantitative justification and methodology for accurately accounting for the impact of HSI on systems engineering.

Future research on adapting the COSYSMO model to HSI will focus on HSI's impact on the other size and cost drivers of the model. In order to better understand level 3 and ROI costs, there needs to be strong agreement on the measurement of benefit and cost savings (Ahram & Karwowski 2009; U.S. Air Force, 2008a). As these areas mature, the management and implementation of HSI on large complex systems will benefit from economic modeling tools in support successful systems.

8.8 ACKNOWLEDGMENTS

The authors gratefully acknowledge support for this research provided through the U.S. Air Force Human Systems Integration Office, the MIT Systems Engineering Advancement Research Initiative, and the MIT Lean Advancement Initiative. The authors would also like to thank the employees at Pratt & Whitney for their participation in the case study.

REFERENCES

Ahram, T.Z., & Karwowski, W. (2009). Measuring human systems integration return on investment, in *INCOSE Spring Conference*, Suffolk, VA.

ANSI/EIA. (1999). *ANSI/EIA-632-1988 Processes for Engineering a System*. New York: American National Standards Institute.

Aronstein, D.C., Hirschberg, M.J., & Piccirillo, A.C. (1998). *Advanced Tactical Fighter to F-22 Raptor: Origins of the 21ˢᵗ Century Air Dominance Fighter*. Reston, VA: American Institute of Aeronautics and Astronautics.

Beaton, R.J. (2008). *Human Systems Engineering Best Practices Guide*. Washington, DC: Naval Sea Systems Command, Human Systems Integration Group.

Blanchard, B.S., & Fabrycky, W.J. (2005). *Systems Engineering and Analysis* (4ᵗʰ Edition). Upper Saddle River, NJ: Prentice Hall.

Boehm, B. (1981). *Software Engineering Economics*. Upper Saddle River, NJ: Prentice Hall.

Boehm, B., Abts, C., Brown A.W., Chulani, S., Clark, B.K., Horowitz, E., Madachy, R., Reifer, D.J., & Steece, B. (2000). *Cost Estimation with COCOMO II*. Upper Saddle River, NJ: Prentice Hall.

Boehm, B.W. & Port, D. (1999). Escaping the software tar pit: Model clashes and how to avoid them. *ACM Software Engineering Notes*, *24*(1), 36–48.

Boehm, B., Valerdi, R., Lane, J., & Brown, A.W. (2005). COCOMO suite methodology and evolution. *CrossTalk—The Journal of Defense Software Engineering*, *18*(4), 20–25.

Boehm, B., Valerdi, R., & Honour, E. (2008). The ROI of systems engineering: Some quantitative results for software-intensive systems. *Systems Engineering*, *11*(3), 221–234.

Bolkcom, C. (2007). *CRS Report for Congress: F-22A Raptor*. Congressional Research Service. Washington, DC.

Booher, H.R. (1997), Human factors integration: Cost of and performance benefits to army systems. *Army Research Laboratory Report ARL-CR341*, Aberdeen Proving Ground, MD.

Booher, H.R. (2003). *Handbook of Human Systems Integration*. Hoboken, NJ: Wiley-Interscience.

Chairman of the Joint Chiefs of Staff (CJCS). (2007). Operation of the joint capabilities integration and development system. In *Chairman of the Joint Chiefs of Staff Manual 3170.01C*. Washington, DC: Government Printing Office.

Cockburn, A. (2001). *Writing Effective Use Cases*. New York: Addison-Wesley.

Dalkey, N.C. (1969). *The Delphi Method: An Experimental Study of Group Opinion*. Santa Monica, CA: The RAND Corporation.

Deskin, W.J., & Yankel, J.J. (2002). Development of the F-22 propulsion system, in *38th AIAA/ASME/SAE/ASEE Joint Propulsion Conference and Exhibit*. Indianapolis, IN.

Drew, C. (2008). A fighter jet's fate poses a quandary for Obama. *New York Times*. http://www.nytimes.com/2008/12/10/us/politics/10jets.html.

Dunaway, A. (2008). Flying high. *US Air Force Photo*. http://www.af.mil/shared/media/photodb/photos/090123-F-2828D-942.JPG.

General Accounting Office. (1981). *Effectiveness of US Forces Can Be Increased Through Improved Systems Design*. *(Rep PSADSI-17)*, Washington, DC: Government Printing Office.

General Accounting Office. (1985). *The Army Can Better Integrate Manpower, Personnel, and Training into the Weapons Systems Acquisition Process*. Washington, DC: Government Printing Office.

Government Accountability Office. (2003a). *Defense Acquisitions Improvements Needed in Space Systems Acquisition Management Policy*. Washington, DC: Government Printing Office.

Government Accountability Office. (2003b). *Best Practices: Setting Requirements Differently Could Reduce Weapon Systems' Total Ownership Costs*. Washington, DC: Government Printing Office.

Government Accountability Office. (2009). *GAO Cost Estimating And Assessment Guide: Best Practices for Developing and Managing Capital Program Costs*. Washington, DC: Government Printing Office.

Gille, W.H. (1988). *Design to Cost Handbook*. St. Louis, MO: US Army Troop Support Command, Directorate for Resource Management, Cost Analysis Division.

Gillette, Jr, F.C. (1994). Engine design for mechanics. *SAE International*.

Gunston, B. (2007). *Jane's Aero-Engines*. Alexandria, VA: Jane's Information Group Incorporated.

Hasselrot, A., & Montgomerie, B. (2005). *An Overview of Propulsion Systems for Flying Vehicles*. Stockholm, Sweden: Swedish Defense Research Agency.

Honour, E.C. (2002). Toward an understanding of the value of systems engineering, in the 1st *Annual Conference on Systems Integration*, Hoboken, NJ.

Human Factors and Ergonomics Society (HFES). (2009). About HFES: HFES History, *Human Factors and Ergonomics Society Website*. http://www.hfes.org/web/AboutHFES/history.html.

International Council on Systems Engineering (INCOSE). (2007). *INCOSE Systems Engineering Handbook, version 3.1*. http://www.incose.org.

International Society of Parametric Analysts (ISPA) and the Society of Cost Estimating and Analysis (SCEA). (2004). *Parametric Estimating Handbook* (4th Edition). Vienna, VA: ISPA/SCEA Joint Office. http://www.ispa-cost.org.

Jones, C. (2007). *Estimating Software Costs*. New York: McGraw-Hill.

Jørgensen, M. (2004). A review of studies on expert estimation of software development effort. *Journal of Systems and Software*, *70*(1–2), 37–60.

Keith, D.R., Williams Jr., J.G., Mullins, J.P., & Marsh, R.T. (1983). Joint agreement on increased R&D for readiness and support. Alexandria, VA: *Department of the Army*; Washington, DC: Department of the Navy; Wright-Patterson AFB, OH: Department of the Air Force.

Liu, K.K. Valerdi, R., & Rhodes, D.H. (2009). Economics of human systems integration: A systems engineering perspective, in *Proceedings of Conference on Systems Engineering Research*, Loughborough, UK.

Liu, K.K., Valerdi, R., Rhodes, D.H., Kimm, L., & Headen, A. (2010). The F119 engine: A success story of human systems integration in acquisition. *Defense Acquisition Review Journal*, *17*(2), 284–301.

Landsburg, A.C., Avery, L., Beaton, R., Bost, J.R., Comperatore, C., Khandpur, R., Malone, T.B., Parker, C. Popkin, S., & Sheridan, T.B. (2008). The art of successfully applying human systems integration. *Naval Engineers Journal*, *120*(1), 77–107.

Mack, D.D., Higgins, L.A., & Shattuck, L.G. (2007). Applying human systems integration to the rapid acquisition process. *Naval Engineers Journal*, *119*(1), 97–108.

Malone, T.B., Baker, C.C., Kirkpatrick, M., Anderson, D.E., Bost, J.R., Williams, C.D., Walker, S.A., & Hu T.H.G. (1998). Payoffs and challenges of human systems integration (HSI) modeling and simulations in a virtual environment. *Naval Engineers Journal*, *110*(4), 21–37.

Malone, T.B., & Carson, F. (2003). HSI top down requirements analysis. *Naval Engineers Journal*, *115*(2), 37–48.

Meister, D. (2000), *The History of Human Factors and Ergonomics*. Mahwah, NJ: Erlbaum.

National Aeronautics and Space Administration (NASA). (2002). *Parametric Cost Estimating Handbook*. http://cost.jsc.nasa.gov/PCEHHTML/pceh.htm.

National Aeronautics and Space Administration NASA. (2008). *2008 NASA Cost Estimating Handbook*. Washington, DC: NASA Headquarters, Cost Analysis Division.

Nemeth, C.P. (2004). *Human Factors Methods for Design: Making Systems Human-Centered*. Boca Raton, FL: CRC Press.

Pratt & Whitney. (2003). F119 cutaway. *Pratt & Whitney's F119 Receives ISR Approval from USAF, Surpasses 4,000 flight Hours, Demonstrates Unprecedented Reliability*. http://www.pw.utc.com/StaticFiles/Pratt%20&%20Whitney/News/Press%20 Releases/Assets/Images/f119_low3.jpg.

Rechtin, E. (1991). *Systems Architecting: Creating & Building Complex Systems*. Upper Saddle River, NJ: Prentice Hall.

Reynolds, J.C. (1983). JAFE field visit. *Air Force Memorandum*. Wright-Patterson AFB, OH.

Stanco, J., & Malesich, M. (1999). Reducing CV life cycle costs through process modeling and simulation. *Naval Engineers Journal*, *111*(3), 359–370.

U.S. Air Force. (2008a). *Air Force Human Systems Integration Handbook*. Washington, DC: 711 Human Performance Wing, Directorate of Human Performance Integration, Human Performance Optimization Division.

U.S. Air Force. (2008b). *Air Force Cost Analysis Handbook*. Arlington, VA: Air Force Cost Analysis Agency.

U.S. Air Force. (2008c). *Fact Sheet: F-22 Raptor*. http://www.af.mil/factsheets/ factsheet.asp?fsID=199.

U.S. Army. (2002). *Department of the Army Cost Analysis Manual*. US Army Cost and Economic Analysis Center. Washington, DC: Government Printing Office.

U.S. Army. (2007). MANPRINT History. *US Army MANPRINT Program*. http://www.manprint.army.mil/manprint/history.html.

U.S. Department of Defense. (1992). *Cost Analysis Guidance and Procedures*. Department of Defense 5000.4-M. Washington, DC: Government Printing Office.

U.S. Department of Defense. (2007). Volume I: Definitions and guidelines. *DoD Architecture Framework Version 1.5*. Washington, DC: Government Printing Office.

U.S. Department of Defense. (2008). Operation of the defense acquisition system. *Department of Defense Instruction 5000.02*. Washington, DC: Government Printing Office.

USCM. (2002). *USCM8 Knowledge Management System*. Goleta, CA: Tecolote Research, Inc.

Valerdi, R., Rieff, J., Roedler, G., & Wheaton, M. (2004). Lessons learned from collecting systems engineering data, in *2nd Conference on Systems Engineering Research*, Los Angeles, CA.

Valerdi, R., Rieff, J., Roedler, G., & Wheaton, M. (2007). Lessons learned from industrial validation of COSYSMO, in *17th INCOSE Symposium*, San Diego, CA.

Valerdi, R. (2008a). Zen in the art of cost estimation, in *2nd Asia-Pacific Conference on Systems Engineering*, Yokohama, Japan.

Valerdi, R. (2008b). *The Constructive Systems Engineering Cost Model (COSYSMO)*. Saarbrücken, Germany: VDM Verlag.

Valerdi, R., & Davidz, H.L. (2009). Empirical research in systems engineering: Challenges and opportunities of a new frontier. *Systems Engineering*, *12*(2).

Wallace, D.F., Bost, J.R., Thurber, J.B., & Hamburger, P.S. (2007). Importance of addressing human systems integration issues early in the science & technology process. *Naval Engineers Journal*, *119*(1), 59–64.

Wright, T. (1936). Factors affecting the cost of airplanes. *Journal of Aeronautical Science*, *3*(4), 122–128.

Yin, R.K. (2003). *Case Study Research: Design and Methods* (3rd *Edition*). Thousand Oaks, CA: Sage Publications.

Young, P. (2003). *Report of the Defense Science Board/Air Force Scientific Advisory Board Joint Task Force on Acquisition of National Security Space Programs*. OMB No. 0704-0188. Washington, DC.

Chapter **9**

A Spreadsheet-Based Tool for Simple Cost–Benefit Analyses of HSI Contributions During Software Application Development

Deborah J. Mayhew

9.1 INTRODUCTION

One fundamental type of human systems integration (HSI) involves providing software applications to individual users to assist them in some aspect of performing their jobs. In this case, the HSI issue may be optimizing the productivity of the trained and experienced user (designing for "ease-of-use" or efficiency) or optimizing the ability of the new or casual user to get up to speed quickly with or without training (designing for "ease-of-learning"), or both.

Examples from the *industrial and commercial context* (see Chapter 2 in this volume) might include:

- Providing a *database application of customer information* (the system) to customer service representatives (the humans) in a credit card or insurance company to help them handle customer queries and requests (in this case, ease-of-use or productivity is the primary HSI issue)
- Providing subscribers of a health plan (the humans) with a secure *Web site* (the system) *for accessing information about their benefits and medical records, as well as other services*, such as e-mail access to health-care providers (in this case, ease-of-learning and remembering is the primary HSI issue)

The Economics of Human Systems Integration: Valuation of Investments in People's Training and Education, Safety and Health, and Work Productivity. Edited By William B. Rouse

- Providing software product users (the humans) with *Web-based services* (the system) for *troubleshooting and updating their software products*, purchasing product-related services, and networking with other users (in this case, the primary HSI issues are both ease-of-learning and ease-of-use)

Examples from the *government and defense context* (see Chapter 3 in this volume) might include:

- Providing a *database application for creating and sharing documents and tracking activities* (the system) to media liaisons (the humans) in a government agency (in this case, ease-of-use or productivity is the primary HSI issue, but ease-of-learning also is a goal)
- Providing an online *"inventory" tracking application* (the system) for arresting officers in a metropolitan police department (the humans) who must keep track of property and evidence taken from prisoners (in this case, ease-of-learning and remembering is the primary HSI issue)
- Providing a *resource management intranet* (the system) to all employees of a government agency (the humans) to help them obtain office supplies and equipment they need to do their jobs (in this case, ease-of-learning is the primary HSI issue)

In each of these cases, automation (the system) *potentially* can increase job performance (of the human). However, it is the *quality* of the *HSI* — in this case, the interface between the application and the user — that will determine to what extent user job performance is improved... or in fact degraded.

Automation itself — that is, building or purchasing software tools — is generally assumed to be cost-justifiable. That is, the assumption is that if you invest in automated tools for workers, it will pay off in increased productivity, decreased training and support, increased effectiveness, or some other quantifiable benefit that justifies the investment. However, it is the quality of the HSI that will determine to what extent — if any — automation will be cost-justifiable in the end. Investing in HSI resources and activities can help ensure — and increase — the return on investment (ROI) of an investment in automation.

The focus of this chapter is to offer and explain a free spreadsheet-based tool to help estimate the *potential ROI* of an investment in *adding HSI resources and activities* to an automation development effort or purchase. The spreadsheet-based tool is provided as a free download on my Web site (Mayhew, 2009). Table and page numbers referred to in this chapter as well as in the tool worksheets are references to table and page numbers in the book *Cost-Justifying Usability* (Bias & Mayhew, 2005) — a reference for much more detail on cost-justifying HSI efforts.

Any cost–benefit analysis must start with estimating the *cost* of an investment, followed by predicting the potential *benefits*, so as to determine whether, and to what extent, the benefits might be expected to exceed the costs, that is, the extent to which an ROI will be realized. To estimate the costs of an HSI investment, one must have a plan for HSI resources and activities as a part of the overall automation project plan. My book *The Usability Engineering Lifecycle* (Mayhew, 1999) describes and explains a generic approach to HSI in the case of developing software applications. It is this life-cycle that is the basis of estimating the cost side in the cost-justification tool described and explained in this chapter. The life-cycle is a general framework that is relevant to cost-justifying an HSI effort when developing any kind of software application, from commodity trading to an e-commerce Web site. The next section provides an overview of the usability engineering life-cycle.

9.2   THE USABILITY ENGINEERING LIFE-CYCLE—AN OVERVIEW

The first step in cost-justifying an HSI effort on a particular automation project is to lay out an HSI plan for that project. This section provides a high-level synopsis of such a plan, based on *The Usability Engineering Lifecycle* (Mayhew, 1999).

The usability engineering life-cycle consists of a set of HSI tasks applied in a particular order at specified points in an overall software application development life-cycle.

Several types of tasks are included in the usability engineering life-cycle:

- Structured usability requirements analysis tasks
- An explicit usability goal-setting task, driven directly from requirements analysis data
- Tasks supporting a structured, top-down approach to user interface design that is driven directly from usability goals and other requirements data
- Objective usability evaluation tasks for iterating design toward usability goals

Figure 9.1 represents in summary, visual form, the usability engineering life-cycle. The life-cycle is cast in three phases: Requirements Analysis, Design/Testing/ Development, and Installation. Specific HSI tasks within each phase are presented in boxes, and arrows show the basic order in which tasks should be carried out. Much sequencing of tasks is iterative, and the specific places where iterations most typically would occur are illustrated by arrows returning to earlier points in the life-cycle. Brief descriptions of each life-cycle task follow.

9.3   PHASE 1: REQUIREMENTS ANALYSIS

9.3.1   User Profile

A description of the specific user characteristics relevant to user interface design (e.g., level of general computer literacy, expected frequency of use, and level of job experience) is obtained for the intended user population. This will drive tailored user interface design decisions and identify major user categories for study in the Task Analysis task.

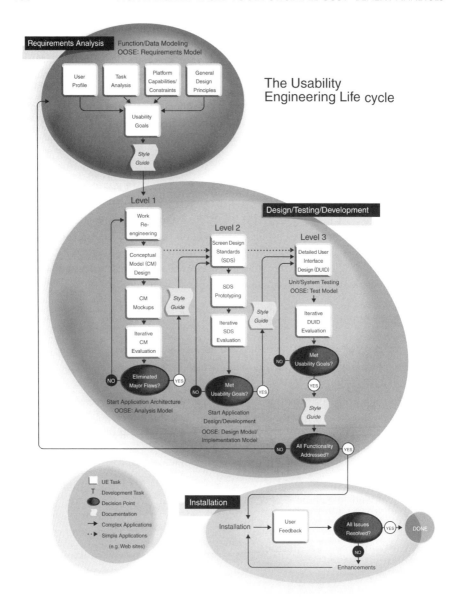

FIGURE 9.1   The usability engineering life cycle, from Mayhew (1999).

9.3.2   Task Analysis

A study of users' current tasks, workflow patterns, and conceptual frameworks is conducted, resulting in a description of current tasks and workflow, and an understanding and specification of underlying user goals. These will be used to set usability goals and drive work reengineering and user interface design.

9.3.3 Platform Capabilities/Constraints

The user interface capabilities and constraints (e.g., windowing, direct manipulation, screen size, color, etc.) inherent in the technology platform chosen for the product (e.g., Apple Macintosh, MS Windows, and product-unique platforms) are determined and documented. These will define the scope of possibilities for user interface design.

9.3.4 General Design Guidelines

Relevant general user interface design guidelines available in the usability engineering literature are gathered and reviewed. They will be applied during the design process to come, along with all other project-specific information gathered in the previous tasks.

9.3.5 Usability Goals

Specific *qualitative* goals reflecting usability requirements are developed, extracted from the User Profile and Task Analysis tasks. In addition, *quantitative* goals (based on a subset of high-priority qualitative goals) are developed, defining minimal acceptable user performance and satisfaction criteria. These usability goals focus later design efforts and form the basis for later iterative usability evaluation.

9.4 PHASE 2: DESIGN/TESTING/DEVELOPMENT

9.4.1 Level 1 Design

9.4.1.1 Work Reengineering. Based on all requirements analysis data and the usability goals extracted from them, user tasks are redesigned at the level of organization and workflow to streamline work and exploit the capabilities of automation. No *visual* user interface design is involved in this task, just abstract organization of functionality and workflow design. The result of this task is sometimes referred to as *information architecture*. It defines how users will navigate through the information and/or functionality of the application.

9.4.1.2 Conceptual Model Design. Based on all the previous tasks, a set of design conventions is generated for visually presenting the different levels in the (usually hierarchical) information architecture and for interactions for navigating through it. Screen design detail is *not* addressed at this design level.

9.4.1.3 Conceptual Model Mockups. Paper-and-pencil or live prototype mockups of high-level design ideas generated in the Conceptual Model Design task are prepared, representing ideas about high-level functional organization and conceptual model design. Detailed screen design and complete functional design are *not* in focus here.

9.4.1.4 Iterative Conceptual Model Evaluation. The mockups are evaluated and modified through iterative evaluation techniques such as formal usability testing, in which real, representative end users attempt to perform real, representative tasks with minimal training and intervention, imagining that the mockups are a real product user interface. This and the previous two tasks are conducted in iterative cycles until all major usability "bugs" are identified and engineered out of level 1 (i.e., conceptual model) design. Once a conceptual model is relatively stable, system architecture design can commence.

9.4.2 Level 2 Design

9.4.2.1 Screen Design Standards. A set of application-specific standards and conventions for all aspects of detailed screen or page design is developed, based on any industry and/or corporate standards that have been mandated (e.g., Microsoft Windows, Apple Macintosh, etc.), the data generated in the Requirements Analysis phase and the product-unique conceptual model design arrived at during level 1 design. Screen design standards will ensure another level of coherence and consistency—the foundations of usability—across the application user interface.

9.4.2.2 Screen Design Standards Prototyping. The screen design standards (as well as the conceptual model design) are applied to design the detailed user interface to selected subsets of application functionality. This design is implemented as a running prototype.

9.4.2.3 Iterative Screen Design Standards Evaluation. An evaluation technique, such as formal usability testing, is carried out on the screen design standards prototype, and then redesign/reevaluation iterations are performed to refine and validate a robust set of screen design standards. Iterations are continued until all major usability bugs are eliminated and usability goals seem within reach.

9.4.2.4 Style Guide. At the end of the design/evaluate iterations in design levels 1 and 2, you have a validated and stabilized conceptual model design, and a validated and stabilized set of standards and conventions for all aspects of detailed screen design. These are captured in the document called the application style guide, which already documents the results of requirements analysis tasks. During detailed user interface design, following the conceptual model design and screen design standards in the application style guide will ensure quality, coherence, and consistency—the foundations of usability.

9.4.3 Level 3 Design

9.4.3.1 Detailed User Interface Design. Detailed design of the complete product user interface is carried out based on the refined and validated information architecture, conceptual model design, and screen design standards documented in the product style guide. This design then drives application development.

9.4.3.2   Iterative Detailed User Interface Design Evaluation. A technique such as formal usability testing is continued during application development to expand evaluation to previously unassessed subsets of functionality and categories of users, and to continue to refine the user interface and validate it against usability goals.

9.5   PHASE 3: INSTALLATION

9.5.1   User Feedback

After the application has been installed and in production for some time, feedback is gathered to feed into design enhancements, design of new releases, and/or design of new but related applications.

9.6   GENERAL APPROACH TO COST–BENEFIT ANALYSIS OF HUMAN SYSTEMS INTEGRATION

To cost-justify a proposed HSI plan, you simply adapt a very generic and widely used cost–benefit analysis technique. Having laid out a detailed HSI project plan based on the usability engineering life-cycle (see previous discussion, and Mayhew, 1999), it is a fairly straightforward matter to calculate the costs of that plan. Then you need to calculate the predicted benefits. This is a little trickier, and it is where the adaptation of the generic cost-justification analysis comes into play. Then, you simply compare costs to benefits to find out if, when, and to what extent the benefits are predicted to outweigh the costs. If they do to a satisfactory extent, then you have cost-justified the planned HSI effort.

More specifically, first an HSI plan is laid out. The plan specifies particular techniques to employ for each HSI task, breaks the techniques down into steps, and specifies the personnel hours and equipment costs for each step. The *cost* of each task is then calculated by multiplying the total number of hours for each type of personnel by their effective hourly wage (fully loaded, that is, including salary, benefits, office space, equipment, utilities, and other facilities), and adding up personnel costs across types. Sometimes it is hard to get data on fully loaded wages for an organization. In this case, I use a rule of thumb I have heard informally to double the before-tax annual salary, and then divide by the typical number of hours a full time worker is paid for in a year, usually approximately 2,000. Even if my audience is unwilling or unable to give me actual figures for fully loaded wages, they can contest—or not—my ballpark figure based on this rule of thumb. Then the costs from all tasks are summed to arrive at a total estimated cost for the plan.

Next, the overall *benefits* of the specific HSI plan are predicted by selecting relevant benefit categories, calculating expected benefits by plugging project-specific parameters and assumptions in to benefit formulas, and summing benefits across categories.

The potential benefit categories relevant to a particular cost–benefit analysis will depend on the basic business model of the application being developed. Benefit categories potentially relevant to different types of software applications are sum-marized in Table 9.1 and in the **Benefits Categories** worksheet of the spreadsheet tool (see also Table 3.1 on p. 58 of Bias & Mayhew, 2005).

Note that the relevant benefit categories for different types of applications/Web sites vary somewhat. In a cost–benefit analysis, one wants to focus attention on the potential benefits that are *of most relevance to the bottomline business goals for the application*, either short term or long term or both.

Note also that these benefits represent just a sample of those that might be relevant for the types of software applications listed. Other benefits relevant to these particular types of automation projects might be included as appropriate, given the business goals of the application sponsors and the primary concerns of the audience, and they could be calculated in a similar fashion within the spreadsheet tool as those described in the following discussion. And of course, very different kinds of automation projects exist that may have very different kinds of benefits, for example, lives and equipment saved and wars won in a military context. The general cost-justification approach can be applied in these latter types of situations, but the tool would have to be modified significantly to address them.

Finally, overall predicted benefits are compared with overall estimated costs to see if, and to what extent, the overall HSI plan is justified.

When HSI practitioners are invited to participate in application development projects already in progress, which is often the case, it may be difficult to include all life-cycle tasks, and to influence overall schedules and budgets. They are likely to have to live within already-committed-to schedules, platforms, and system archi-tectures; use shortcut techniques for life-cycle tasks; and impact budgets minimally. Nevertheless, it is almost always possible to create an HSI plan that will make a sig-nificant contribution to an application development project, even when one comes in relatively late. And, you can use the cost–benefit analysis technique to prepare and support even plans that involve only parts of the overall life-cycle and only shortcut techniques for tasks within it.

9.7   EXAMPLE COST–BENEFIT ANALYSIS

In this section, I will both provide a concrete example of conducting a cost–benefit analysis of an HSI plan as well as introduce and explain the use of my spreadsheet-based cost-justification tool, which is available as a free download from my Web site (Mayhew, 2009). The example I will use involves the development of a software application to support internal customer service representatives at a credit card company.

9.7.1   An Application for Internal Users—Customer Service

First, let us look at the overall results of the cost–benefit analysis for this example, both in the spreadsheet tool and in the book, if you have it (Bias & Mayhew, 2005). We will assume the HSI project plan with its associated cost that is shown

TABLE 9.1 Benefit Categories by Application/Site Type

APPLICATION/ SITE TYPE BENEFITS	Internal Application	Commercial Product	E-Commerce Site	E-Services Site	Site Funded by Advertising	Product Information Site	Customer Service Site	Intranets
Increased buy-to-look ratios			✓					
Decreased abandoned shopping carts			✓					
Increased number of visits			✓		✓			
Increased return visits					✓			
Increased length of visits					✓			
Decreased failed searches			✓		✓			
Decreased costs of other sales channels			✓					
Decreased use of "Call Back" button (i.e., live customer service)			✓	✓			✓	✓
Savings due to making changes earlier in development life cycle	✓	✓	✓	✓	✓	✓	✓	✓
Increased "click though" on ads					✓			
Increased sales leads		✓				✓		
Increased sales		✓						
Decreased costs of traditional customer service channels		✓		✓			✓	
Decreased training costs	✓							✓
Increased user productivity	✓							✓
Decreased user errors	✓							✓

in Table 9.2 and presented in the **Total Costs** worksheet of the spreadsheet tool (see also Table 3.2 on p. 61 of Bias & Mayhew, 2005).

In this example, the project human systems integrator estimated that in the case of this project, the HSI plan would produce a customer service application with the *expected benefits* summarized in Table 9.3 and in the **Internal** worksheet in the spreadsheet tool (see also Table 3.3 on p. 62 of Bias & Mayhew, 2005).

TABLE 9.2   Cost of a Usability Engineering Plan

						Your Data	Formulas
COST CALCULATIONS							
PHASE	TASK (Technique)	Usability Engineers Hours @	Developers Hours @	Managers Hours @	Users Hours @		TOTAL COST
		$175	$175	$200	$25		
Requirements Analysis	User Profile (Questionnaire)	62	0	4	33		$12,475
	Contextual Task Analysis	138	8	8	60		$28,650
	Platform Capabilities and Constraints	16	6	0	0		$3,850
	Usability Goals	20	0	4	2		$4,350
Design/Testing/ Development	Work Reengineering (Information Architecture)	80	0	0	16		$14,400
	Conceptual Model Design	80	8	0	8		$15,600
	Conceptual Model Mockups (Paper Prototype)	36	0	0	0		$6,300
	Iterative Conceptual Model Evaluation (Usability Test)	142	0	0	22		$25,400
	Screen Design Standards	80	8	0	8		$15,600
	Screen Design Standards Prototyping (Live Prototype)	28	80	0	8		$18,900
	Iterative Screen Design Standards Evaluation (Usability Test)	142	40	0	22		$32,400
	Detailed User Interface Design	80	8	0	8		$15,600
	Iterative Detailed User Interface Design Evaluation (Usability Test)	142	40	0	22		$32,400
	TOTALS	1046	198	16	201		$225,925

TABLE 9.3   Expected First-Year and Lifetime Benefits for an Application for Internal Users

BENEFIT CATEGORY	BENEFIT VALUE-*FIRST YEAR*
Increased Productivity	$199,652.78
Decreased Errors	$47,916.67
Decreased Training	$62,500.00
Decreased Late Design Changes	$84,000.00
TOTAL BENEFIT	$394,069.44

BENEFIT CATEGORY	BENEFIT VALUE-*LIFETIME (5 yrs)*
Increased Productivity × 5 years	$998,263.89
Decreased Errors × 5 years	$239,583.33
Decreased Training × 1 years	$62,500.00
Decreased Late Design Changes × 1 year	$84,000.00
TOTAL BENEFIT	$1,384,347.22

Comparing these benefits and costs, the human systems integrator argued that the proposed HSI plan would more than pay for itself in the first year after launch, as shown in Table 9.4 and in the **Product Info** worksheet in the spreadsheet tool (see also Table 3.4 on p. 62 of Bias and Mayhew, 2005), and that during an anticipated system lifetime of five years, the benefits would continue to accrue to a total of $1,158,422.22.

Note that the simple analyses offered in this example and in the spreadsheet-based tool do not consider the time value of money—that is, that the money for the costs is spent at one point in time, whereas the benefits come later in time, and this is *not* taken into account in this simple analysis. Also, if the money were *not* spent on the costs, but instead were invested in some other way, this money would likely increase in value. In my experience, the predicted benefits of usability engineering are usually so dramatic that these more sophisticated financial considerations are

TABLE 9.4   Net Benefit Calculations for an Application For Internal Users

NET BENEFIT CALCULATIONS	
First Year	
Benefit =	$394,069.44
Cost =	$225,925.00
NetBenefit =	$168,144.44
Lifetime (5 yrs)	
Benefit =	$1,384,347.22
Cost =	$225,925.00
NetBenefit =	$1,158,422.22

not necessary to convince the audience for the analysis. However, if needed, these calculations based on the time value of money are explained in Chapter 7 of Bias and Mayhew (2005).

In addition, note that the application of the cost-justification analysis assumes that *potential* benefits *actually* can be realized. For example, in this example, one could conclude that to realize the potential benefits, one would have to reduce personnel, and this may or may not be possible. Another way to phrase the benefit, however, could be that instead of saving money by maintaining current transaction rates while cutting the cost of personnel, one could save money by increasing the transaction rate without increasing personnel costs. In any case, all the cost–benefit analysis really can do is identify ways in which employing an HSI program *potentially* could pay off. Then it needs to be determined whether potential benefits can be taken advantage of practically given the full context of the organization and situation.

The project human systems integrator expected the HSI plan to be approved based on the cost justification in this example. Now we will see exactly how this net benefit was calculated, using the spreadsheet tool.

9.7.2   Start with the Human Systems Integration Plan

This is the first step in conducting a cost–benefit analysis. The HSI plan identifies which usability engineering life-cycle tasks and techniques (see previous discussion and Mayhew, 1999) will be employed and breaks them down into required staff and hours. Costs can then be computed for these tasks in the next two steps that follow.

In this scenario, we start with the assumed plan represented in Table 9.2 and in the **Total Costs** worksheet of the spreadsheet tool (see also Table 3.2 on p. 61 of Bias and Mayhew, 2005). Note that almost all the cells in this worksheet have a *dark gray* background. This indicates that the values in these cells are computed according to *formulas* that reference cells in other worksheets. Thus, when using the tool, you should *not* directly edit any of the patterned cells in the worksheets. Cells you *should* edit to reflect your particular project will appear with a *light gray* background in worksheets. For this stage in the analysis process, really we are just focused on the first two columns of this table, which layout the plan.

It is important to note that there is not one correct HSI plan. This—as much else—is something that will vary across projects. The choice of technique for carrying out each task in the usability engineering life-cycle will depend on project budgets, schedules, and complexity. Thus, the example plan presented here should not be assumed. A project-unique plan must be designed around the parameters of a specific project.

9.7.3   Establish Analysis Parameters

Most calculations for both estimated costs and predicted benefits are based on project-specific parameters. These should be researched, established, and documented before proceeding with the analysis. Analysis parameters for this example

TABLE 9.5 Analysis Parameters for an Application for Internal Users

Analysis is Parameters	Values
Application type:	for Internal Users
Number of end users:	250
User work days per year:	230
User fully loaded hourly wage:	$25
Developer fully loaded hourly wage:	$175
Usability Engineer fully loaded hourly wage:	$175
Manager fully loaded hourly wage:	$200
Expected system lifetime (years):	5
Current transactions per day:	100
Current recovery time per error (2 minutes expressed as hours):	0.033333333
Time per early design change (hours):	8
Ratio of late to early design changes:	4
Usability Lab:	In place

are presented in Table 9.5 and in the **Total Costs** worksheet of the spreadsheet tool (see also Table 3.5 on p. 63 of Bias and Mayhew 2005).

In this example, we assume that there is an existing application with known parameters and that the project involves a redesign, which is a common scenario for application development today.

It should be emphasized that when using the general cost–benefit analysis technique illustrated here, the particular parameters and parameter *values* used in this example should *not* be assumed. Both the particular parameters themselves and the parameter values of *your* project and organization should be substituted for those in Table 9.5 and in the **Total Costs** worksheet (as well as all other worksheets) of the spreadsheet tool (see also Table 3.5 on p. 63 of Bias and Mayhew, 2005). For example, your application may be intended for many more—or less—than 250 users, and the fully loaded hourly wage (the costs of salary plus benefits, office space, equipment, utilities, and other facilities) of your personnel may be significantly lower or higher than those assumed in these sample analyses. See also other worksheets in the spreadsheet tool for examples of relevant parameters for other types of development projects.

Note that in general, certain parameters in a cost–benefit analysis have a major impact on the magnitude of potential benefits. For example, when considering *user productivity*—of primary interest to internal development organizations such as in this example—the critical parameters are the *number of users* and the *volume of transactions*, and to some extent, the *users' fully loaded hourly wage*. When there is a large number of users and/or a high volume of transactions, even very small performance advantages (and low hourly wages) in an optimized application user interface will add up quickly to significant overall benefits. On the other hand, where there is a small number of potential users, and/or a low volume of transactions, benefits may not add up to much even when the potential per transaction performance advantage seems significant and the user hourly wage is higher.

For example, consider the following two scenarios. First, imagine a case where there are 5,000 users and 120 transactions per day per user. Even a half-second advantage per transaction in this case adds up:

$$5,000 \text{ users } \times 120 \text{ transactions } \times 230 \text{ days } \times 1/2 \text{ second } = 19,167 \text{ hours}$$

If the users' hourly rate is $25, the annual savings are as follows:

$$19,167 \text{ hours } \times \$25 = \$479,175$$

This is a pretty dramatic benefit for a tiny improvement on a per-transaction basis.

On the other hand, if there were only 25 users, and they were infrequent users, with only 12 transactions per day, even if a per-transaction benefit of one minute could be realized, the overall benefit would only be

$$25 \text{ users } \times 12 \text{ transactions } \times 230 \text{ days } \times 1 \text{ minute } = 1,150 \text{ hours}$$

At $25 per hour, the overall annual productivity benefit will only be

$$1,150 \text{ hours } \times \$25 = \$28,750$$

Thus, in the case of productivity benefits, the costs associated with optimizing the user interface are more likely to pay off when there are many users and many transactions.

9.7.4 Calculate the Cost of each Usability Engineering Life-cycle Task in the Human Systems Integration Plan

The cost of each task/technique listed in Table 9.2 and in the **Total Costs** worksheet of the spreadsheet tool (see also Table 3.2 on p. 61 of Bias and Mayhew, 2005) was estimated by breaking the task/technique down into small steps, estimating the number of hours required for each step by different types of personnel, and multiplying these hours by the known fully loaded hourly wage of each type of personnel (if outside consultants or contractors are used, their simple hourly rate plus travel expenses would apply, and if external users are recruited to participate, they would be paid at some simple hourly rate or flat fee.)

In our example, the project human systems integrator used the task cost calculations as shown in the **User Profile** through **DUID** worksheets of the spreadsheet tool. One example of these task cost calculations can be observed in Table 9.6 (see also Tables 3.6 to 3.18 on pp. 66–72 in Bias and Mayhew, 2005).

These task cost calculations show the derivation of the numbers summarized in Table 9.2 and in the **Total Costs** worksheet of the spreadsheet tool (see also Table 3.2 on p. 61 of Bias and Mayhew, 2005).

One parameter used in the calculations of cost is the fully loaded hourly wage of involved personnel. Fully loaded hourly wages are calculated by adding together the

TABLE 9.6 Cost of User Profile (Questionnaire)

User Profile (Questionnaire)	Usability Engineers	Developers	Managers	Users
STEP	HOURS	HOURS	HOURS	HOURS
Needs Finding	4		2	2
Draft Questionnaire	6			
Management Feedback	2		2	2
Revise Questionnaire	4			
Pilot Questionnaire	4			4
Revise Questionnaire	2			
Select User Sample	4			
Distribute Questionnaire/Respond	8			25
Data Analysis	8			
Data Interpretation/Presentation	20			
Total Hours	62	0	4	33
Times Hourly Rate	$175	$175	$200	$25
Equals	$10,850 plus	$0 plus	$800 plus	$825 = $12,475

cost of salary, benefits, office space, equipment, and any other relevant overhead for a type of personnel, and by dividing this by the number of hours paid for each year by that personnel type. The hourly rate used here for human systems integrator staff is based on an informal average of typical current salaries of senior-level internal usability engineering staff and external consultants in my recent experience (see also UPA, 2009), for the most recent salary survey of usability practitioners.) The hourly rate of developers was similarly estimated (see, for example, Payscale, 2009.) However, the fully loaded hourly rate figures used to generate this and the other sample cost–benefit analyses in the spreadsheet tool and Bias and Mayhew (2005) are just examples, and you would have to substitute the actual hourly rates of personnel in your own organization in an actual analysis. Additional costs, such as equipment and supplies, also could be estimated and added into the total cost of each task/technique, although that was not done in this example for simplicity's sake.

In the spreadsheet tool, you would start by plugging in your fully loaded hourly wage parameters in the **Fully Loaded Hourly Rate** worksheet, as in Table 9.7.

Then, for each table in the **User Profile** through the **DUID Evaluation** worksheets, you would enter your planned level of effort for each task, as for example

TABLE 9.7 Fully Loaded Hourly Rates

Developer	$175
Usability Engineer (UE)	$175
User	$25
Manager	$200
Customer support	$50
Trainer	$50

in Table 9.6 (see also Tables 3.6 through 3.18 on pp. 66–72 in Bias and Mayhew, 2005). *For any task that is not included in your plan, it is important to enter zeroes in all light gray cells.* Similarly, if you plan to conduct a given task but not every listed step in it, simply enter zeros in the cells of any steps you do not plan to conduct.

Once you have entered in your unique project parameters in the hourly wage and tasks worksheets, you can look back at the **Total Costs** worksheet and Table 9.2 to see the total estimated cost of your plan.

9.7.5   Select Relevant Benefit Categories

Because our example is a redesign of an *application for internal use*, only certain benefit categories are of relevance to the business goals of the project, as shown in Table 9.1 and the **Benefits Categories** worksheet in the spreadsheet tool (see also Table 3.1 on p. 58 in Bias and Mayhew, 2005). These include:

- Increased productivity
- Decreased errors
- Decreased training
- Decreased late design changes

Others might have been included—for example, decreased cost of user support time—but just these four were selected to keep the analysis simple and conservative. As discussed, the best benefit categories to include in a cost–benefit analysis will depend on the type of project and the intended audience for the analysis.

In this example, the project human systems integrator expected to achieve *increased productivity* by focusing on streamlining across-screen navigation within tasks, by minimizing typing and mouse clicks on individual screens, and by designing to facilitate scanning and interpreting displays. S/he is expected to *decrease errors* both by following well-established design principles during design and by detecting and eliminating common errors through usability testing. S/he is expected to *decrease training* time by designing a consistent, rule-based user interface that matches users' knowledge and expectations and in which the smallest number of design conventions accounts for the widest scope of functionality. Finally, *late design changes* would be *minimized* and replaced by less expensive early design changes by following an iterative design process that incorporated usability inspection and testing.

Table 9.3 and the **Internal** worksheet in the spreadsheet tool (see also Table 3.3 on p. 62 in Bias and Mayhew, 2005) summarize the predicted magnitude of each of these benefit categories and then sums across them to predict a total benefit.

When selecting benefit categories to use in your own cost-justification analysis of an HSI plan for an internal application development project, you can use the benefits assumptions table in the **Internal** worksheet in the spreadsheet tool (also shown in Table 9.8) to make your benefit assumptions. You can use all four benefit categories in that table or choose to use any combination of them. Simply enter

**TABLE 9.8   Benefits Assumptions for an Application
of Internal Users**

BENEFIT ASSUMPTIONS	
Increased Productivity	
Decreased time per transaction (5 seconds expressed as hours):	0.001389
Decreased Errors	
Number errors eliminated per day:	1
Decreased Training	
Hours saved off current 1 week training:	10
Decreased Late Design Changes	
Number of changes made early:	20

zeros in the cells representing assumption values for benefit categories you do not wish to include.

What follows is an explanation of how benefit predictions in each category were derived from project-specific parameters and assumptions.

9.7.6   Predict Benefits

In this step, the project human systems integrator predicted the magnitude of the benefits that would be realized—*relative to the current application*, which is being redesigned—*if* the HSI plan (with its associated costs) were implemented. Benefits were predicted in each selected benefit category by doing some simple arithmetic based on project-specific analysis parameters and some simple project-specific assumptions.

Note that although at this point in the process of your own analysis, you have already filled in your project-unique parameters, you now need to consider your benefits assumptions, and modify the values for them in the benefits assumptions table in the spreadsheet tool. The project *parameters* for this example are laid out in Table 9.5 and in the **Internal** worksheet in the spreadsheet tool (see also Table 3.5 on p. 63 in Bias and Mayhew, 2005). The benefits *assumptions* are given in Table 9.8 and in the **Internal** worksheet in the spreadsheet tool (see also Table 3.19 on p. 73 in Bias and Mayhew, 2005).

In the case of calculating predicted productivity benefits, the relevant *parameters* are from Table 9.5 and in the **Internal** worksheet in the spreadsheet tool (see also Table 3.5 on p. 63 in Bias and Mayhew, 2005):

- The total number of users
- The number of days each user works per year
- The number of transactions each user currently performs each working day
- The users' fully loaded hourly wage

The *assumptions* made regarding increased productivity in our example (see Table 9.8, and the **Internal** worksheet in the spreadsheet tool) are that:

- Each transaction will take five seconds (0.001389 hours) less on a user interface developed by incorporating the HSI plan, as compared with a user interface developed without the HSI plan
- Each user will make one less error per day
- Each user will require ten hours less of training to get up to speed
- Twenty design changes will be made early in the design process, rather than after implementation

Benefit assumptions are the crux of the whole cost–benefit analysis. Although *costs* can be calculated with a high degree of confidence based on past experience, and all the *parameters* fed into the analysis are known facts, the *assumptions* made are just that—assumptions and predictions, rather than known facts or guaranteed outcomes. The audience for the analysis is asked to accept that these assumptions are reasonable ones, and they must to be convinced by the overall analysis.

It should be pointed out that *any* cost–benefit analysis for *any* purpose must ultimately include some assumptions that are really only predictions of the likely outcome of investments of various sorts. The whole point of a cost–benefit analysis is to try to evaluate in advance, in a situation in which there is some element of uncertainty, the likelihood that an investment will pay off. The trick is basing the prediction of ROI on a firm foundation of known facts. In the case of a cost–benefit analysis of an HSI effort, there are several foundations on which to formulate sound assumptions regarding predicted benefits, including:

- References to the general usability literature documenting impacts of certain types of design decisions on user performance
- References to after-the-fact case studies of benefits achieved through HSI
- Anecdotes from colleagues
- One's past experiences as a human systems integrator
- One's past experience working with a particular design/development organization

See Bias and Mayhew (2005) for more in-depth discussion of making and supporting analysis assumptions.

In general, usually it is wise to make *very conservative* benefit assumptions, for several reasons. First, any cost–benefit analysis has an intended audience, who must be convinced that benefits will in fact outweigh costs. Assumptions that are very conservative are less likely to be challenged by the relevant audience, thus increasing the likelihood of acceptance of the analysis conclusions. In addition, conservative benefits assumptions help to manage expectations. It is always better to achieve a greater benefit than was predicted in the cost–benefit analysis, than to achieve a lesser benefit, even if it still outweighs the costs. Having underestimated benefits will likely make future cost–benefit analyses more credible and more readily accepted. Also, it is important to realize that some validly predicted benefits may be canceled out by other nonusability-related changes, such as decreases in user morale and motivation, decreased system reliability or response time, and so on. Having made conservative benefits predictions decreases the possibility that other factors will wipe out completely any benefits from improved usability.

Returning to the explanation of the derivation of benefit predictions in our example, we see the benefit assumptions for each benefit category given in Table 9.8 and in the **Internal** worksheet in the spreadsheet tool (see also Table 3.19 on p. 73 in Bias and Mayhew, 2005). The project human systems integrator selected these assumptions believing they were very conservative ones. S/he referenced some literature, which showed up to 20% to 30% savings in task time on one design approach relative to another, and pointed out that s/he was assuming only a modest 4% increase in productivity. S/he pointed out the assumption of eliminating one error per day per user is extremely conservative and cited internal statistics showing high typical user error rates on current internal applications. S/he noted that the reason most current internal applications took a week to train was from the lack of consistency in the user interfaces of those applications and from the need to memorize many cryptic codes and unclear error messages. S/he made the case that an interface in which a small number of rules explained a wide scope of functionality, and in which the user needed to memorize less, would be teachable in a significantly shorter period of time. Finally, to defend the assumption about the relative cost of early versus late design changes, s/he cited a classic paper in the literature (Mantei & Teorey, 1988).

Table 9.9 and the **Internal** worksheet in the spreadsheet tool (see also Table 3.20 on p. 78 in Bias and Mayhew, 2005) shows the calculation of the total predicted benefit in each benefit category, based on parameters and assumptions in Tables 9.5 and 9.8.

In the case of increased productivity, multiplying the number of users by days per user, by transactions per day, by hours saved per transaction, and by hourly rate results in the total benefit given in Table 9.3 and in the **Internal** worksheet in the spreadsheet tool for this benefit category (see also Table 3.3 on p. 62 in Bias and Mayhew, 2005): $199,652.78. The total benefits per category are summed in Table 9.3 to a total benefit in the first year alone and to a lifetime benefit during an assumed five year application lifetime.

When conducting your own project-specific cost–benefit analysis, note that although you need to enter values for some parameters and for assumptions in the project type worksheet in the spreadsheet tool, all tables to the right of those two are based on calculations involving previously entered values in other tables, which are indicated by the dark gray background in those cells. You will not edit those tables at all. Also recall that you will want to enter zeros for any parameter or assumption you do not wish to use in your analysis.

TABLE 9.9   Benefits Calculations for an Application for Internal Users

INDIVIDUAL BENEFIT CALCULATIONS

Increased Productivity

Users ×	Days ×	Transactions ×	Transaction ×	Hourly Rate =	
250	230	100	0.001389	$25	$199,652.78

Decreased Errors

# Users ×	# Days ×	# Eliminated Errors ×	Hours Saved per Error ×	Hours Rage =	
250	230	1.0	0.033333	$25	$47,916.67

Decreased Training

# Users ×	per Users ×		Hours Rage =	
250	10		$25	$62,500.00

Decreased Late Design Changes Cost of Early Changes:

# Changes ×	Hours per Change ×	Hours Rage =	
20	8	$175	$28,000.00

Cost of Late Changes:

Cost of Early Changes ×	Ratio of Late of Early Changes =	
$28,000.00	4	$112,000.00

Savings of Early Changes Relative to Late Changes:

Cost of Late Changes -	Cost of Early Changes =	
$112,000.00	$28,000.00	$84,000.00

9.7.7   Compare Costs with Benefits

Having calculated the costs of a particular HSI plan and having predicted the total benefits to result from executing that plan as compared with not executing it, the next step is simply to subtract the total costs from the total benefits to arrive at a net benefit. In this example, this calculation is shown in Table 9.4 and in the **Internal** worksheet in the spreadsheet tool (see also Table 3.4 on p. 62 in Bias and Mayhew, 2005). The analysis predicts a clear and substantial net benefit ($168,144.44) in the first year alone and a dramatic net benefit ($1,158,422.22) during the expected application lifetime.

Our project human system integrator's initial usability engineering plan seemed to be well justified. It was a fairly aggressive plan, in that it included all life-cycle tasks, and the most reliable and thorough techniques for each task. Given the very clear net benefit, the human system integrator would have been wise to stick with this aggressive plan and to submit it to project management for approval and funding.

If the net benefit had been marginal, or if there had in fact been a net *cost*, then it would have been well advised to go back and rethink the proposed human system integration plan, scaling back to shortcut techniques for some tasks. Perhaps, for example, the human system integrator should have planned to do only a shortcut User Profile by interviewing user management, a shortcut Task Analysis consisting of just a few rounds of contextual observations/interviews with users, and just one iterative cycle of usability testing on a complete detailed design, to catch major flaws and be sure the predicted benefits had been achieved. Of course, this would make the predictions more risky and call for an even more conservative analysis.

As explained, to plan the budget for a usability engineering program, it makes sense to start out by calculating the costs of the most aggressive HSI plan that you would like to implement, including the more reliable and thorough techniques for most, if not all, life-cycle tasks. If predicted benefits outweigh costs dramatically, as they usually will when critical parameters are favorable, then you easily can make a good argument for even the most aggressive usability engineering program, because only the most conservative claims concerning potential benefits have been made and as such can be defended easily.

If, however, costs and benefits in the initial calculation seem to match up fairly closely, then you might want to consider scaling back the planned HSI program, maybe even to just a bare-bones plan, with more shortcut techniques applied for each life-cycle task.

To illustrate this planning strategy, consider the following two scenarios. First, revisit our example analysis, which involved building a system for 250 internal users. Fairly conservative assumptions were made concerning benefits: task time reduced by five seconds, training time reduced by ten hours, and one error eliminated per day per user at two minutes saved per error. Even with these fairly conservative assumptions, the fairly aggressive HSI plan was predicted to pay off in the first year, with net benefits continuing to accrue dramatically after that.

In fact, if you had made the more aggressive and yet still realistic benefits assumptions of training time reduced by 20 hours (rather than by 10), two errors eliminated per user per day (rather than just one), and task time reduced by 15 seconds (rather than just by 5 seconds), the benefits would have summed to \$903,839.58 in the first year alone, outweighing the costs of \$225,925 by \$677,914.58, and to \$3,683,197.92 across five years, outweighing the costs by \$3,457,272.90. Thus, one could argue that although even the most conservative assumptions predict a fairly dramatic payoff of a comprehensive usability engineering program, the likelihood is that the payoff will be higher still.

In contrast, suppose you again started out by costing out a comprehensive HSI program at \$225,925. In this case, however, suppose that there are only 50 intended users (instead of 250) performing 50 transactions per user per day (instead of 100.) In this case, calculations using the original more conservative benefits assumptions would show a loss until well into the second year, and a five-year lifetime net benefit of only \$18,318.06.

Even though the benefits assumptions were conservative, although a first-year loss is not necessarily a bad thing, it still seems risky to make an aggressive investment that, based on conservative assumptions, really does not show a significant payoff even during the course of five years. In this case, one would want to scale back the planned usability engineering program and its associated costs. Because the benefits assumptions made were so conservative, it is likely that they will be achieved even with a minimal usability effort. In this way, you can use the spreadsheet-based cost–benefit analysis tool to "what if" in order to plan a level of HSI effort that is most likely to pay off.

9.8 SUMMARY

The cost–benefit analysis example offered in this chapter is based on a simple subset of all actual costs and potential benefits and on very simple and basic assumptions regarding the value of money over time. More complex and sophisticated analyses can be performed—see Karat, Chapter 4 in Bias and Mayhew (2005.) However, often a simple and straightforward analysis of the type offered in the preceding example will be sufficient for the purpose of winning funding for HSI investments during software application development in general, or planning appropriate HSI programs for specific development projects.

The example analysis offered here suggests that it usually is fairly easy to justify a significant investment of time and money in HSI during the development of software applications. The framework and example presented in this chapter, along with the free spreadsheet tool available from my Web site, should help you demonstrate that this is the case for your development projects.

REFERENCES

Bias, R. G., & Mayhew, D. J. (2005). *Cost Justifying Usability—An Update for the Internet Age*. San Francisco, CA: Morgan Kaufmann Publishers.

Mantei, M. M., & Teorey, T. T. J. (1988). Cost/benefit for incorporating human factors in the software lifecycle. *ACM Communications*, *31*(4), 428–439.

Mayhew, D. J. (1999). *The Usability Engineering Lifecycle*. San Francisco, CA: Morgan Kaufmann Publishers.

Mayhew, D. J. (2009). Deborah J. Mayhew & Associates web site, http://drdeb.vineyard.net/index.php?loc=12&nloc=1.

Payscale. (2009). Career Planning page. http://www.payscale.com/research/CN/Job=Software_Engineer_%2f_Developer_%2f_Programmer/Salary.

UPA. (2009). Resources, Salary Surveys page. http://www.usabilityprofessionals.org/usability_resources/surveys/SalarySurveys.html.

Chapter 10

Multistage Real Options

MICHAEL J. PENNOCK

10.1 INTRODUCTION

Chapter 7 introduced the notion that investments in human systems integration (HSI) can be evaluated as real options. In short, real options analysis is based on the recognition that many real-world business investment opportunities resemble financial options. Consequently, they can be evaluated using the powerful tools developed to value financial options.

This chapter explores the real options approach to valuing investments, in particular, the multistage investments typical of technology development efforts. To motivate the exposition, a notional example of an HSI investment opportunity is presented. The example highlights the fundamental investment valuation issues that real options analysis attempts to address. This leads to a discussion of the fundamentals of investment analysis and to a discussion of the evolution of real options analysis.

With a basic understanding of the principles of options analysis, the chapter then covers how to value a real option. First, a conceptual approach to valuing an option is presented. This is followed by a discussion of the pros and cons of the most common methods for solving real options with particular emphasis on lattice methods. Next, the example problem is solved using the binomial lattice method. Finally, the chapter concludes with a brief discussion of the practical issues involved in the application of real options analysis.

The Economics of Human Systems Integration: Valuation of Investments in People's Training and Education, Safety and Health, and Work Productivity. Edited By William B. Rouse
Copyright © 2010 John Wiley & Sons, Inc.

10.2   A MOTIVATING EXAMPLE

To understand how multistage real options can be applied to investments in human systems integration, we will motivate the discussion with an example. Imagine that an engineer at the ACME Computer Corporation has developed a concept for a new piece of software that will aid the company's technical support personnel. In particular, it will aid them in diagnosing problems that customers experience when setting up their newly purchased computers. It is expected that if the idea proves successful, it could reduce significantly the duration of these technical support calls. Of course, it will take a certain amount of investment by the company to develop the software, deploy it to technical support centers, and train personnel in its use. ACME estimates that it will cost approximately $11 million to do so.

To determine the value of this research and development (R&D) investment opportunity, we need to have some estimate of how much money the company will save in the long run. ACME estimates that 1% of new computer sales result in a call to tech support for help with setup, and the average duration of these calls is one hour. Although the new piece of software will not reduce the frequency of calls, it is expected to reduce the duration of the calls by 50%. Because tech support calls at ACME cost $50 per hour, the company stands to save $25 per call if the new software proves successful. Thus, if we know the expected number of computers sold each year, we can determine the stream of expected cost savings. ACME sold 5 million new computers last year. Given that sales rate, the new software would lead to a cost savings of $1.25 million per year.

Using traditional investment analysis, evaluating this opportunity is relatively trivial. To calculate the net present value (NPV), we need to determine the timing and amount of future cash flows, discount them, and sum them. The negative cash flows would be the costs to develop and deploy the new software. The positive cash flows would be the savings on technical support that can be derived from forecasts of future computer sales. As we would demand that our returns exceed the cost to finance the investment, we would discount the cash flow streams by the company's cost of capital. If the NPV is positive, we move forward with the investment. Otherwise, we drop it.

In financial markets, many factors can affect the expected rate of return of an investment, but two of the principle contributors are the time value of money and the perceived riskiness of the investment. The time value of money is essentially rent on capital. Investors must be compensated for postponing the use of their money, and this is the rate of return that they would demand from a risk-free investment. Virtually all investments entail some degree of risk, however. As the riskiness of an investment increases, investors demand larger premiums to compensate for bearing the additional risk. This takes the form of an increase in the required rate of return.

Thus, using the cost of capital as a discount rate implies that the investment opportunity is about as risky as the company overall. This seems unlikely. In fact, we are concerned with two kinds of risk, technical and market. Technical risk is related to uncertainty in the performance of the end product. It may turn out that the software, once developed, does not save any time at all. Market risk pertains

TABLE 10.1   Staging Parameters

Stage	Cost ($ million)	Probability of Success	Duration (years)
Prototype	0.1	0.5	1
Pilot	1	0.7	1
Deployment	10	0.8	N/A

to the random fluctuations in the marketplace. Earlier we indicated that our cost savings were dependent on the number of computers sold. That quantity is certainly going to fluctuate with the state of computer markets. If computer sales decline sufficiently, the cost savings may not justify development costs.

Faced with both market and technical risk, many decision makers would simply increase the discount rate in the NPV calculation to a level that they feel is commensurate with level of risk. As was discussed in Chapter 7, this approach to handling risk is fundamentally flawed. The reason is that many real business investments are staged. For example, we might divide our technical support software investment into three stages: prototype development and testing, pilot testing, and full enterprise deployment. Table 10.1 lists the costs and technical risks that ACME estimated for its staging scheme. Each of these stages provides management the opportunity to terminate the project early, thus limiting downside risk exposure. The NPV approach implicitly assumes that the project will proceed regardless of failure, a very unlikely scenario. Even worse, it exacerbates the penalty by using a high hurdle rate.

To value this investment correctly, we must account for the downside risk mitigation provided by staging the investment, and this requires dynamic programming. Dynamic programming is a decision analysis technique that handles situations where a series of decisions must be made over time. The well-known decision tree is a simple form of dynamic programming. For the moment, let us assume that our investment problem is a decision tree. Each of the three stages constitutes a decision node where the decision maker can decide either to continue the project or to terminate it. Technical and market uncertainty can be represented by probability nodes in the tree. Figure 10.1 constitutes a notional representation of our R&D investment.

In standard decision tree notation, decision nodes are squares and probability nodes are circles. Decision nodes are associated with each stage of the project. In each case, management may choose to continue or terminate. Following each decision node is a probability node that determines whether the technical effort in the stage is successful. If the stage fails, obviously, the project is over. Even if the technical effort succeeds, however, the cost effectiveness of continuing the project will hinge on the state of the market. Consequently, each technical probability node is followed by a market probability node that determines the change in the computer market over the course of the stage. For simplicity of illustration, the movement of the market is limited to an up or a down movement, but as we will see in subsequent sections, we will not be limited to only two market outcomes.

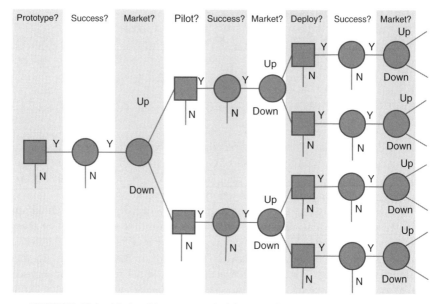

FIGURE 10.1   Notional investment decision tree for technical support project.

It would seem then that valuing our R&D investment should be straightforward. Unfortunately, there is one key piece of information that we do not know. What should the discount rate be? Any fixed discount rate implies that the riskiness of the investment is the same regardless of how the project evolves over time. That does not seem very likely. What if over the course of developing the technical support software, the market for computers expanded rapidly? That would lead to a major reduction in market risk and imply that the discount rate should decline. Thus, it would be economically incorrect to use a fixed discount rate. So how do we know what the discount rate should be? It turns out that we can get around this problem if we employ options analysis. To understand how, it is necessary to present a brief explanation of options theory.

10.3   OPTIONS THEORY

Options are an integral part of modern investment analysis, and consequently, understanding options theory requires at least a cursory understanding of investment analysis in general. To that end, what follows is a very brief review of the evolution of options analysis.

10.3.1   Traditional Investment Analysis

Traditionally, corporate investment decisions have been made using discounted cash flow methods such as the NPV criterion. To calculate NPV, one simply discounts all expected cash flows, both positive and negative, and sums them. If the value is positive then investment is worthwhile. Of course, the perennial question regarding the practical use of NPV is what should the discount rate be? Modigliani and Miller established the convention of using the corporate cost of capital as the discount rate (Modigliani & Miller, 1958). Their rationale is simple; a capital investment should generate a rate of return greater than the cost to finance it. Even so, in a risky world, it is fairly common practice to adjust the hurdle rate so that it is commensurate with the perceived level of investment risk. As many critics have noted, however, this approach tends to result in an overly conservative investment portfolio (Hayes & Abernathy 1980; Hayes & Garvin, 1982). High-risk, high-return projects are rejected in favor of safer but low-return investments.

The flaw with NPV as an investment criterion is that it fails to account for managerial flexibility. In reality, managers have the flexibility to delay, expand, contract, or abandon investment projects. This flexibility allows a manager to limit downside risk exposure. For example, high-risk, high-return projects are typically staged. As each stage proceeds, more information about the nature of the investment is gathered, and uncertainty is reduced. If new information reveals that the investment is no longer desirable, the project can be terminated without incurring additional losses. NPV, on the other hand, implicitly assumes that a failing investment project will continue regardless of the losses incurred. This shortcoming led to the search for alternative methods to value corporate investments.

Of course, dynamic programming is the obvious candidate for handling such dynamic decision problems. Dynamic programming is characterized by starting with a terminal condition and working backward in time to find the optimal policy for every possible decision point (a decision tree is a type of dynamic program). For investment problems, there is the added complication of risk, and thus, there is a stochastic component to any dynamic program used to evaluate them. If the decision makers involved were risk neutral, this would not be a major complication. A risk-neutral decision maker is one who is indifferent between the expected value of a risky return and the equivalent lump sum. Thus, traditional stochastic dynamic programming methods such as decision trees could be used with the discount rate set to the risk-free rate to account for the time value of money.

Unfortunately, most investors are risk averse and would like to be compensated for bearing additional risk. Thus, we find that although dynamic programming solves the managerial flexibility problem, we are still left with the problem of finding an appropriate discount rate. Decision analysis would suggest that utility theory could be used to assess the degree of risk aversion, but with publicly traded companies, doing so is clearly futile because of the inability to query investors in such companies. It turns out that to solve this problem, we must turn to financial options theory.

10.3.2 Financial Options

The impetus for developing financial options methods was born out of the desire to determine a fair price for financial instruments that provide a contingent claim on a traded asset (e.g., stock options). Before discussing the pricing of options, however, it is instructive to review the pricing of traded assets.

Asset pricing has its roots in mean-variance portfolio theory. Mean-variance portfolio theory was developed by Markowitz (Markowitz, 1952, 1956, 1987a, 1987b) and postulates that investors must trade off between expected return and risk (i.e., price volatility). Given two assets with the same expected return, an investor would prefer the one with lower volatility. As traded stocks are not correlated perfectly, an investor actually can reduce volatility by combining stocks into a portfolio, hence, the value of a diversified portfolio. In fact, when considering the entire market, there is an efficient frontier of portfolios.

A portfolio on the efficient frontier is Pareto optimal in that if an investor wished to create a portfolio with a higher expected return, he would have to incur additional risk. Conversely, if he wanted to reduce risk, he would have to accept a lower expected return. Thus, an investor who wants to earn a particular expected return always would prefer to choose the appropriate portfolio from the efficient frontier. It is an interesting property of the efficient frontier that the entire frontier can be generated through a linear combination of any two efficient portfolios. Furthermore, the efficient frontier for a market of risky assets is concave and increasing with risk so investors see diminishing returns as they bear additional risk. If we introduce a risk-free asset into the decision space, the efficient frontier becomes linear and is tangent to the efficient frontier for risky assets at a single point. In theory, this point should be the market portfolio (i.e., a portfolio made up of every asset weighted according to its market capitalization). Thus, every investor should only hold a combination of the risk-free asset and the market portfolio. To do otherwise would be inefficient. This principle forms the basis for the capital asset pricing model (CAPM).

CAPM was developed by Sharpe (1964), Lintner (1965), Mossin (1966), and Treynor (1961), and its premise is that the price and, consequently, the expected return of a traded asset are determined by its level of systematic risk. Essentially CAPM divides the risk inherent in an asset into two components, systematic and nonsystematic. Nonsystematic risk is specific to the asset and can be diversified away by holding additional assets. Systematic risk, however, is the underlying risk of participating in the market, and it cannot be reduced through any more diversification. More generally, nonsystematic risk is firm or industry specific, whereas systematic risk is inherent in the economy. Consequently, when investors price a particular asset, they are only concerned with its level of systematic risk because any other risks can be eliminated through diversification. The remarkable result of CAPM is that the expected return of a security is a linear function of the security's covariance with the market portfolio (see Chapter 7).

Despite the intuitive appeal of CAPM, theoretical and empirical studies have revealed that it is not an entirely accurate model of reality (Black et al., 1972; Merton & Samuelson, 1974). Consequently, several other pricing models have been developed to remedy some of its shortfalls. Despite its shortcomings, CAPM provides an intuitive foundation for understanding the properties of financial options, and all one needs to assume is that some CAPM-like model holds.

Although pricing financial options was an active field of research for some time, it was not until Black and Scholes's seminal paper that combined stochastic calculus with the idea of a replicating portfolio that a closed-form solution was achieved (Black & Scholes, 1973). In particular, Black and Scholes developed a partial differential equation that could be solved to find the price of a European call option. A European call option is a contract that provides the right, but not the obligation, to buy an asset at a prespecified date and price. Perhaps most striking was that this equation depended only on a few observable values: the current price and volatility of the market-traded asset and the risk-free interest rate. What Black and Scholes recognized is that the payoff of a contingent claim such as a call option could be replicated via a dynamic trading strategy involving the underlying asset and risk-free bonds. As the payoffs of the call option and the replicating portfolio are the same, they must have the same value. If that were not the case, then there would be an arbitrage opportunity. This approach solved two problems in investment analysis. It accounted for managerial flexibility in terms of exercising the option, and it made the discount rate endogenous to the model. As price of the replicating portfolio is governed by the dynamics of the traded asset, the implication is that there is an implicit expected return for the portfolio that is based on its level of systematic risk. Thus, the Black–Scholes approach means that a risk-adjusted discount rate need not be specified exogenously. Instead, the pricing of the replicating portfolio reveals investors' risk attitudes.

Black and Scholes's results were enhanced quickly by Merton (Merton, 1973), who developed a more rigorous method of deriving the Black–Scholes result as well as extending the result to account for dividends and some other option formulations. In a subsequent paper, Merton reinforced the Black–Scholes equation by reducing the number of assumptions required to derive it (Merton, 1977).

With the basic approach for option pricing established, other researchers produced a plethora of papers that provide analytic solutions to several different types of options. Some of these include the European compound call option valued by Geske (1977, 1979), the American put option valued by Geske and Johnson (1984), the European exchange option valued by Margrabe (1978), and the European sequential exchange option valued by Carr (1988). There was also interest in valuing options over different price processes. In their original paper, Black and Scholes assumed that the price of the underlying asset followed a diffusion process called geometric Brownian motion (GBM). GBM presumes an underlying exponential growth trend such that asset prices are distributed lognormally. Although this is a good first-order approximation of stock prices, it is not always the most appropriate. Cox and Ross developed properties and pricing methods for options written on assets governed by jump processes (e.g., a Poisson process) (Cox & Ross, 1976), and Merton took this one step further and developed option pricing methods for jump diffusion processes (i.e., a diffusion process such as GBM with randomly occurring jumps) that can be a more accurate representation of stock prices (Merton, 1976).

Perhaps more important than the special-case analytic solutions was the discovery of the existence of an equivalent martingale measure that allows for risk-neutral pricing. The concept was developed over a series of papers by Cox and Ross (1976), Ross (1976), Harrison and Kreps (1979), and Harrison and Pliska (1981; 1983). The basic idea is that if no arbitrage opportunities exist in a market, then an equivalent probability measure to the real probability measure exists (i.e, both probability measures agree on which outcomes have zero probability). Essentially, this equivalent probability measure allows one to generate an alternative probability distribution. Using the alternative probability distribution in place of the real probability distribution allows options to be solved as dynamic programs where the discount rate is the risk-free rate. Thus, the alternative probability distribution is known as the risk-neutral probability distribution. It turns out that this method of solving options yields exactly the same price as the replicating portfolio method, but in many cases, it simplifies the solution procedure.

10.3.3 Real Options

Returning to the topic of corporate investment decisions, it was realized quickly following the Black–Scholes result that many real investment opportunities were analogous to financial options. For example, the construction of a factory provides the owner the option to produce a product. This could be considered a call option. In fact, Myers suggested that corporate discretionary investments could be considered growth options (Myers, 1977). Consequently, the powerful tools developed to price financial options, risk-neutral pricing in particular, could be leveraged to value real investment opportunities.

Option pricing methods remedy two of NPV's flaws. They can account for managerial flexibility, and they obviate the need to specify an exogenous discount rate. In fact, the discount process implicit in an option price is commensurate with the risk attitudes of a firm's shareholders regarding the systematic risk of the investment opportunity. The key implication here is that a firm's shareholders are indifferent to any risks specific to the investment under consideration. As shareholders can diversify away these risks, they can be evaluated as expectations and discounted at the risk-free rate. Any systematic risk introduced by the investment will require additional compensation for shareholders in the form of an increased expected return. Thus, in theory, the implied discount rate for the investment opportunity should be consistent with the discount rate predicted by CAPM given the level of systematic risk.

Under the banner of real options or contingent claims analysis, several researchers developed solutions for valuing many real investment opportunities. Bhattacharya valued a project under a mean reverting cash flow (Bhattacharya, 1978). McDonald and Siegel provided a means for valuing options on assets that earn below the equilibrium rate of return (McDonald & Siegel, 1984) as well as the value of a firm with the option to shut down (McDonald & Siegel, 1985) and the value of deferring investments (McDonald & Siegel, 1986). Brennan and Schwartz considered natural resource investments when valuing a mine that could be shut down and restarted (Brennan & Schwartz, 1985), and Paddock, Siegel, and Smith valued offshore oil leases (Paddock, et al., 1988). Majd and Pindyck valued the option to defer a sequential construction project when there are limits on how fast the project can proceed (Majd & Pindyck, 1987), and Trigeorgis and Mason explored the options to defer, expand, or contract (Trigeorgis & Mason, 1987). Also, Pindyck explored a firm's capacity choice (Pindyck, 1988). Ultimately, Kulatilaka and Marcus claim that most corporate investment options are really switching or flexibility options (Kulatilaka & Marcus, 1988). Of course, options can be generalized even more as Trigeorgis and Mason suggest that real options are just economically corrected versions of decision trees (Trigeorgis & Mason, 1987). The real options literature is now relatively mature, and the key concepts have been collected and distilled in several books, including notably *Investment Under Uncertainty* by Dixit and Pindyck (1994) and *Real Options* by Trigeorgis (1996).

Although it might seem that the sole benefit of options research is an economically corrected form of dynamic programming, its true benefits have been to provide a conceptual framework for corporate investment decisions as well as a set of solution methods for stochastic dynamic decision problems in economic contexts.

10.3.4   Solving a Simple Call Option

To illustrate the theoretical concepts just discussed, let us consider a very simple option. Imagine that we have a call option to buy a share of stock for a particular company at a future time. The price of the stock, S, is governed by a very simple probability process depicted in Figure 10.2. The current price is $10, but it can only assume two possible values in the future. It can either go up to $15 with a probability of 0.6 or it can go down to $5 with a probability of 0.4.

According to the terms of our call option, we have the right, but not the obligation, to buy the share of stock at a price of $10 at the future time. The question, then, is how much is our option worth now? We will denote this as C. If we consider the potential payoffs of the option, we get Figure 10.3.

If the price of the stock goes up to $15, we get a payoff of $5 (since we paid $10 to buy the share). If the price of the stock goes down, exercising the option would result in a loss of $5. As we are not obligated to exercise the option, our best course of action is to forego exercising and walk away with nothing.

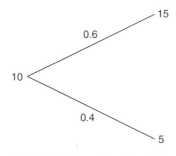

FIGURE 10.2   Price of stock S.

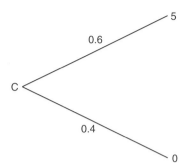

FIGURE 10.3   Call option payoffs.

At this point we have a complete decision tree, and if we knew the appropriate discount rate, we could determine the value of the option, C, by taking the discounted expectation of the payoffs. As discussed, however, we do not know what the discount rate should be. Instead we will employ the replicating portfolio method to determine the value of the call option. The idea is that if we can create a portfolio of market-traded assets that exactly replicates the payoff structure of the option, then the value of the option must be equal to the value of the portfolio. If that were not the case, there would be an arbitrage opportunity in which an investor could make a profit at no risk. Let us examine this concept in more detail.

First, for the purposes of comparison, we note that the effective one period growth rate for the stock is 10%. What if we were to use this discount rate to value the option? As the call option is a derivative of the stock, it seems reasonable that the risk of the option might be similar to the stock. This is loosely analogous to using the cost of capital to evaluate a business investment. When we use 10% to discount the expectation, we get $C = \$2.73$.

Now, let us see whether we can replicate the payoff of the option using a portfolio of traded assets. According to options theory, we can accomplish this by using the underlying stock and a risk-free asset such as a government bond. For sake of this example, we will assume that the risk-free bond earns 5% interest. So, we need to determine the quantities of the stock and the bond that we need to hold in order to replicate the payoff of the call option. Let x be the number of shares of stock we must buy and y be the dollar amount of the risk-free bond we must buy. We must ensure that the value of the portfolio matches the payoff of the option regardless of whether the stock goes up or down. Thus, we require two equations, one for each possible outcome. These are listed as follows:

$$15x + 1.05y = 5$$
$$5x + 1.05y = 0$$

The first equation accounts for the case where the stock goes up. Consequently, the value of the portfolio will be the number of shares purchased times the price ($15) plus the value of the bonds that we purchased (which always earn 5% interest).

This must equal the payoff of the option, which is $5. The equation for the case where the stock declines is structured similarly. When we solve for x and y in this system of equations, we obtain $x = 0.5$ and $y = -2.38$. What this means is that to build our replicating portfolio we must purchase half a share of stock and borrow $2.38. If we do that, then no matter what happens, our portfolio will always have the same payoff as the option. Of course, we cannot really purchase half a share of stock, but if we were to replicate the call option on a large number of shares, this would not be a problem. Regardless, the value of our replicating portfolio at the starting time is $2.62.

This means that the value of the call option also must be $2.62. Note that this value for the option differs from the value of $2.73 that we calculated using the discount rate of the underlying stock. Thus, if we were to price the option at $2.73, we would create an arbitrage opportunity. We could buy the replicating portfolio for $2.62 and then sell the call option for $2.73. As they both have the same payoffs, our positions offset each other, and we always end up with a profit of $0.11 no matter what the underlying stock does. All we would need to do is sell a large number of options and offset it with a replicating portfolio and we could make an extremely large profit, risk free. If this situation existed in the marketplace, we could imagine that the opportunity would not last very long because, once discovered, no one would be willing to pay us $2.73 for the option. The price must be $2.62. Although the replicating portfolio pricing approach was demonstrated here for a trivial example, the concept can be generalized to a dynamic trading strategy to replicate the payoff of an actual stock option. This concept is the basis for the Black–Scholes equation.

We can demonstrate a few other interesting concepts with our simple option example. First, it should be noted that the no-arbitrage price in no way depended on the probabilities of the stock going up or down. We could calculate the option value without them. Second, the no-arbitrage price implies the discount rate for the option. For this example, the discount rate for the option is 14.5%, which suggests that the option carries more systematic risk than the underlying stock. Considering that the replicating portfolio involves borrowing money to buy stock, a higher discount rate certainly seems like a reasonable finding.

There is one final concept that is crucial to understanding how options are valued. It was mentioned earlier that options could be solved using something called risk-neutral probability. The risk-neutral probability distribution is an alternative probability distribution that allows us to dispense with the replicating portfolio and price an option using discounted expectation. It is not a "real" probability distribution in the traditional sense, but it is a convenient mathematical device to simplify pricing options. It turns out that if we replace the real probability distribution governing a traded asset with the risk-neutral probability distribution, we can evaluate an option as if we were risk-neutral decision makers. That means we can discount at the risk-free rate. In essence, we have solved the discount rate problem, and we can solve the option using dynamic programming.

Fortunately, the risk-neutral probability distribution is easy to obtain and will be discussed in greater detail in the next section, but first, let us show how we

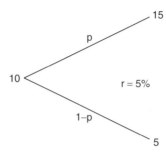

FIGURE 10.4 Solving the risk-neutral probabilities.

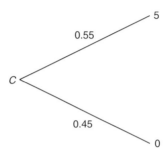

FIGURE 10.5 Call option with risk-neutral probabilities.

can price our simple option using risk-neutral probabilities. To find the risk-neutral probability distribution, we return to our stock and assume that it grows at the risk-free rate. The question then becomes what probabilities would justify a current price of $10 given that we assume that it grows at 5% (Figure 10.4)?

To solve for p, we need to use the discounted expectation:

$$\frac{15p + 5(1-p)}{1.05} = 10$$

This results in $p = 0.55$, and this is the risk-neutral probability that the price of the stock will increase to $15 in the future time period. We can now revisit our option pricing problem but replace the original probabilities with the risk-neutral probabilities (Figure 10.5).

If we apply discounted expectation with the risk-free discount rate, we get $C = \$2.62$, the same price for the option that we obtained using the replicating portfolio.

10.4 SOLUTION METHODS

Although the example from the previous section priced a trivial example of an option, pricing a more complex option requires the application of exactly the same principles. The challenge lies in handling the price process of the underlying asset.

Obviously, the future price of a market-traded asset such as a stock has more than two possible values. Instead, plotting the price history of any stock usually results in a volatile and fairly jagged graph. Practical options analysis requires that we find some way to model the stochastic behavior of such a price process.

10.4.1 Geometric Brownian Motion

The most common price model used for stocks is the previously mentioned GBM (Figure 10.6). It is essentially noisy exponential growth. Although it is not the most accurate model of a stock price, it is a good first-order approximation and is by far the easiest to work with. Unfortunately, even though GBM is the easiest price model, it still poses some challenges.

Let us designate $X(t)$ as the price of stock at time t and assume that this stock has historically grown at an annual rate of μ with a volatility of σ. If the stock price is governed by geometric Brownian motion, then the price process $X(t)$ can be described with the following stochastic differential equation:

$$dX = \mu X \, dt + \sigma X \, dZ$$

This equation defines how much the price changes, dX, over a very short period of time, dt. The first term on the right-hand side provides the exponential growth we typically expect from a stock price. Imagine that our current price is $X(t) = \$10$ and that our stock grows at an annual rate of $\mu = 10\%$. That means that $\mu X = \$1$. Thus, we would expect our stock price to increase by \$1 during the course of a year. The dt term scales the price increase by time. So if our time increment were half a year, we would expect \$0.50 of a price increase. Of course, dt represents an infinitely small increment of time.

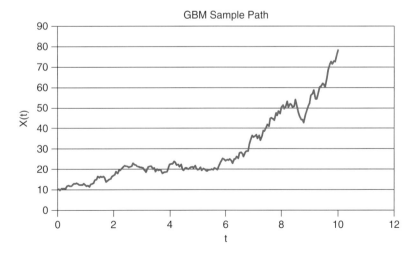

FIGURE 10.6 Sample geometric Brownian motion path.

The second term on the right-hand side is the noise term. This is where stochastic behavior of the stock price comes into play. The term dZ is akin to a random draw from a normal distribution (although we are skipping some details here). This draw is then scaled by the volatility and the current stock price. Thus, we can see that the change in stock price is driven by the expected growth plus a noise term.

On the surface, it would seem that this price process should be relatively easy to work with. Unfortunately, the term dZ is not particularly well behaved. It is an increment of Brownian motion and cannot be addressed using "normal" calculus. Therefore, if one would like to describe the behavior of such a price process, stochastic calculus is required. It is beyond the scope of this chapter to delve into the finer points of stochastic calculus. Suffice it to say that it is often challenging to solve integrals over these types of stochastic processes. This is the chief reason for the difficulty in evaluating options.

10.4.2 Solution Approaches

Assuming that we have selected a process such as geometric Brownian motion to model the stochastic behavior of our underlying asset, there are four main approaches to solving for the value of an option: analytic methods, partial differential equation (PDE) methods, Monte Carlo simulation, and lattice methods.

These four approaches can be divided into two pairs based on their approach. The first pair, analytic and PDE methods, both involve the development of a PDE that describes the option value. The option value PDE is derived from the stochastic process that describes the price of the underlying asset and the dynamic trading strategy that leads to the replicating portfolio. Boundary conditions such as the value of the option at exercise are identified, and if an analytic solution to the differential equation exists, then there is an analytic solution for the value of the option. The Black–Scholes option pricing equation was derived in this manner. If no analytic solution exists, then numerical methods for solving the PDE may be employed. Of course, this is not a general solution, and the option only can be valued over a subset of the state space.

Although many analytic solutions have been found for financial options, most real options do not lend themselves to these approaches, and they require some knowledge of stochastic calculus. The interested reader is directed to Dixit and Pindyck's *Investment Under Uncertainty* (1994) for a fairly accessible treatment of analytic solutions to real options.

The remaining pair of approaches both attempt to model the underlying stochastic price process. Monte Carlo methods simulate the stochastic price process by repeatedly performing random draws and observing the outcome. Unfortunately, Monte Carlo simulation is only feasible for extremely simple options. As stated earlier, solving an option is essentially solving a dynamic program, but the forward simulation essentially precludes the backward induction required to solve all but the simplest options.

Lattice methods are based on the principle that geometric Brownian motion can be approximated using a random walk, and one of the most popular is the binomial lattice method developed by Cox et al. (1979). It functions by employing a random walk in which the state variable only can move discretely up or down. The moves are multiplicative, and the down move is the reciprocal of the up move. Thus, the resulting achievable state space forms a lattice, hence the name. With a discrete state space, the option is effectively a decision tree and can be solved using backward induction. The binomial lattice method can achieve an arbitrary level of accuracy by reducing the size of the time step.

For the nonexpert, lattice methods are the most practical and flexible of these approaches. Consequently, we will focus on describing the lattice approach in detail, in particular the binomial lattice approach.

10.4.3 Binomial Lattice Approach

The binomial lattice approach effectively operates by extending the solution method employed to solve the simple call option example. If we take our simple, discrete up/down model and repeat it in a particular way, we have the binomial lattice method.

Figure 10.7 illustrates a lattice for a price process. The lattice is a discrete approximation of the continuous geometric Brownian motion process. The smaller the time step and the longer the time horizon, the better the lattice approximates GBM. Of course, the sizes and probabilities of the up and down price movements must be selected in a very particular way. In particular, the up movement is defined as

$$u = e^{\sigma\sqrt{\Delta t}}$$

where σ is the volatility and Δt is the size of the time step. The lattice is multiplicative, so the current price is multiplied by the up movement to get the price

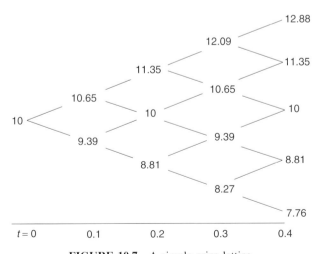

FIGURE 10.7 A simple price lattice.

increase for the next time step. The current price is multiplied by the reciprocal of the up movement to get the price decrease. It is in this manner that the set of possible prices over time form a lattice. The lattice depicted in Figure 10.7 was generated assuming an annual volatility of 0.2 and a time step of 0.1 years.

The probability of an up move is determined by

$$p_{up} = \frac{1}{2}\left(1 + \frac{\mu - \frac{\sigma^2}{2}}{\sigma}\sqrt{\Delta t}\right)$$

where μ is the growth rate. If we assume that $\mu = 0.1$ per year, then we obtain a probability of 0.56 for an up move. Thus, we have a discrete and finite representation of our stochastic process. That means that any decisions based on the value of this price process can be evaluated just as we would a decision tree. Decision trees are evaluated using backward induction. Thus, we start with the last decision and determine the optimal course of action for every possible state. Then we roll back and discount the values obtained from the optimal policy to the next-to-last decision. The process repeats until we reach the first decision.

The lattice method could be used to evaluate any dynamic decision problem over a GBM random variable, but because we are considering options, we want the no-arbitrage solution. We certainly could use the replicating portfolio approach, but this would require us to calculate the replicating portfolio for every possible price and time combination. Instead, if we replace the up and down probabilities with the risk-neutral probabilities, we can evaluate the option as a decision tree with the risk-free discount rate. Fortunately, the risk-neutral probabilities are trivial to calculate for the binomial lattice.

$$p_{up} = \frac{(1 + r\Delta t) - d}{u - d}$$

where r is the risk-free rate, u is the up move, and d is the down move.

The power of the lattice method is that it easily accommodates any number of decisions. Imagine that we wanted to evaluate the price of an American call option with a decision horizon of one year. An American option is an option that can be exercised at any time. Thus, we would build a lattice with a one year horizon and consider whether or not we should exercise the option at each possible time step. This also means that we can use the lattice method to evaluate a multi-stage option like the one presented in the example. In the next section, we will use the binomial lattice method to do just that.

10.5 SOLVING THE EXAMPLE

Now that we have a method for solving a multistage option, we can return to our tech support example. We determined earlier that the value of deploying the new software will come in the form of labor savings on tech support calls. Thus, the asset we are seeking to acquire is the expected net present value of future cost savings from shorter tech support calls. As the number of tech support calls is linked to the number of computers sold by the company, we must have a model for computer sales. Let $S(t)$ be the annual sales rate for new computers. Historically, computer sales for this company have grown at an annual rate of $\alpha = 10\%$ but have been fairly volatile with an annual log-volatility of $\sigma = 30\%$. We will assume that we can model the sales rate using geometric Brownian motion. Thus,

$$dS = \alpha\, S dt + \sigma S dZ.$$

So if 1% of new computer sales require tech support and we expect the new software to save an average of $25 per call, then the instantaneous cash flow of savings would be 0.25 $S(t)$.

To find the net present value of this cash flow, we need only discount and integrate. Let $X(t)$ be the expected net present value of the cash flow stream:

$$X(t) = E\left[\int_t^\infty 0.25\, S(s)e^{-\mu s} ds\right]$$

As discussed, the expected net present value of future cash flows can be considered analogous to a stock price. Thus, $X(t)$ is like the price of the stock that we have an option to acquire. But there is a remaining complication. Our equation contains a discount rate, μ. Where do we get μ? The real options approach assumes that we can find a twin-security that is traded in the marketplace. The price of the twin security, or portfolio of securities, must move up and down with the value of our underlying asset, in this case, the cost savings from our tech support software.

Identifying or justifying the existence of the twin security for a real option on a nontraded asset is one of the trickiest aspects of real options analysis. As a practical matter, we may not be able to find an exact match, and it is probably not worth the effort regardless. Probably the most common approach in the literature is to identify a traded asset that can be assumed reasonably to correlate highly with the underlying asset. Often this is assumed to be the stock price of the company or industry in question.

For our example, we will assume that the stock price of the computer company correlates strongly with the company's computer sales, and consequently, the company stock will serve as our twin security. The stock price has the same volatility of 30% and an annual growth rate of 20%. Because our twin security and our underlying asset should have the same level of systematic risk, investors in the marketplace would demand a 20% rate of return on our asset if it were traded. Therefore, our discount rate, μ, must be 20%.

Simplifying the equation for $X(t)$, we get[1]

$$X(t) = \int_t^\infty 0.25\, E[S(s)]e^{-\mu s} ds$$

---

[1] As the integrand is non-negative we can apply Tonelli's Theorem and move the expectation inside the integral. The expected value of a geometric Brownian motion process, S, with growth rate α is simply $S_o e^{\alpha t}$. Because $\mu > \alpha$, we can solve the resulting integral.

$$X(t) = \int_t^\infty 0.25\, S(t) e^{-(\mu-\alpha)s}\, ds$$

$$X(t) = \frac{0.25}{\mu - \alpha} S(t)$$

Those familiar with traditional investment analysis may note that $X(t)$ is basically a growing annuity. Of course, it is dependent on a stochastic process, $S(t)$, and, thus, is stochastic itself. What kind of stochastic process is $X(t)$? It turns out $X(t)$ is also a geometric Brownian motion process with the same growth rate and volatility as $S(t)$.[2] Thus, we have the option to acquire an asset, $X(t)$, that behaves just like a stock price.

We are almost ready to apply the binomial lattice method to solve our option, but first, we must discuss one more issue. Let $\delta = \mu - \alpha$, where δ is the difference between the growth rate of our underlying asset and the twin security, and it constitutes an opportunity cost. It can be considered equivalent to a dividend payment on a traded stock. Imagine that we had a call option on a share of stock that pays a dividend. Until we exercise the call option, we are missing out on those dividend payments. Thus, the dividend is an opportunity cost when we hold the option over the stock, and it must be accounted for when we price the option.

Fortunately, pricing an option with a continuous proportional dividend requires only a minor modification to our binomial lattice method. When we calculate our risk-neutral probabilities, we simply replace r with r-δ.

We are now ready to apply the binomial lattice method to our tech support software option. All parameters we need are listed in Table 10.1 and 10.2. Table 10.1 lists the cost and duration of each stage of the project as well as the probability that the technical effort will succeed. Table 10.2 lists all parameters necessary to construct the binomial lattice. The lattice essentially describes the market risk. Note that the time step selected for our lattice is one month and that the risk-free discount rate is assumed to be 5%.

The first step in applying the binomial lattice method is to generate the price lattice. The price lattice is the set of all possible values of the underlying asset, $X(t)$, that are achievable in each time period. The price lattice for the first six months is shown in Figure 10.8. We start with the initial value of $12.5 million at the starting time and multiple it by the up and down moves to obtain the possible values of $X(t)$ for the first month. This process is repeated to generate the remainder of the lattice. As the development of our tech support software will take 2 years, we must extend our price lattice out to 24 months.

Now that we have a model for the behavior of the underlying asset over the development period, we can solve the multistage option using backward induction. First, we create a second parallel lattice with an empty slot for each element of the price lattice. For convenience, we will call this the option lattice. The option lattice serves to identify the value of the multistage option for every possible value of the underlying asset. To populate the option lattice, we start with the last decision first.

---

[2] Applying Ito's Lemma to the definition of $X(t)$ will verify this.

TABLE 10.2   Option Model Parameter Values

Parameter	Value	Description
S_0	5	Computer sales in million units per year at time 0
X_0	12.5	Expected NPV of cost savings in $ million per year at time 0
μ	0.2	Annual growth rate of the twin security
σ	0.3	Annual log volatility of the of underlying asset
δ	0.1	The opportunity cost of holding the underlying asset
ΔT	0.083333	The time increment of the binomial lattice in years
r	0.05	The risk-free growth rate
u	1.090463	The up movement for the binomial lattice
d	0.917042	The down movement for the binomial lattice
p_{up}	0.454337	The probability of an up movement for the binomial lattice
p_{down}	0.545663	The probability of a down movement for the binomial lattice

Months	0	1	2	3	4	5	6
Price	12.5	13.63079	14.86387	16.20851	17.67478	19.2737	21.01726
		11.46302	12.5	13.63079	14.86387	16.20851	17.67478
			10.51206	11.46302	12.5	13.63079	14.86387
				9.639999	10.51206	11.46302	12.5
					8.840279	9.639999	10.51206
						8.106903	8.840279
							7.434367

FIGURE 10.8   Price lattice for the first six months.

The last decision is the deployment decision and that will occur in two years. If ACME successfully deploys the new tech support software, it will obtain the next present value of the cost savings, $X(t)$, minus the cost of deployment. Of course, Table 10.1, indicated that there is only an 80% chance that the deployment effort will be successful. Thus, the value of ACME exercising its option to deploy the tech support software is determined by $0.8X(t) - 10$. According to our price lattice, there are 25 possible values for $X(t)$ at the two-year mark. Thus, we must evaluate the value of exercising the option for all 25 values. If the value of exercising the option is negative, ACME will choose not to exercise its deployment option, and consequently, the option will be worth nothing. All 25 option values for the deployment decision are listed in Table 10.3 in two sets of two columns.

We can see from the table that when expected net present value of future savings is greater than or equal to $14.86 million, it is optimal for AMCE to exercise the deployment option. If it is less than $14.86 million, ACME will drop the project.

Now we can continue working backward. We must roll back the option values along the lattice using discounted expectation with the risk-neutral probabilities and the risk-free rate. Each time we roll back, we fill in another slot in the option lattice. We do this until we reach the pilot testing option one year from now. This

TABLE 10.3 Option Values for Deployment

X(t)	Option Value	X(t)	Option Value
99.90	69.92	10.51	0
84.01	57.21	8.84	0
70.65	46.52	7.43	0
59.42	37.53	6.25	0
49.97	29.97	5.26	0
42.02	23.62	4.42	0
35.34	18.27	3.72	0
29.72	13.77	3.13	0
24.99	9.99	2.63	0
21.02	6.81	2.21	0
17.67	4.14	1.86	0
14.86	1.89	1.56	0
12.50	0		

option is evaluated just like the previous, only now we have the option to acquire the discounted expected value of the deployment option rather than $X(t)$. We note that the pilot stage will cost \$1 million and has a 70% chance of succeeding. The option values for this stage are listed in Table 10.4. Note that this table provides the optimal exercise policy for ACME based on the value of $X(t)$ at the decision point for the pilot testing stage.

We repeat the rollback process until we reach the first decision point, prototype development. As that is the first decision, there is only one option value to compute, and this also happens to be the value of the multistage option to develop the tech

TABLE 10.4 Option Values for Pilot Testing

X(t)	Option Value
35.34	10.25
29.72	7.40
24.99	5.01
21.02	3.04
17.67	1.47
14.86	0.33
12.50	0
10.51	0
8.84	0
7.43	0
6.25	0
5.26	0
4.42	0

support software. For this example, the net option value is approximately \$45,000. As it is positive, it is worthwhile for ACME to invest in this project.

One other important point to note is that using the binomial lattice method also yields the optimal policy for each stage in terms of $X(t)$. As $X(t)$ is determined by the level of sales, we can also express the optimal policy in terms of sales, a much more observable and meaningful quantity.

Earlier in this chapter we discussed the shortcomings of the traditional NPV method. The problem with NPV can be illustrated most clearly by calculating it for this example and by comparing it with the net option value. To do so, we will assume that there is no staging, so the development costs are incurred to obtain the expected value of the cost savings. For a discount rate, we will use $\mu = 0.2$, the growth rate for the twin security, since it has the most comparable risk characteristics. Under these assumptions, we find the NPV for investing in this project is $-\$4.77$ million. If ACME were to use NPV as its decision criteria, it would reject an otherwise profitable opportunity.

Although this illustrative example was relatively simple, the advantages of the real options approach should be clear. Managerial flexibility can have a significant value, especially for high-risk, high-return investments. A real options approach captures this value, whereas traditional approaches to investment analysis do not.

10.6 PRACTICAL ISSUES

It is important to note that although real options analysis does remedy some shortcomings of traditional investment analysis, it is still subject to many of the same limitations of other discounted cash flow (DCF) methods, including NPV. Perhaps the most important of these is that both real options and NPV are predicated on the existence of a twin security or portfolio of traded assets that replicates the stochastic behavior of the underlying asset. It is through these market-traded assets that we can determine the rate of return that investors demand from an investment. Depending on the nature of the investment under consideration, it can be challenging to identify such market traded assets. Imagine if someone had tried to value the option to introduce the first personal computer into the marketplace. There was simply no comparable market to use as a proxy for risk and expected return. This situation violates the assumption of market completeness implicit in DCF methods. More specifically, the investment cannot expand investors' decision set.

A related issue is that of parameter estimation. If the identification of a replicating portfolio of market-traded assets is difficult, then estimating parameters such as volatility is likely to be difficult as well. Particularly challenging is the assessment of the δ parameter like the one used in the example problem. The use of δ originates because the expected return on non-market-traded assets may differ from that of the twin security traded in the marketplace. In the options literature, δ may fill various roles such as representing a dividend, a convenience yield, or competition. In all cases, however, it constitutes an opportunity cost incurred from holding versus exercising the option. Unfortunately, in many cases, it is difficult to estimate, and all too often, it is simply ignored.

Another challenging aspect of real options is the incorporation of competition. Imagine that two firms both have the option to develop competing products. It would be incorrect to evaluate either firm's option to develop the product as exclusive. Competitive effects must be considered because each firm must consider the possibility that its competitor may beat it to the market. Typically this is accomplished by introducing game theory into the real options framework. Although this has been accomplished in the academic literature for special cases, solving options that incorporate competition is often challenging without extreme simplification.

Finally, when a real option is used to evaluate an investment opportunity, the result is how the marketplace would value the option, not necessarily its value to the option holder. As the marketplace only demands compensation for systematic risk, it implies risk neutrality toward any risks specific to the investment (e.g., technical risk). It is unlikely, however, that many decision makers faced with multistage real investments would be indifferent to technical risk, especially when their jobs are on the line. Real options analysis, by its very nature, does not account for the risk aversion that may be of critical importance to many real decision makers.

10.7 CONCLUSIONS

Real options remedy some shortcomings of traditional investment analysis techniques such as NPV. In particular, real options account for the value of flexibility that originates in investments that occur over time. The ability to make course corrections based on emerging information significantly reduces downside risk and increases the value of an investment. This chapter, in particular, considered the application of real options analysis to an investment opportunity that is broken into multiple stages. A notional example was presented and served as a vehicle for explaining both the theory and the practice of solving a multistage real option.

Although in some sense real options are simply economically corrected decision trees, the benefit of the financial options analogy is two-fold. First, financial options are well understood by many business decision makers. Framing investment decisions in terms of options makes the results of an investment analysis readily accessible to those decision makers. In particular, it highlights contingent opportunities that one would like the right, but not the obligation, to pursue. Second, many powerful tools such as the binomial lattice method have been developed to value financial options. Extending the analogy allows those tools to be applied to evaluate real investment decisions.

REFERENCES

Bhattacharya, S. (1978). Project valuation with mean reverting cash flow streams. *Journal of Finance*, *33*(5), 1317–1331.

Black, F., Jensen, M., & Scholes, M. (1972). The capital asset pricing model: Some empirical tests. In M. Jensen, Ed., *Studies in the Theory of Capital Markets*. New York: Praeger.

Black, F., & M. Scholes, M. (1973). The pricing of options and corporate liabilities. *Journal of Political Economy*, *81*(3), 637–654.

Brennan, M.J., & Schwartz, E.S. (1985). Evaluating natural resource investments. *Journal of Business*, *58*(2), 135–157.

Carr, P. (1988). The valuation of sequential exchange opportunities. *Journal of Finance*, *43*(5), 1235–1256.

Cox, J.C., & Ross, S.A. (1976). The valuation of options for alternative stochastic processes. *Journal of Financial Economics*, *3*, 145–166.

Cox, J.C., Ross, S.A., & Rubinstein, M. (1979). Option pricing: A simplified approach. *Journal of Financial Economics*, *7*(3), 229–263.

Dixit, A.K., & Pindyck, R.S. (1994). *Investment Under Uncertainty*. Princeton, NJ: Princeton University Press.

Geske, R. (1977). The valuation of corporate liabilities. *Journal of Financial and Quantitative Analysis*, *12*, 541–552.

Geske, R. (1979). The valuation of compound options. *Journal of Financial Economics*, *7*, 63–81.

Geske, R., & Johnson, H.E. (1984). The American put option valued analytically. *The Journal of Finance*, *39*(5), 1511–1524.

Harrison, J.M., & Kreps, D.M. (1979). Martingales and arbitrage in multiperiod securities markets. *Journal of Economic Theory*, *20*(3), 381–408.

Harrison, J.M., & Pliska, S.R. (1981). Martingales and stochastic integrals in the theory of continuous trading. *Stochastic Processes and their Applications*, *11*(3), 215–260.

Harrison, J.M., & Pliska, S.R. (1983). A stochastic calculus model of continuous trading: Complete markets. *Stochastic Processes and their Applications*, *15*(3), 313–316.

Hayes, R.H., & Abernathy, W.J. (1980). Managing our way to economic decline. *Harvard Business Review*, *58*(4), 67–77.

Hayes, R.H., & Garvin, D.A. (1982). Managing as if tomorrow mattered. *Harvard Business Review*, *60*(3), 70–79.

Kulatilaka, N., & Marcus, A.J. (1988). General formulation of corporate real options. *Research in Finance*, *7*, 183–199.

Lintner, J. (1965). The valuation of risk assets and the selection of risky investments in stock portfolios and capital budgets. *Review of Economics and Statistics*, *47*, 13–37.

Majd, S., & Pindyck, R.S. (1987). Time to build, option value, and investment decisions. *Journal of Financial Economics*, *18*, 7–27.

Margrabe, W. (1978). The value of an option to exchange one asset for another. *Journal of Finance*, *33*(1), 177–186.

Markowitz, H.M. (1952). Portfolio selection. *Journal of Finance*, *7*(1), 77–91.

Markowitz, H.M. (1956). The optimization of a quadratic function subject to linear constraints. *Naval Research Logistics Quarterly*, *3*(1–2), 111–133.

Markowitz, H.M. (1987a). *Mean-Variance Analysis in Portfolio Choice and Capital Markets*. New York: Basil Blackwell.

Markowitz, H.M. (1987b). *Portfolio Selection*. New York: Wiley.

McDonald, R.L. & Siegel, D.R. (1984). Option pricing when the underlying asset earns a below-equilibrium rate of return: A note. *Journal of Finance*, *39*(1), 261–265.

McDonald, R.L., & Siegel, D.R. (1985). Investment and the valuation of firms when there is an option to shut down. *International Economic Review*, 26(2), 331–349.

McDonald, R.L., & Siegel, D.R. (1986). The value of waiting to invest. *Quarterly Journal of Economics*, 101(4).

Merton, R.C. (1973). Theory of rational option pricing. *Bell Journal of Economics and Management Science*, 4(1), 141–183.

Merton, R.C. (1976). Option pricing when underlying stock returns are discontinuous. *Journal of Financial Economics*, 3, 125–144.

Merton, R.C. (1977). On the pricing of contingent claims and the Modigliani-Miller theorem. *Journal of Financial Economics*, 5, 241–249.

Merton, R.C., & Samuelson, P.A. (1974). Fallacy of the log-normal approximation to optional portfolio decision-making over many periods. *Journal of Financial Economics*, 1, 67–94.

Myers, S.C. (1977). Determinants of corporate borrowing. *Journal of Financial Economics*, 5, 147–175.

Modigliani, F., & Miller, M.H. (1958). The cost of capital, corporate finance, and the theory of investment. *American Economic Review*, 48(3), 261–297.

Mossin, J. (1966). Equilibrium in capital asset markets. *Econometrica*, 34(4), 768–783.

Paddock, J.L., Siegel, D.R., & Smith, J.L. (1988). Option valuation of claims on real assets: The case of offshore petroleum leases. *Quarterly Journal of Economics*, 103(3), 479–508.

Pindyck, R.S. (1988). Irreversible investment, capacity choice, and the value of the firm. *American Economic Review*, 78, 969–985.

Ross, S.A. (1976). The arbitrage theory of capital asset pricing. *Journal of Economic Theory*, 13(3), 341–360.

Sharpe, W.F. (1964). Capital asset prices: A theory of market equilibrium under conditions of risk. *Journal of Finance*, 19, 425–442.

Treynor, J.L. (1961). *Towards a Theory of Market Value of Risky Assets*. Unpublished manuscript.

Trigeorgis, L. (1996). *Real Options: Managerial Flexibility and Strategy in Resource Allocation*. Cambridge, MA: MIT Press.

Trigeorgis, L., & Mason, S.P. (1987). Valuing managerial flexibility. *Midland Corporate Finance Journal*, 5, 14–21.

Chapter **11**

Organizational Simulation for Economic Assessment

Douglas A. Bodner

11.1 INTRODUCTION

This chapter focuses on using organizational simulation as a method for assessing the economics of human systems integration (HSI), particularly from an investment decision-making perspective, and within the context of systems or product development. In other words, what investments in HSI should be made, when in the system or product life-cycle process should they be made, and what outcomes result from these investments. Organizational simulation is proposed as one means of evaluating such investments before they are made. Simulation is an appropriate method for analyzing investment decision making because it enables modeling of complex organizations and systems that are not amenable to purely analytic approaches, and it incorporates the uncertainty inherent in investing. Organizational simulation is an emerging methodology that incorporates computer simulation technology with organizational modeling, human behavior modeling, and potentially immersive organizational experiences (Carley, 2002; Prietula et al., 1998; Rouse & Boff, 2005). As HSI investment decisions occur within the context of organizations, organizational simulation has the potential for a richer assessment of HSI investments than traditional approaches to simulation.

This chapter considers the system or product life cycle as a sequence of activities ranging from research and development, to system design and integration, to

The Economics of Human Systems Integration: Valuation of Investments in People's Training and Education, Safety and Health, and Work Productivity. Edited By William B. Rouse
Copyright © 2010 John Wiley & Sons, Inc.

production and deployment, to operation and sustainment, and finally to retirement (Cochrane & Hagan, 2005; Saaksvuori & Immonen, 2003). (As the focus is on human systems integration, the term "system" generally will be used to refer to both systems and products that are organizational outputs.) Typically, in framing HSI investments, the investment is considered as an upstream cost (e.g., in design and integration), whereas the outcomes are realized downstream (e.g., in operation). These outcomes may be in the form of increased revenue, reduced cost, or improved system performance. Of course, this assumes positive outcomes from the investment, which may not necessarily be the case. Fundamentally, though, the two outcome types discussed in this chapter are financial and performance outcomes. Of course, there may be multiple ways to measure performance for a particular system.

Earlier chapters discuss several relevant topics in detail. First, because outcomes are realized downstream of investments, the time value of money comes into play. Thus, discount rates and cost of capital are of concern when making investments. These can be used to compute net present values of investment opportunities. Alternatively, investments can be posed in an options framework, using real options analysis. Some investment opportunities can be posed as multistage options. This is especially true in complex systems development, which involves multiple stages of research, development, design, integration, and so on, potentially with a decision at each stage as to whether to continue funding. The dual nature of outcomes (cost and revenue vs. performance) lends itself to multiattribute utility analysis, cost/benefit analysis, or cost effectiveness analysis, depending on the particular type of organization involved and system being developed. Finally, organizations have competitors and collaborators, making concepts from game theory relevant.

The cash flows and performance outcomes tell only part of the story, though. The type of organization involved and its structure have a major impact on the investment analysis. For a complex system, these functions of the system life cycle may involve separate organizations within a networked enterprise. Thus, organizational dynamics arise. Consider an organization charged only with design and integration, for example. It may view an HSI investment as purely a cost, if the downstream payoff is realized in a separate organization that addresses production and deployment. Thus, there is little incentive for it to invest in HSI.

This chapter starts by addressing simulation in general and organizational simulation in particular. It then discusses use of organizational simulation in analyzing HSI investments and provides examples of how the economics of HSI can be assessed. As organizational simulation is an emerging field, thoughts are provided on future directions.

11.2 SIMULATION

Simulation has a rich history of use in the domain of systems engineering for design and analysis (Banks et al., 2005; Law & Kelton, 2000). It is important to understand basic simulation terminology and how that terminology is modified slightly for organizational simulation.

11.2.1   Terminology

In the traditional simulation context, a *system* is the set of entities and interactions between them that is under study. Here, a system has a broad connotation, as opposed to usage in this chapter for organizational simulation, and it can mean anything from a galaxy, to an economy, to a factory, to an airplane cockpit, to a bacterial colony. A simulation *model* is a computational representation of the system and its behavior over time, where behavior typically is considered as a set of state changes. The model is used to analyze the system's behavior and performance under different circumstances. Such analysis can explore outcomes of new systems or new system configurations. It also can be framed as a simulation *experiment*, in which alternative system designs are compared statistically to see which performs best, or in which the effects of various factors on performance are determined. Thus, simulation can support system design and improvement.

Although simple models can be studied by manual simulation, simulation models typically are executed as software programs. A simulation *run* is a single execution of the model, which simulates the system's behavior over a set time horizon. Statistics are then collected during the run on various phenomena of interest (performance metrics, costs incurred, etc.). Some simulation models need a *warmup period* prior to collection of statistics so that the model reaches steady-state behavior. Most simulation models employ *random number generators* used to sample values from various probability distributions so as to model randomness or uncertainty. This means that the statistics collected during one run tend not to be representative of system behavior in general. Thus, analysts typically execute multiple *replications* of a run, varying the random number streams in each, so as to generate a set of statistics across the replications that represent general system behavior and can be analyzed for statistical significance. Multiple replications are especially needed in experimental analysis.

In the context of organizational simulation, it should be noted that the system under consideration is an *organization*, or potentially an *enterprise* consisting of multiple organizations. The assumption here is that this organization or enterprise is engaged in systems development. Thus, in this chapter, the term "system" refers to the products that it develops. These products might be military systems (e.g., planes or ships), consumer products (e.g., cell phones or computers) or infrastructure (e.g., transportation systems or power plants). The remainder of this chapter uses the concepts and terminology depicted in Figure 11.1. The systems of concern here involve human systems integration as shown. A system undergoes a life cycle within the context of an enterprise. It should be noted that the enterprise may consist of different organizations responsible for different aspects of the life cycle. The organizational simulation model represents both the enterprise and the system(s) produced. Depending on the modeling and analysis purpose, of course, it may represent only a subset of the enterprise and/or systems produced. It may also focus on the enterprise and its behavior (i.e., does the enterprise make good HSI investments), or on the systems produced and theirs (i.e., did a particular HSI investment result in reduced cost or improved performance in the organization

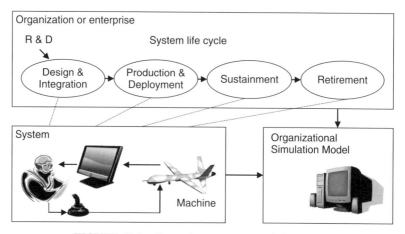

FIGURE 11.1 Enterprises, systems, and simulation.

where it is deployed). In this chapter, the focus of the model is called the *simulated world* (i.e., specific elements that are modeled) or the *application* (i.e., the general class of simulated worlds to which the model might be applied).

11.2.2 Modeling Human Behavior

In economic assessment of HSI, simulations can be called on to model investment decision making within an organization. Although organizational decision making typically is based on rules, it also includes a component of human behavior. Similarly, in assessing HSI outcomes in particular systems, simulations are called on to model human behavior and interaction with a system. There are two primary methods for modeling human behavior. First, a simulation model can be designed to interact with a human subject, so that a human-in-the-loop simulation results. The simulation model then addresses the "rest of the simulated world" (e.g., sensors, machines, physical phenomena, etc.). Human-in-the-loop simulations have been used extensively and often are used for training in such environments as airplane cockpits, nuclear power plants, and military command centers. A human-in-the-loop simulation requires two important elements—interfaces to enable interaction (e.g., computer screen interfaces and/or controls) and pauses between state changes in the simulation that reflect the delays that someone would experience interacting with the real organization or system. Simulations that do not address HSI typically are executed without such delays, so as to reduce the time needed to execute the simulation.

The other approach is to model human behavior as an embedded part of the simulation, relying on programmed decision rules and other such constructs. Traditional simulation approaches have treated humans as resources that perform physical tasks such as loading a part onto a machine in a factory and have modeled human decision making and cognition in very limited detail. Recent advances in agent-based simulation (Hillebrand & Stender, 1994) and cognitive frameworks (Gluck & Pew, 2005) have brought more realistic human decision-making and cognition modeling capabilities to simulation. In particular, it is of interest to model how humans respond to incentives, and how they respond to the availability of information.

Clearly, there are tradeoffs between these two approaches. Human-in-the-loop simulation provides a more accurate representation of human behavior but is expensive in terms of personnel time required and development time (primarily for interface development). This is especially true if experimentation is desired (which requires multiple replications). It is also potentially more limited in the types of human behavior that can be modeled because of limitations on the availability of human subjects. Also, the human-in-the-loop approach generally has not been practical when analyzing systems with many humans. Hence, one alternative is to meld these two approaches such that a human-in-the-loop interacts with simulated humans in the simulated world. Another alternative is enabled by the advent of massively multiplayer online role playing games (MMORPGs), in which multiple humans-in-the-loop interact with simulated humans in the context of game-play (Castronova, 2006).

Finally, human interaction with simulation models is not confined to modeling human behavior within the simulated world. A simulation may be designed so that a user can interact with the model without being part of the simulated world. For instance, the user might be an observer who navigates through the simulated world, seeing the effects of various scenarios play out. The user also might take a more active role, if the capability is provided to allow the user to change features of the simulated world while the simulation model is executing. Thus, the user can experiment and perform what-if analysis based on what currently is happening in the simulation. Figure 11.2 illustrates the various ways in which humans can be modeled and can interact with a simulation model.

11.2.3 Paradigms for Modeling

In general, behavior can be viewed from multiple perspectives, and indeed different paradigms of simulation have been developed to accommodate different perspectives. One common distinction between behavioral perspectives involves whether

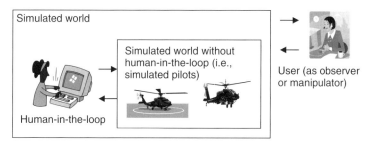

FIGURE 11.2 Human modeling and interaction.

behavior is viewed as a continuous state change process or as a discrete state change process. Continuous state change models are appropriate for fluid, mechanical, and physiological applications, where continuous motions are important to the analysis of behavior. Examples of such applications include chemical processing plants (fluid flows), robotic systems (kinematics), and ergonomics applications (human motion).

Discrete state change models, however, abstract behavior to a series of events or instantaneous state changes. Events often represent the start and completion of activities. The set of events then is used to characterize behavior rather using than continuous state changes. Examples of applications typically represented using discrete state change models include discrete-parts manufacturing factories (start and finish of material processing on machines), airports (takeoffs and landings), and call centers (arriving calls and service events). The focus of these models is on the events such as plane landings, rather than on the aerodynamics of plane flight. One common theme underlying many of these applications is the need to schedule events or to analyze the effectiveness of different ways to schedule events.

Based on these concepts, three common paradigms have evolved that are particularly relevant for organizational simulation: discrete-event simulation, system dynamics simulation, and agent-based simulation. Discrete-event simulation, as implied by the name, focuses on modeling the events in the simulated world. Discrete-event simulation supplies three perspectives for representing events—the process-interaction perspective, the event-scheduling perspective, and the activity-scanning perspective (Law & Kelton, 2000). The process-interaction perspective is perhaps the most commonly used. Process-interaction simulations use threads (also called processes, entities, or transactions) that flow through a modeler-specified set of blocks representing the processes in the simulated world. The typical set of blocks exhibits a seize-hold-release behavior typical of many resource-processing organizations. Hence, the process-interaction perspective has come to predominate in simulations of discrete manufacturing and service organizations, where the flow of material and work through resources is an important world feature. Event-scheduling, however, provides a more detailed representation by modeling individual events. Event-scheduling typically has been used for detailed, application-specific simulators. Activity-scanning simulations continuously scan availability of resources in the model. When resources become available, the simulation assigns them to activities that have requested them. Activity scanning is computationally intensive, and is not widely used, except for applications in construction organizations, where it is well suited for modeling infrequent activities that require multiple resources (Martinez & Ioannou, 1999).

System dynamics takes a different approach to modeling the world than discrete-event simulation. Derived from Forrester (1961), a system dynamics representation consists of a set of stocks, flows between stocks, and feedback mechanisms. Stocks represent a quantity of items (e.g., funds, inventory, and people), and flows represent rates of change in those quantities. Feedback mechanisms represent nonlinear relationships between different elements in the system. A system dynamics model is a continuous representation focusing on rates of change and time delays, rather than on a discrete one focusing on events. System dynamics has been applied to model businesses (Sterman, 2000), the environment (Ford, 1999), populations (BenDor & Metcalf, 2005), and human–technology relationships (Pavlov & Saeed, 2004). The power of system dynamics comes from its ability to explain and experiment with complex, nonlinear phenomena through these feedback mechanisms.

Neither discrete-event simulation nor system dynamics simulation is particularly effective at modeling humans for the purposes of HSI. To represent humans more effectively, one promising technology is agent-based simulation (Hillebrand & Stender, 1994). As such, agent-based simulation has gained significant attention in the social sciences (e.g., Saam & Schmidt, 2001) and in military studies (e.g., Ilachinski, 2004). An agent-based simulation consists of several agents representing various real-world elements. Each agent has its own set of behaviors, and at each step in the simulation, an agent may execute behaviors based on its internal state or the state of the simulated system. Thus, agent-based simulation can give rise to emergent behavior originating from the actions and interactions of independent actors. Although an agent need not be a human, there is potential for using agents to represent human behavior and interactions.

11.3 ORGANIZATIONAL SIMULATION FRAMEWORK

Organizational simulation involves a variety of concepts illustrated by the conceptual framework in Figure 11.3. This figure does not illustrate an architecture but focuses on individual layers that apply to organizational simulation. The discussion in this chapter focuses on the three shaded levels after an initial overview of the others. As organizational simulation is an emerging methodology, this section concludes with an overview of the current state of the art relative to representation and technology.

Of course, organizational simulation is based on hardware platforms that execute the simulation models. This hardware could involve a single computer or multiple networked computers. Basic simulation functionality is provided by simulation

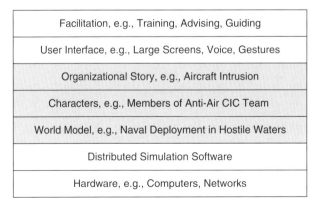

FIGURE 11.3 Organizational simulation framework.

software. This functionality includes the state change mechanisms needed to simulate behavior, random number generators, and statistics collection capability. If the simulation model executes over multiple networked computers, then the simulation software must be distributed in nature. The software must then synchronize the different submodels running on each computer to maintain logically correct behavior and proper temporal sequence (Fujimoto, 2000).

Advanced organizational simulation capabilities include facilitation and rich user interfaces. Facilitation provides assistance for users that are interacting with the organizational simulation, whether as human-in-the-loop simulation participants or as users seeking to manipulate or observe the simulation. This is especially relevant for complex simulation models that require navigation and decision aids. Users visualize and interact with the simulation through interfaces that might provide three-dimensional graphics, animation of the simulation's unfolding behavior, and visual analytics that display complex statistics of enterprise performance.

11.3.1 World Models

The world model fundamentally captures the entities in the enterprise being modeled, their states (including allowable states and allowable state transitions), and their relationships with one another. For instance, in a factory simulation, the world model consists of the set of material processing and transport resources, the various materials consumed and products produced (related through a bill of materials), and the relationship between resources and material (i.e., a process plan that dictates the sequence in which material is transformed from raw materials into finished goods). In a battlefield simulation, the world model consists of the war-fighting resources (e.g., tanks and planes), their capabilities with respect to other types of resources, and the command-and-control structure that relates the various resources to one another.

In an organizational simulation model addressing the economics of HSI, the world model consists of the following important elements:

- The organizational structure of the enterprise (including incentives and information availability)
- The processes by which systems are evolved within an enterprise over their life cycles
- The research and development processes that provide technologies for systems
- The finances of the enterprise with respect to funding systems across their life cycles (including the structure of investments and payoffs)
- The enterprise rules that govern how finances are applied to funding systems
- Metrics that are used to judge system and enterprise performance
- The systems themselves and their characteristics (costs, performance, modularity, etc.)
- Representations of exogenous entities and phenomena that affect the enterprise but are not modeled explicitly within the simulation

Discrete-event simulation and system dynamics simulation are both well suited to modeling different aspects of the world model, such as business and technical processes, population growth and decline, resource usage, and investment cash flows.

11.3.2 Character Models

Humans exist as part of the world model, certainly, whether they are real or simulated. However, the character model is considered separately to model traits and behavior of people in the simulation.

A workshop on organizational simulation was held in 2003 to bring together approximately 50 experts to assess the current state of the art and identify research needs (Boff & Rouse, 2004). Participants came from the areas of modeling and simulation, behavioral and social sciences, computing, artificial intelligence, gaming, and entertainment. Participants also included people with extensive experience in business and military operations. They identified the following representational elements that would be desirable to include in character models.

- Emotions
- Personalities
- Interaction with one another
- Ability to learn from experience and training
- Ability to respond appropriately
- Believability
- Being members of formal organizations
- Ability to serve as coaches

In the context of investments and human systems integration, the following are relevant, as well:

- Biases (for HSI investment decision makers)
- Skills (for system users, i.e., HSI effectiveness)
- Human–technology interaction (for system users, i.e., HSI effectiveness)
- Economic utilities (for HSI investment decision makers, system users, and other system stakeholders)

Of course, having realistic representations of these elements requires significant advances in theory (e.g., decision making, social networks, cognition, etc). The required advances were summarized by the National Research Council (2008). From a modeling and simulation perspective, there is also the need for tools to allow a modeler to design a character with specific traits and behaviors. Although agent-based simulation provides a framework for representing human behavior, it is not a character design environment. Rather, it requires extensive customization to create character traits and behaviors as identified in the preceding discussion. However, tools such as character programming frameworks (e.g., Mateas & Stern, 2002) are providing capability for character design.

11.3.3 Organizational Stories

The organizational story tells how a set of characters interact in the world model to form a scenario of interest for exploration or experimentation. Any particular organizational story is constrained by the world model (e.g., allowable states and state transitions) and also by the set of character models (e.g., allowable behaviors and reactions). Within these constraints, the goal is to examine the possible outcomes originating from a scenario.

Discrete-event simulations operate with an event calendar that dictates the order of events happening in the simulation. This certainly is one way to construct a story. Typically, one event starts the simulation, others are prescheduled, and these events schedule others for future occurrence. For instance, in a factory simulation, a machine start processing event might schedule a machine finish processing event to execute after the processing time duration. Certain events, such as preventive maintenance, might be prescheduled. Thus, the simulation unfolds as a combination of these types of events.

In systems dynamics, the story unfolds as a set of state changes in stocks that occur as a result of flows and feedbacks. This results in a potentially unpredictable story, which is of interest because system dynamics seeks to model nonlinear and unpredictable behavior in systems.

Finally, stories in agent-based simulation occur as the result of interacting agents. At each time step in the simulation, the agents react to the state of the simulated world and execute a behavior in response. The collection of such behaviors over time forms the story line. Although agents often represent people, they can be used to represent other elements in the world model. Thus, agent-based simulation models interaction between characters and other elements. This interaction often results in interesting emergent behaviors for the simulated world. Similar to system dynamics, this emergent behavior can be unpredictable, as when a combination of simple agent behaviors results in complex overall behavior.

In organizational simulation, one goal is to provide the modeler or analyst more control over the story line, rather than have it result purely from an unplanned sequence of events, interactions among flows and feedback loops, or emergent agent behavior. This applies to situations where the intent is to study a more narrowly posed scenario. Of interest in this regard are the emerging fields of interactive drama (Bates, 1992; Weyhrauch, 1997) and drama management (Nelson et al., 2006). An interactive drama functions similarly to a game, except that the goal is for the user simply to interact with a set of simulated characters that can react realistically to the user's conversation and actions, perhaps without any scoring. Drama management, however, provides guidance through a network of events comprising a story line according to certain criteria (e.g., most likely story, the most likely given a certain starting condition, or potentially bad outcomes). This guidance is provided by restricting the set of possible state transitions from one place in the story (or by specifying probabilities for the transitions).

11.3.4 State of the Art

The representations underlying discrete-event and system dynamics simulation are relatively mature. The representations underlying agent-based simulation, however, are not as mature but are in the process of evolving to maturity. Organizational simulation, however, is still an emerging methodology whose representations are not yet specified fully. In large part, this results from limitations in modeling human behavior in computational form. A recent report from the National Research Council (2008) outlines several challenge areas in which significant additional research is needed to provide a capability for behavioral simulation (ranging from individuals to societies):

- *Theory development*. Better theories of individual and group human behaviors are needed to support theory-based modeling representations for organizational simulation. These need to be in quantitative form to support implementation in software.
- *Uncertainty, dynamic adaptability, and rational behavior*. A fundamental and unresolved question is how to model human behavior with respect to nondeterminism, learning over time and rationality versus irrationality.
- *Data collection methods*. One limitation on organizational simulation is that many organizations simply do not collect the types of data needed for simulation models (e.g., fully specified work processes or value-add at each process step). Traditional approaches to addressing this problem include expert opinion (interview an expert for his or her opinion on values for data elements), sensitivity analysis (select parameter values that seem reasonable and experiment to determine how sensitive model outcomes are to changes in parameter values), and Bayesian analysis (select an initial parameter value and update it as new information becomes available).
- *Federated models*. As the simulated world scales up from a team, to a department, to an organization, and finally to an enterprise, it becomes appealing to develop submodels for the different groups within the simulated model. These submodels may execute as separate software programs. Some may be modeled in more detail than others. Some submodels may have a human-in-the-loop element, whereas others rely on simulated humans. Making such decisions typically is done on an ad hoc basis. In addition, there are technical issues involving interoperability of the submodels.
- *Validation and usefulness*. Model validation and usefulness are fundamental and unresolved issues in simulation. These are compounded in organizational simulation with the difficulties in representing human behavior.
- *Tools and infrastructure for model building*. The lack of tools and infrastructure for model building is largely a function of the gaps in representation and underlying theory.

Consequently, there is not yet a software package that implements the features described in this chapter for organizational simulation. This leaves the modeler with two possible approaches. The first approach is to develop a new modeling language and software implementation for organizational simulation modeling language. The second is to use currently available modeling and simulation tools. Several commercially and freely available software packages implement the major simulation paradigms. Table 11.1 lists several widely used discrete-event simulation software packages.

Table 11.2 lists the two most widely used software packages for system dynamics simulation.

Table 11.3 lists widely used agent-based simulation software.

TABLE 11.1 Discrete-Event Simulation Software

Software	Description/Applications
ARENA (Kelton et al., 2004)	Commercial package with modeling GUI and animation built on underlying SIMAN language. Applications include manufacturing, business processes, military operations, and call centers.
AutoMod (Rohrer, 1997)	Commercial package with modeling GUI and animation, with primary application in semiconductor manufacturing.
Delmia/QUEST (Kim et al., 2006)	Simulation engine with three-dimensional, CAD-based graphics used primarily for product design and manufacturing applications.
DSOL (Jacobs, 2005)	Open-source Java-based simulation library.
MicroSaint (Dahn & Laughery, 1997)	Commercial package integrated with tools for modeling impact of environmental and workspace factors on human tasks.
SIMLIB (Law & Kelton, 2000)	Open-source simulation library implemented in C or Fortran. Applications in queueing systems.
WITNESS (Markt & Mayer, 1997)	Commercial package with modeling GUI and animation and interactive simulation (breakpoints). Applications include manufacturing, energy and project management.

Java is a trademark of Sun Microsystems, Inc. (Santa Clara, CA).

TABLE 11.2 System Dynamics Simulation Software

Software	Description/Applications
Stella/iThink (High Performance Systems, 2001)	System dynamics simulator with hierarchical modeling, model-building GUI, and animated charting of state variables. Applications in business process modeling, manufacturing, and strategy.
Vensim (Garcia, 2006)	System dynamics simulator with causal tracing and model-building GUI. Applications include plant design, team skills analysis, and aerodynamics.

TABLE 11.3 Agent-Based Simulation Software

Software	Description/Applications
Netlogo (Wilensky, 1999)	ABM environment based on Logo language and having a modeling GUI, animation, and model library for various application domains. Applications include social and physical sciences.
Repast (North et al., 2006)	Java library with animation. Based on Swarm. Applications include social science and physical sciences.
Swarm (Minar et al., 1996)	One of the first agent-based offerings. Objective-C library with capability to integrate with Java. Applications include environment and neighborhood segregation.

A recent trend in simulation is the concept of integration platforms (i.e., software frameworks that allow integration of multiple simulation paradigms into a single model). An early example of such an integration platform is the SIMAN discrete-event simulation language (precursor to ARENA), a primarily process-interaction, discrete-event language that provides integration with continuous models. Other examples include Repast (an agent-based package that allows discrete-event modeling), NetLogo (an agent-based package that allows system dynamics modeling), and AnyLogic (which combines agent-based, discrete-event, and systems dynamic modeling constructs) (Wartha et al., 2002). Finally, simulation tools have evolved in terms of data integration, where model data can be stored in spreadsheets or databases, allowing for improved data management.

11.4 ECONOMIC ASSESSMENT

Here, as elsewhere in this book, the focus is on assessing the economics of specific investment decisions related to HSI. Investments are considered as costs at various points in the system life cycle. In research and development, HSI investments may include developing new technologies for displays or decision support. In system design and integration, HSI investments may include replacing old technologies with newer technologies or performing such tasks as user-centered design. In production and deployment, HSI investments may include training personnel. Training often is considered an operational cost related to personnel costs, but it can be considered an investment related to future benefits. In operation and sustainment, investments may include subsystem upgrades, plus new training programs.

Returns from successful HSI investments relate to revenue, reduced cost, or increased performance. An automobile with better HSI-based design is likely to outsell automobiles with poor design. A well-designed ship command center, from an HSI perspective, may require fewer personnel, thus reducing costs. A well-designed air traffic control station, from an HSI perspective, may result in fewer accidents and near-accidents caused by reduced human error (in other words, improved performance). In some systems, it is difficult to value improved performance. This may occur when the system has certain set requirements to be met, and there is no reward for exceeding the requirements. One way to address this situation is to seek and measure cost reduction in meeting the requirements (Rouse & Boff, 2003).

The following economic assessment methods are considered:

- Net present value (NPV), which discounts the value of a future cash flow stream to the present using a discount rate.
- Internal rate of return (IRR), which computes the rate of return, or discount rate such that the NPV is zero.
- Cost–benefit ratio (CBR), which computes a ratio of the cost component of NPV to a similarly computed "net present benefit" expressed in units of the benefit (e.g., system performance metric).
- Net option value (NOV), which computes the value of an option minus the initial outlay to purchase the option.

Let T be the time horizon over which returns are measured, expressed in an appropriate time unit (e.g., months or years). Let r be the discount rate, expressed per unit of time used for T. For $i = 0, 1, \ldots T$, let c_i be the cost per time unit i, f_i be the financial return per time period i, and b_i be a performance benefit accrued in time period i. That is, c_1 is the cost incurred in the first time period; c_0, f_0, and b_0 are accrued initially at time zero, as for example c_0 being the initial outlay for an investment. Then these metrics can be expressed as in Equations 11.1–11.3 (adapted from Rouse and Boff, 2003). In Equation 11.3, the assumption is that the same discount rate is used for performance and cost; this need not be the case.

$$\text{NPV} = \sum_{i=0}^{T} \frac{f_i - c_i}{(1 + r)^i} \tag{11.1}$$

$$\text{IRR} = r \text{ such that } \sum_{i=0}^{T} \frac{f_i - c_i}{(1 + r)^i} = 0 \tag{11.2}$$

$$\text{CBR} = \frac{\sum_{i=0}^{T} c_i / (1 + r)^i}{\sum_{i=0}^{T} b_i / (1 + r)^i} \tag{11.3}$$

These metrics depend on cash flows and/or performance outcomes. NPV, IRR, and CBR all assume that future cash flows and performance outcomes are known and are deterministic (in both magnitude and timing). This assumption rarely holds in reality. Even when cash flows are specified contractually as to magnitude and timing, for instance, there is a risk of lateness or default. New systems typically have significant uncertainty in their costs, performance outcomes, and revenues (if applicable). This situation can be addressed to some extent in NPV, IRR, and CBV by specifying different outcome scenarios and assigning probabilities to each outcome. For instance, if three outcomes are judged equally likely, the NPV can be stated as the average of the NPVs from each outcome. In more general terms, an expected value for NPV, IRR, and CBR can be derived from a set of outcomes and associated probabilities. In addition, a discrete probability distribution can be specified for the outcomes of each, with a standard deviation and range. One fundamental question is which probabilities to assign to different outcomes.

Incorporating uncertainty is one issue; another issue with these three metrics is that they do not capture the value of flexibility in future decision making. One method that captures uncertainty and flexibility is option models. The option model is based on the concept of financial options, in particular the call option that a party can purchase, granting the future right to buy a stock at a guaranteed price within a certain timeframe. Options are attractive because they incorporate flexibility in future decision making—one can buy the stock or choose not to buy it before the expiration. Buying the stock is called exercising the option. In making investments in systems, the analogous concept is called real options (Trigeorgis, 1996). In real options, the analogy for the option purchase price is the amount to be invested initially, whereas the analogy for the stock price is the expected future system value discounted to the present. The exercise price is the purchase (or full development) of the system. The net option value is the value of the option minus the initial investment, or purchase price of the option (Equation 11.4):

$$\text{NOV} = \text{Option Value} - c_0 \qquad (11.4)$$

The option value is not necessarily straightforward in computation because there may be a complex structure of flexibility in future decision making for any particular situation. In fact, few closed-form solutions exist to computing values. One exception is the Black–Scholes formula for European options, where exercise must occur on the expiration date (Black & Scholes, 1973). Another approach is to use backward computations involving a lattice structure decision tree that models up and down movements in future system value (Cox et al., 1979). A third method, for complex options, is to use Monte Carlo simulation (Glasserman, 2004). Monte Carlo simulation differs from other forms of simulation in that it often is used to derive values for parameters via probabilistic sampling methods. In addition, real options rely on several assumptions that may not be met in practice, and they are sometimes difficult for managers to understand because of their complexity (Lander & Pinches, 1998).

11.5 ASSESSING HSI INVESTMENT USING ORGANIZATIONAL SIMULATION

Simulation addresses the uncertainty inherent in investing by assigning probabilities to various events and behaviors that occur during execution of a simulation model. Simulation models also must provide decision rules whereby certain investments are selected and others are not. This implies that models must have some representation of the possible future returns from these investments, as well as present funds available to make investments. This section addresses the use of organizational simulation to aid in economic assessment of HSI investments.

11.5.1 Simulation-Based Analysis of a Single HSI Investment

As a start, it is useful to examine how simulation can be used to assess the economics of making a single HSI investment. Below is an example of using simulation for two types of economic assessment.

An agency may wish to assess whether it should invest in a new software system with improved human systems integration for its call center. This new system requires new training for call center personnel. The exact cost of the new system and training may not be known with certainty. For the purposes of the simulation, this cost can be modeled as a random variable with a triangular distribution. Such a distribution has a lower bound, an upper bound, and a mode between the two bounds. This distribution often is used for unknown parameter values that do not have a known distribution (Law & Kelton, 2000). The performance outcome, however, is an output of the simulation run. Assume that performance is measured as average hold time for callers. If the simulation is modeled, for example, as an agent-based simulation, then the system and training may affect performance of each call handler differently, because of the handler's characteristics (represented by character models). In addition, performance improvement may vary across different types of calls. Finally, the improvement may not be the same for each different call, because of random variation. Thus, the improvement would be a function of the call handlers, the mixture of call types received, and random variation. If call handlers interact with one another (i.e., collaborate to handle particular calls), then performance improvement likely depends on the interaction of call handlers.

A simulation model can be developed to test the effect of this new system and training on performance by incorporating these factors and modeling the random variations in performance improvement as samplings from a probability distribution. The model run starts once system installation and training are complete. The run length needs to be specified as a time horizon during which a realistic picture of call center behavior can be established. Let this run length be denoted as T, in appropriate units (e.g., months or years), and let r be the discount rate, reported as the rate per time unit of T. Because of the randomness, multiple replications are needed. Let n be the number of replications (assumed to be determined so that there are statistically significant results). Let c_{0j} be the initial new system and training investment cost associated with replication j. (The values of c_{0j} differ because of random variation across replications.) The simulation needs to be designed such that it collects the hold times for each caller during the simulation run. Let h_{ij} be the average hold time for all callers during time unit i ($i = 1, 2 \ldots T$) of replication j. Then h_{ij} are statistics collected by the simulation based on the individual hold time observations. Assuming there is historical data for hold times, let H_b be the historical average hold time. Then let $p_{ij} = H_b - h_{ij}$ be the observed performance improvement for time unit i of replication j. Finally, let H_a be a parameter denoting the post-installation average hold time.

First, it is of interest to know whether the improvement is positive. This can be done using statistics [a review of statistics for simulation is provided by Law and Kelton (2000)]. H_a can be estimated by the average of h_{ij} over the T time units and n replications (Equation 11.5). Note that there is no h_{0j} term, because this average is computed starting in the first period.

$$\hat{H}_a = \sum_{i=1}^{T} \sum_{j=1}^{n} h_{ij} \qquad (11.5)$$

Assuming that H_b is the true pretraining average hold time, the issue is to test the hypothesis that $H_a < H_b$. In general, this can be done with a t test (a statistical test used to compare two values) that uses the h_{ij} data. The test is more accurate with additional replications. However, one issue in a call center or other applications where customers queue for service is the autocorrelation of hold times between adjacent customers. That is, there is a correlation between the hold times of customers who call into the call center within a short time of one another. This causes statistical issues because of an understated variance in the observed values. Techniques for conducting the t test, determining the appropriate number of replications, and addressing autocorrelations are discussed in Banks et al. (2005) and Law and Kelton (2000).

If it can be established statistically that $H_a < H_b$, then it makes sense to perform an economic assessment to determine whether this improvement is worth the investment cost. A cost–benefit analysis for each replication then can be constructed as in Equation 11.6, using p_{ij} as the benefit (adapted from Equation 11.4). Note that this example considers only a one-time initial system and training cost and assumes all other costs of the as-is call center and call center with the new system are the same. If subsequent training is required, these costs could be expressed as c_{ij} ($i = 1, 2 \ldots T$), and a numerator identical to Equation 11.4 would result.

$$\text{CBR}_j = \frac{c_{0j}}{\sum_{i=0}^{T} p_{ij} \big/ (1 + r)^i} \qquad (11.6)$$

Taking the average of the cost–benefit ratio outcomes over all n replications then results in a statistic for the cost–benefit ratio for the new system and training, and this statistic can be compared with the organization's benchmarks to see whether the investment should be made.

Although improving performance may be the organizational goal, it also may be that the goal is to achieve the same performance but to reduce the personnel cost. In this case, a similar analysis can be conducted. The model must be adjusted to test the effect not only of the new system and training but also of reducing personnel. The question then becomes at what reduced level of personnel does $H_a = H_b$? To determine what staffing level achieves the same performance may require some trial-and-error analysis in which different levels are tested (each level having several different replications).

This analysis can be complicated by the existence of different types of call handlers, represented by different character models in the simulation. As the type of call handler has an impact on performance, changing the mixture by reducing the number of callers can have a performance impact in addition to the actual reduction. Thus, if substantial differences in call handlers are modeled, considerable trial-and-error analysis may be needed.

Once a staffing level (and mixture of call handler types) has been identified such that its average hold time approximately equals H_b, then a t test can be conducted to determine whether this is the case statistically. If the two are identical statistically, then a net present value analysis can be conducted using Equation 11.1. Here, only the cost differences between the as-is call center versus the call center with the new system and training are considered (i.e., revenues are not considered in this example). Let k_i denote the estimated future costs associated with the as-is call center for $i = 1, 2 \ldots T$. Because the as-is call center is not simulated, it does not have it does not have replications for k_i. Note that k_0 is considered to be zero because the as-is system does not experience an investment. Let c_{ij} be the future estimated costs derived from the simulation model for the call center with the new system and training for period i and replication j. Note that c_{ij} here contains personnel costs and the costs of any additional on-going training required for the new system. Then the net present value of the savings from the to-be system is expressed in Equation 11.7:

$$\text{NPV} = \sum_{j=1}^{n} \sum_{i=0}^{T} \frac{k_i - c_{ij}}{(1 + r)^i} \tag{11.7}$$

11.5.2 HSI Investment across an Organization

Viewing HSI investments as applied to a particular system is one perspective. Organizations and enterprises, however, typically have many such investments to make over multiple systems. These HSI investments compete for limited funds, along with other types of investments and costs. In addition, these decisions and the associated returns may be spread over different organizations within an enterprise. Organizational simulation can be used to examine these phenomena, as well.

Consider a research and development organization where R&D in HSI is pursued, along with other technologies. R&D often is considered as an investment because it yields future benefits and is conducted in an uncertain environment. Thus, it can and should be analyzed for its economic benefit to the firm. R&D organizations typically use a staged system for projects, dividing them into basic research, applied research, development, and so on. This mitigates risk because a project need not be continued through all stages if it does not meet certain thresholds at each one. Figure 11.4 depicts a typical R&D organization.

Each stage of the R&D organization receives a budget that it can use to invest in R&D projects. Thus, there are two levels of economic decisions—how to allocate

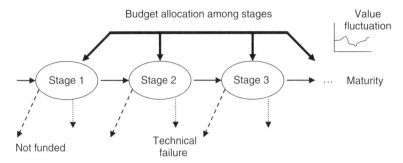

FIGURE 11.4 R&D organization.

the budget over stages and which projects each stage should select. In addition, different types of risk are associated with these investments (Boer, 2000). Technical risk refers to the possibility that the R&D may not yield the desired result. At the level of basic research, it may simply fail for scientific reasons. At later stages, results may fail to scale up or to integrate with other technologies needed for a successful system. Market risk, however, occurs when a successful result is fielded, but the market for it has decreased or may no longer exist. In military systems, this often is called mission risk (i.e., the intended mission for the system no longer exists). As a project progresses, it has an estimated value that would result from maturity and usage (i.e., deployment). This value fluctuates because of market risk.

Considering the two types of economic investment decisions for an R&D system, two questions can be asked:

- How should the overall R&D budget be allocated over the various stages?
- On what basis should R&D projects be selected at each stage?

It can be argued that budget allocation across stages should be done by considering the expected budget requests at each stage, factoring in the expected rates of technical failure (Hansen et al., 1999). However, this approach does not account for the effect of market risk. Likewise, traditional economic assessment of R&D has been done using such measures as net present value and internal rate of return. In recent years, it has been argued that the staged nature of R&D lends itself to valuation by real options (Faulkner, 1996; Myers, 1984). In addressing the funding of individual projects, the question becomes as to whether NPV or NOV is the better metric to use for project selection.

An organizational simulation model has been used to address these questions in the context of R&D (Bodner & Rouse, 2007). This model addresses only revenues and costs, not system performance, and it does not specifically address HSI. Nevertheless, it is useful in framing economic assessment questions that can be addressed via organizational simulation.

In the simulation, the world model focuses on the R&D stages, processes, decision making, and risks internal to an organization. The external effects are modeled indirectly by a probabilistic market risk that causes fluctuation of the value of the organization's R&D. The basic organizational story focuses on creating the most value from a limited R&D budget. Characters in the simulation are program managers or similar personnel who make funding decisions for R&D.

Thus, the world model includes a process-based workflow system, in which R&D projects traverse processes that represent funding decisions, R&D work, possibility of technical failure, and adjustment of anticipated deployed value (fluctuation from market risk) as the simulation progresses. There are three R&D stages, plus a fourth representing deployment. Prior to each stage, a selection process determines which projects are funded for that stage, based on the available budget. Thus, the economic assessment computations are implemented within the simulation to support the selection process. The selection seeks to maximize the value of those projects selected, subject to the budget constraint. Those not selected are discarded. Each project has the following attributes used for the selection:

- A funding request for each stage that grows with progressive stages
- An estimated postdeployment cash flow (revenues minus operating expenses) that varies over simulated time

Those funded proceed to the technical work associated with the R&D stage. At each stage, a project may fail for technical reasons, based on probabilistic failure rates. If the project is successful at a particular stage, it progresses for consideration at the next stage. The budget cycle is one year, as are project stage durations. Market risk is modeled as a geometric Brownian motion process, where the cash flow value discounted to present time incrementally moves either up or down (Trigeorgis, 1996). The magnitude of the movement is governed by volatility of the cash flow. This volatility is the standard deviation of the return on the cash flow asset, and it is a surrogate for the market risk (i.e., higher volatility implies higher risk, both upside and downside). In practice, a revised value is computed only when an R&D project is to be valuated for selection.

This simulation model, implemented in the ARENA simulation environment, has been used to address the two economic assessment questions in an experimental study that also includes the level of market risk (volatility) and the initial investment outlook from the perspective of technical risk. Initial investment outlook is modeled as the probability that the initial NPV of the project result, based on the estimated future cash flows of the project, is negative. As pointed out in Herath and Park (1999), there are examples of highly successful R&D investments that have had an initially negative NPV but a positive initial NOV. The four factors are considered at two levels each:

- Budget allocation is compared between the approach of Hansen et al. (1999), which will be referenced as the baseline method versus an alternative that shifts funding to upstream stages from downstream ones.
- Valuation metric is compared between NPV and NOV, where NOV is computed using the Black–Scholes formula.

- Market risk is compared between high and low volatilities.
- Initial investment outlook is compared between high and low probabilities of initial negative NPV for R&D projects.

Consideration of these four factors as independent variables, at two levels each, results in a 2^4 factorial experiment. This yields 16 scenarios for simulation runs.

Value creation from R&D is the primary concern. Two dependent variables are used to represented value creation—the total deployed value from R&D results over the time horizon and the ratio of the total deployed value to the R&D expenses incurred over the time horizon. Let V_1 denote total deployed value and V_2 denote total deployed value per R&D expenditure.

To test the effects and interactions of the four independent variables, each scenario was executed over a time horizon of $T = 25$ years, with a prior warm-up period to reach a steady state of five years. Statistics from the warm-up period were not used. Ten replications of each scenario were performed for statistical significance. The results then were analyzed separately for each dependent variable. The experimental analysis and assumptions underlying the model are described more fully in Bodner and Rouse (2007). However, the summary results yield interesting insights into R&D decision making. The focus here is on the two questions of interest and their interaction with market risk and initial investment outlook, both of which are considered as environmental factors.

- *Valuation method.* For V_1, using NOV outperforms using NPV as the valuation metric, especially when the average initial investment outlook is more negative. However, the reverse is the case for V_2. No significant interaction effect exists between V_2 and either environmental variable. The difference between the two is that with NOV, a higher percentage of the R&D budget is expended. Fundamentally, NPV is the more conservative metric, emphasizing return on R&D investment dollar spent. NOV, however, emphasizes total value creation from a given budget.
- *Budget allocation.* With a high level of volatility, it is better to shift funds to upstream R&D stages, in terms of V_1. With lower volatility, it is slightly better to use the baseline method. Considering V_2, the baseline method has better performance, especially when volatility is low. As volatility increases, it becomes a better strategy to shift funds upstream, so that a larger portfolio of possibilities is created. The upside of the winners can be exploited with continued funding, whereas those that lose value over time can be eliminated.

Although this example does not specifically address HSI investments, it does provide insight into economic assessment issues in investments that include HSI. It has been argued that HSI issues should be addressed earlier in the military R&D process than they currently are (Wallace et al., 2007). Thus, an application of the model could consider the effect of HSI investments that are integrated into the deployment process with other non-HSI technologies. This would require the joining of technologies into systems in the deployment stage of the model, which is a straightforward adaption of the existing model. There the technical risk rate models the risk that integration may not be successful. Particular systems have particular HSI issues. Thus, the model could be elaborated to address the details of particular types of systems. This type of complexity can be addressed through a relational database model integrated with the simulation. Preliminary work in modeling this approach for R&D is presented in Bodner et al. (2005). Finally, the model could be used to compare investing in HSI early in the R&D stages versus addressing HSI issues as they in integration and deployment. The performance outcomes from poor HSI and the costs of additional work to correct HSI shortcomings during integration are the key issues for study.

11.6 FUTURE WORK

This section discusses the future work needed in the field of economic assessment with organizational simulation. To put such work in context, enhancements to the R&D model are discussed in parallel. Topics discussed include multiorganization effects, economic modeling, organizational realism, as well as immersion and manipulation.

11.6.1 Multiorganization Effects

This particular model addresses only the R&D aspects of system development. The deployment part of the enterprise is modeled simply as another stage. In many situations, R&D and deployment operate as two separate organizations within an enterprise. Thus, a more elaborate model of the deployment organization can be used to demonstrate organizational effects.

Consider the military acquisition system, for example. It operates as a set of programs, each of which is seeking to develop a system (e.g., ship, plane, missile, etc.). These programs typically contract work to companies. Usually, new systems require new technologies. The acquisition organization may seek out commercially available technologies or may rely on the military R&D organization (which consists of multiple service-specific agencies that contract work to companies and universities). Thus, there is a complex acquisition and R&D enterprise. For this discussion, the level of analysis is set at the level of R&D as an organization and acquisition as a collaborating organization.

In effect, the R&D organization is a producer of technologies, and the acquisition organization consumes them. Traditional acquisition approaches tend to emphasize large leaps in system capability. This is achieved by using relatively immature technologies and maturing them within the system development process. This can result in delays and cost overruns, because of the technical risks assumed by the acquisition organization in maturing the technologies. However, recent reforms in the acquisition process call for an evolutionary approach that emphasizes incremental capability increases, relying on more mature technologies (Lorell et al., 2006).

The question becomes which approach is superior. One key consideration is that acquisition programs have multiple stakeholders (e.g., military personnel across different services, defense contractors, Congress, etc.). It can be demonstrated via game theory that pursuing an overly aggressive technology policy (i.e., selecting immature technologies for system development) results from a "tragedy of the commons" effect in the acquisition process (Pennock, 2008). That is, the stakeholders are incentivized to seek larger capability leaps for their own objectives than they would otherwise because the program is funded publicly. This results in a more aggressive technology policy than is optimal.

Extending this notion to organizational simulation, a model that represents both the R&D organization and the acquisition organization has been used to study the joint effectiveness of both, using traditional acquisition versus evolutionary acquisition, for example (Pennock, 2008). Similarly, the goal of the acquisition organization is to field capability, whereas the goal of the R&D organization is to develop technologies (within the context of the military systems mission). A joint simulation model could be used to study the effectiveness of methods to align the incentives of these two organizations in terms of capabilities and cost. Finally, viewing the acquisition enterprise as a whole, organizational simulation could be used to study the multistakeholder nature of the whole enterprise, ranging from military fighters, to command personnel, to defense contractors, to the legislative and executive branches.

Of course, such models can become enormously complex from a computational perspective. Different organizations within the enterprise may be modeled using different software (e.g., different commercial off-the-self packages), potentially causing interoperability issues. Continued research in distributed simulation is needed to help ensure computational feasibility and interoperability. In addition, however, such models can become difficult and time-consuming to develop and maintain. Tools are needed to facilitate model building and maintenance.

11.6.2 Economic Modeling

The example considers a relatively simple selection process focusing on the financial merit of individual projects relative to one another. The option models are relatively simple and rely on the closed-form Black–Scholes formula, which assumes a European options structure (i.e., exercise time is known). To incorporate other types of options may require the backward computations associated with a lattice approach or Monte Carlo simulation. Integrating these assessment methods to value options within an organizational simulation may pose computational challenges.

Of course, performance could be factored into the model, and a cost–benefit approach could then be used to judge project merits, computed in a manner similar to the NPV or NOV used currently.

However, economic assessment can be more complex than these single-metric approaches. In R&D, as in other investment organizations, the organization has a portfolio that combines risk and potential reward, weighted among different asset categories (e.g., types of technologies). In a more general sense, investment decisions can be framed in the context of R&D portfolio management and optimization. In portfolio optimization, the goal may be to minimize risk subject to ensuring a certain expected return, or to maximize return subject to a limit on the amount of risk exposure. Quantitative methods for portfolio management include simulation-optimization frameworks (Better & Glover, 2006), efficient frontier analysis using risk-reward (Graves et al., 2000), screening by stochastic risk-reward dominance (Ringuest et al., 2004), data envelopment analysis (Linton et al., 2002), dynamic programming (Childs & Triantis, 1999), and strategic percentage allocation among differing levels of market risk versus technical risk (MacMillan & McGrath, 2002). Integrating these methods in an organizational simulation may pose computational challenges similar to those of Monte Carlo evaluations. It should be noted that some interactions between different types of investments must be considered in portfolio management (Childs & Triantis, 1999). For instance, HSI investments often interact with new technology investments so that system performance is increased or cost is decreased with successful integration.

11.6.3 Organizational Realism

Two major needs in the area of organizational realism are reliable data on which to base models and realistic characters to populate them. These issues were discussed in detail by the National Research Council (2008). The basic R&D organization model has been applied to a major forest products company's R&D organization (Bodner & Rouse, 2007). Although basic financial data were available to model the organization at a high-level of detail, it is clear that a low level of model detail would require significant effort. For example, data on individual R&D projects may not be kept.

In addition, the characters in the R&D organization model behave rationally, selecting those projects for funding based on quantitative economic assessment metrics. As such, they are not representative of actual human characters that would populate a real organization. However, it would be of interest to compare the outcomes from realistic human decision making with those from pure use of economic assessment metrics.

The personnel who perform the R&D work are not modeled explicitly. If such detailed character models were developed, they could be used to study the effect of creativity, knowledge networks, and incentives on R&D outcomes.

11.6.4 Immersion and Manipulation

Finally, there is the issue of user interaction with the model. The existing R&D model has an animated user interface that illustrates the flow of projects through the stages of R&D. The user can run the simulation with delays to experience simulated time or without delays so that the simulation runs to completion without user interaction. The model supports stopping at breakpoints so that the user can set new values for parameters (e.g., technical risk rates).

What is missing, however, are tools to facilitate richer manipulation of the model during execution. For instance, it may be of interest to select particular projects for funding, regardless of their economic valuation, to see what the effect would be. Expanding on this, the model could be posed so that the user selects what markets the organization is to enter and constructs R&D portfolios to support entering the market. Additional interaction tools could support scenario generation for competition in the market.

Also, visualization can provide powerful user interaction experience via immersion. Visualization can be accomplished via three-dimensional renderings of organizational facilities or by visual analytics that express complex data in graphical form. Three-dimensional renderings may be of more interest in an organizational simulation that studies the HSI effectiveness within a particular system, whereas visual analytics may be of more interest to display large amounts of data originating from an R&D organization (e.g., project success rates by technology category and year, estimated future cash flows from deployed projects, or technology trend forecasts).

11.7 CONCLUSION

This chapter has addressed the use of organizational simulation in economic assessment of human systems integration. Although simulation has enjoyed a rich history in modeling organizations, the traditional focus has been on such aspects as process flow, scheduling, and efficiency. Recent work in agent-based simulation and cognitive frameworks has brought progress in the difficult problem of modeling human behavior. Organizational simulation seeks to model processes, performance, human behavior, organizational dynamics, and other features that comprise organizations. In the context of HSI, it can be used to assess the effectiveness of HSI investments in a single system. It can also be used to assess how HSI investments are made across the organization.

REFERENCES

Banks, J., Carson, J. S. II, Nelson, B. L., & Nicol, D. M. (2005). *Discrete-Event System Simulation* (4th Edition). Upper Saddle River, NJ: Pearson Prentice Hall.

Bates, J. (1992). Virtual reality, art, and entertainment. *Presence: The Journal of Teleoperators and Virtual Environments*, *2*(1), 133–138.

BenDor, T. K., & Metcalf, S. S. (2005). The spatial dynamics of invasive species spread. *System Dynamics Review*, *22*(1), 27–50.

Better, M., & Glover, F. (2006). Selecting project portfolios by optimizing simulations. *The Engineering Economist*, *51*(2), 81–98.

Black, F., & Scholes, M. (1973). The pricing of options and corporate liabilities. *Journal of Political Economy*, *81*, 637–659.

Bodner, D. A., Rouse, W. B., & Pennock, M. J. (2005). Using simulation to analyze R&D value creation. In M. E. Kuhl, N. M. Steiger, F. B. Armstrong, & J. A. Joines, Eds., *Proceedings of the 2005 Winter Simulation Conference*, Piscataway, NJ: Institute of Electrical and Electronics Engineers.

Bodner, D. A., & Rouse, W. B. (2007). Understanding R&D value creation through organizational simulation. *Systems Engineering*, *10*(1), 64–82.

Boer, F. P. (2000). Valuation of technology using "real options." *Research-Technology Management*, *43*(4), 26–30.

Boff, K. R., & Rouse, W. B. (2004). *Organizational Simulation: Workshop Report*. Riverside, OH: Wright-Patterson Air Force Base, Air Force Research Laboratory, Human Effectiveness Directorate.

Carley, K. M. (2002). Smart agents and organizations of the future. In L. Lievrouw & S. Livingstone, Eds., *The Handbook of New Media*. Thousand Oaks, CA: Sage.

Castronova, E. (2006). *Synthetic Worlds: The Business and Culture of Online Games*. Chicago, IL: University of Chicago Press.

Childs, P. D., & Triantis, A. J. (1999). Dynamic R&D investment policies. *Management Science*, *45*, 1359–1377.

Cochrane, C. B., & Hagan, G. J. (2005). *Introduction to Defense Acquisition Management* (7th Edition). Ft. Belvoir, VA: Defense Acquisition University Press.

Cox, J. C., Ross, S. A., & Rubinstein, M. (1979). Option pricing: A simplified approach. *Journal of Financial Economics*, *7*(3), 229–263.

Dahn, D., & Laughery Jr., K. R. (1997). The integrated performance modeling environment—Simulating human-system performance. In S. Andradóttir, K. J. Healy, D. H. Withers, & B. L. Nelson, Eds., *Proceedings of the 1997 Winter Simulation Conference*. Piscataway, NJ: Institute of Electrical and Electronics Engineers.

Faulkner, T. W. (1996). Applying "options thinking" to R&D valuation. *Research-Technology Management*, *39*(3), 50–56.

Ford, A. (1999). *Modeling the Environment: An Introduction to System Dynamics Modeling of Environmental Systems*. Washington, DC: Island Press

Forrester, J. (1961). *Industrial Dynamics*. Waltham, MA: Pegasus Communications.

Fujimoto, R. (2000). *Parallel and Distributed Simulation Systems*. New York: Wiley.

Garcia, J. M. (2006). *Theory and Practical Exercises of Systems Dynamics*. E-book. ISBN 84-609-9804-5.

Glasserman, P. (2004). *Monte Carlo Methods in Financial Engineering*. New York: Springer.

Gluck, K. A., & Pew R. W., Eds. (2005). *Modeling Human Behavior with Integrated Cognitive Architectures: Comparison, Evaluation, and Validation*. Mahweh, NJ: Erlbaum.

Graves, S. B., Ringuest, J. L., & Case, R. H. (2000). Formulating optimal R&D portfolios, *Research · Technology Management*, *43*, 47–51.

Hansen, K. F., Weiss, M. A., & Kwak, S. (1999). Allocating R&D resources: A quantitative aid to management insight. *Research-Technology Management*, *42*(4), 44–50.

Herath, H. S. B., & Park, C. S. (1999). Economic analysis of R&D projects: An options approach. *The Engineering Economist*, *44*(1), 1–35.

High Performance Systems. (2001). *An Introduction to Systems Thinking—iThink*. Watkinsville, GA: High Performance Systems.

Hillebrand, E., & Stender, J. (1994). *Many-Agent Simulation and Artificial Life*. Amsterdam, The Netherlands: IOS Press.

Ilachinski, A. (2004). *Artificial War: Multiagent-Based Simulation of Combat*. Hadansack, NJ: World Scientific Publishing Company.

Jacobs, P. H. M. (2005). The DSOL Simulation Suite. Enabling Multi-Formalism Modeling in a Distributed Context, Ph.D. dissertation, Delft University of Technology, Delft, the Netherlands.

Kelton, W. D., Sadowski, R. P., & Sturrock, D. T. (2004). *Simulation with ARENA*. Boston, MA: McGraw-Hill.

Kim, Y.-S., Yang, J., & Han, S. (2006). A multichannel visualization module for virtual manufacturing, *Computers in Industry*, *57*(7), 653–662.

Lander, D. M., & Pinches, G. E. (1998). Challenges to the practical implementation of modeling and valuing real options. *The Quarterly Review of Economics and Finance*, *38*, 537–567.

Law, A. M., & Kelton, W. D. (2000). *Simulation Modeling and Analysis* (2nd Edition). Boston, MA: McGraw-Hill.

Linton, J. D., Walsh, S. T., & Morabito, J. (2002). Analysis, ranking and selection of R&D projects in a portfolio. *R&D Management*, *32*, 139–148.

Lorell, M. A., Lowell, J. F., & Younossi, O. (2006). *Evolutionary Acquisition: Implementation Challenges for Defense Space Programs*. Santa Monica, CA: RAND.

MacMillan, I. C., & McGrath, R. G. (2002). Crafting R&D project portfolios. *Research-Technology Management*, *45*(5), 48–59.

Markt, P. W., & Mayer, M. H. (1997). WITNESS simulation software: A flexible suite of simulation tools. In S. Andradóttir, K. J. Healy, D. H. Withers & B. L. Nelson, Eds., *Proceedings of the 1997 Winter Simulation Conference*. Piscataway, NJ: Institute of Electrical and Electronics Engineers.

Martinez, J. C., & Ioannou, P. G. (1999). General purpose systems for effective construction simulation. *Journal of Construction Engineering and Management*, *125*(4), 265–276.

Mateas, M., & Stern, A. (2002). A behavior language for story-based believable agents. *IEEE Intelligent Systems*, *17*(4), 39–47.

Minar, N., Burkhart, R., Langton, C., & Askenazi, M. (1996). The Swarm simulation system: A toolkit for building multi-agent simulations. *Working Paper 96-06-042*. Santa Fe, NM: Santa Fe Institute.

Myers, S. C. (1984). Finance theory and financial strategy. *Interfaces*, *14*(1), 126–137.

National Research Council. (2008). *Behavioral Modeling and Simulation: From Individuals to Societies*. Washington, DC: National Academies Press.

Nelson, M. J., Mateas, M., Roberts, D. L., & Isbell, C. L. (2006). Declarative optimization-based management in interactive fiction. *IEEE Computer Graphics & Applications*, *26*(3), 32–41.

North, M. J., Collier, N. T., & Vos, J. R. (2006). Experiences creating three implementations of the Repast agent modeling toolkit. *ACM Transactions on Modeling and Computer Simulation*, *16*(1), 1–25.

Pavlov, O. V., & Saeed, K. (2004). A resource-based analysis of peer-to-peer technology. *System Dynamics Review*, *20*(3), 237–262.

Pennock, M. J. (2008). The Economics of Enterprise Transformation: An Analysis of the Defense Acquisition System, Ph.D. dissertation, School of Industrial & Systems Engineering, Georgia Institute of Technology, Atlanta, GA.

Prietula, M., Carley, K., & Gasser, L. (1998). *Simulating Organizations: Computational Models of Institutions and Groups*. Menlo Park, CA: AAAI Press.

Ringuest, J. L, Graves, S. B., & Case, R. H. (2004). Mean-gini analysis in R&D portfolio selection. *European Journal of Operational Research*, *154*, 157–169.

Rohrer, M. W. (1997). Automod tutorial. In S. Andradóttir, K. J. Healy, D. H. Withers, & B. L. Nelson, Eds., *Proceedings of the 1997 Winter Simulation Conference*. Piscataway, NJ: Institute of Electrical and Electronics Engineers.

Rouse, W. B., & Boff, K. R. (2003). Cost-benefit analysis for human systems integration. In H. R. Booher, Ed., *Handbook of Human Systems Integration*. New York: Wiley-Interscience.

Rouse, W. B., & Boff, K. R., Eds. (2005). *Organizational Simulation: From Modeling & Simulation to Games & Entertainment*. New York: Wiley.

Saaksvuori, A., & Immonen, A. (2003). *Product Lifecycle Management*. New York: Springer-Verlag.

Saam, N. J., & Schmidt, B., eds. (2001). *Cooperative Agents: Applications in the Social Sciences*. New York: Springer.

Sterman, J. D. (2000). *Business Dynamics: Systems Thinking and Modeling for a Complex World*. Boston, MA: McGraw-Hill.

Trigeorgis, L. (1996). *Real Options: Managerial Flexibility and Strategy in Resource Allocation*. Cambridge, MA: MIT Press.

Wallace, D. F., Bost, J. R., Thurber, J. B., & Hamburger, P. S. (2007). Importance of addressing human systems integration issues early in the science and technology process. *Naval Engineers Journal*, *119*(1), 59–64.

Wartha, C., Peev, M., Borshchev, A, & Filippov, A. (2002). Decision support tool—Supply chain. In E. Yücesan, C.-H. Chen, J. L Snowdon, & J. M. Charnes, Eds., *Proceedings of the 2002 Winter Simulation Conference*. Piscataway, NJ: Institute of Electrical and Electronics Engineers.

Weyhrauch, P. (1997). Guiding Interactive Drama, Ph.D. dissertation, School of Computer Science, Carnegie Mellon University, Pittsburgh, PA.

Wilensky, U. (1999). *NetLogo*. Evanston, IL: Center for Connected Learning and Computer-Based Modeling, Northwestern University. http://ccl.northwestern.edu/net/logo/.

PART IV

Case Studies

Chapter **12**

HSI Practices in Program Management: Case Studies of Aegis

Aruna Apte

12.1 INTRODUCTION

On Armed Forces Day, May 16, 1981, the first of a revolutionary new class of ships was launched at Ingalls Shipyard in Pascagoula, Mississippi. Then First Lady Nancy Reagan broke a champagne bottle against her bow and christened her *TICON-DEROGA*, the first AEGIS cruiser. On January 22, 1983, after 20 months of the most extensive and carefully watched trials, TICONDEROGA was commissioned in the U.S. Navy.

AEGIS, named after the mythological armor shield of Zeus, with its state-of-the-art radar and missile-launching systems is the Navy's most capable surface-launched missile system ever put to sea. Its computer programs and displays detect incoming missile or aircraft threats, sort them by assigning a threat value, assign on-board standard surface-to-air missiles, and guide them to their targets. This makes the AEGIS system a fully integrated combat system capable of simultaneous warfare against air, surface, subsurface, and strike threats.

The US Navy's defense against these threats has continued to rely on the winning strategy of defense-in-depth. In the late 1950s, on Navy ships were replaced by the first generation of guided missiles. By the late 1960s, these missiles continued to perform well, but the U.S. Department of Defense (DoD) recognized that reaction time, firepower, and operational availability in all environments would not match the impending threat. To counter this, an operational requirement for an

The Economics of Human Systems Integration: Valuation of Investments in People's Training and Education, Safety and Health, and Work Productivity. Edited By William B. Rouse
Copyright © 2010 John Wiley & Sons, Inc.

Advanced Surface Missile System (ASMS) was officially declared, and a comprehensive engineering development program was initiated to meet that requirement. ASMS was renamed AEGIS in December 1969 (Jane's.com, 2006). In 1974, the USS NORTON SOUND (AVM 1) was fitted with the AEGIS Engineering Development Model (EDM-1), including one SPY-1 Phased-array Radar. On May 17, 1974, the AEGIS Weapon System, manned by the crew of NORTON SOUND, successfully detected, tracked, engaged, and intercepted a BQM-34A aerial target on the Pacific Missile Test Range with the first firing of the Standard-1 Missile. Later, a second non-warhead Standard-1 Missile was fired that destroyed the target at a range of 15 miles. Rear Admiral Wayne E. Meyer, AEGIS/SM-2/AEGIS Ship Acquisition Manager (considered "Father of AEGIS"), termed this performance "A 7 league advance in our Navy's ability to go once more in harm's way" (USS NORTON SOUND, 2006). Thus go the milestones achieved by the AEGIS program.

All ships of the AEGIS fleet require Microwave Tubes (MWTs) (Hoffer, 2003) and radar phase shifters. Early in the development of the AEGIS program, one such Microwave Tube, the Cross Field Amplifier (CFA) proved to be a substantial cost driver resulting from a low mean time between failure (MTBF). However, after isolating the root causes and removing those using tools like total quality management (TQM), human systems integration (HSI), and process management, the MTBF of the tubes increased from 6,000 hours to 40,000–45,000 hours (Apte & Dutkowski, 2006). A radar phase shifter is a two-port device whose basic function is to provide a change in phase of radio-frequency (RF) signal with minimal attenuation. Figures 12.1 and 12.2 show travelling wave tubes (TWTs) and phase shifters used in the AEGIS weapons system. There are two types of phase shifters: mechanical and electronic. Initial contractor, RCA, was able to reduce the cost per unit of the electronic phase shifters to $200 from $2,000 (Bridger & Ruiz, 2006), again using tools such as HSI, Lean and Six Sigma, and process management. This section researches the contributing issues in both these instances by providing the background and the analysis. In the second section, we offer background in management practices in the AEGIS program. In the third and the fourth sections, we describe total ownership cost reduction in microwave tubes and radar phase shifters in the AEGIS program that are based on case studies conducted by Apte and Dutkowski (2006) and Bridger and Ruiz (2006). In the final (fifth) section, we offer conclusions.

The first case study documents the identification of the root causes of CFA low operational availability and the processes that not only eliminated this problem but also yielded an increase in tube MTBF and a much lower total ownership cost (TOC). It chronicles the methods used to reduce TOC in a program that serves as an example of early evolutionary acquisition. The objective of this case study is to understand the process and recognize the business issues within it that are essential to maintaining system combat capability, enhancing system affordability, and reducing TOC.

The second study captures the production and design processes and program management solutions used to reduce the TOC of AEGIS radar phase shifters. Specifically, it focuses on the design and redesign of the SPY-1 radar phase shifter:

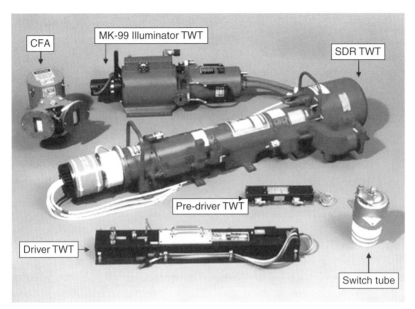

FIGURE 12.1   Tubes used in the AEGIS weapons system. *Source*: Dutkowski, E. J., Jr. (2004–2005).

FIGURE 12.2   Drawing of a Spy-1B/D phase shifter. *Source*: Used with permission from Lockheed Martin, 2006.

a redesign that dramatically improved performance without increasing average procurement unit costs (APUCs). It analyzes various process-improvement projects (PIPs) used to reduce touch-labor and improve production process yield, assess the percentage of manufactured items that are defect free, review programs that improved phase shifter production either directly or indirectly, and determine HSI concepts that helped achieve the implementation.

To date, AEGIS Weapon System capabilities have been or are being installed on 89 U.S. Navy cruisers and destroyers with at least 10 additional destroyer installations being considered by Congress. The SPY-1D (V) littoral radar upgrade superseded the SPY-1D in new-construction ships beginning in FY 1998 and first deployed in 2005. AEGIS is the primary naval weapon system for Japan and is part of two European ship construction programs—the Spanish F-100 and the Norwegian New Frigate. Additionally, Australia and the Republic of Korea recently selected AEGIS for its newest platforms (Lockheed Martin, 2006).

12.2   THE AEGIS PROGRAM

AEGIS, the first fully integrated shipboard combat system is capable of simultaneous warfare against air, surface, subsurface, and strike threats. After success with the AEGIS EDM-1 shipboard application, the decision was made to construct the first AEGIS ships based on the hull and machinery designs of Spruance class destroyers. The sophistication and complexity of the AEGIS combat system were such that the combination of combat system engineering with AEGIS ship acquisition demanded "special management treatment." This combination was affected by the establishment of the AEGIS shipbuilding project at the Naval Sea Systems Command (NAVSEA PMS-400) in 1977 (Jane's.com, 2006). The special management treatment combined the oversight of structural hull mechanical and electrical systems, combat systems, computer programs, repair parts, personnel maintenance documentation, and tactical operation documentation into one unified organization to create the highly capable, multimission surface combatants that are today's AEGIS cruisers and destroyers. The charter for NAVSEA PMS-400 represented a significant Navy management decision, one that had far-reaching impacts on acquisition management, design, and life-cycle support of modern Navy ships. For the first time in the history of surface combatants, PMS-400 introduced an organization that had both responsibility and authority to manage simultaneously development, acquisition, systems integration, and life-cycle support.

The AEGIS team included engineers, designers, operators, shipbuilders, military personnel, and civilians in industry, government, and laboratories. This integrated product team (IPT) was united, productive, and harmonious with a "can-do" attitude. The team consisted of a triad—Navy, industry (e.g., Varian Associates and Vishay Intertechnology), and shipbuilders (e.g., Ingalls Shipbuilding Division of Litton Industries and Bath Iron Works). Admiral Meyer constantly emphasized that success depended on the cooperation among team members and on the significance of each team member. He believed that there are three requirements for creating a successful team: The members must have a collective vision, a collective dedication, and a collective endurance (Truver, 2002).

Human systems integration, as it is called today, was termed human factors engineering (HFE) in the early days of the AEGIS program, which later was called IPT. In the process of HFE, IPT or HSI, the end user, in this case the sailor, was extensively involved. Sailors were brought into the manufacturing plants so that they could share their knowledge and experience with the designers. The AEGIS EDM was rigorously tested on a designated ship at sea so that the sailor could give feedback. Engineers were sent to live aboard warships at sea to gain first-hand knowledge of the shipboard environment and the conditions in which AEGIS would be operating. Combat system engineers were assigned to work with the shipbuilders, and ship designers were assigned to work with the Combat System Integration Agent. A land-based Combat System Engineering Development Site was constructed to serve a dual role. It was a test facility as well as a training center for the officers and crew scheduled to serve in the ships.

A new management information system was devised that was extremely success-ful. The assignment of a single contractor to integrate the entire combat system was an important innovation. Engineering personnel from Naval Warfare Centers were involved extensively in these processes although ultimate project control remained with the Project Manager (PM) because the PM was responsible for the success of the entire project. In addition, every weapon system was shipped for shipboard installation similar to the Just-In-Time concept immediately after production testing, of current operations management.

We now describe the case studies that will illustrate how these management concepts and HSI considerations helped reduce the total ownership cost for two key elements of the AEGIS Project: MWTs and radar phase shifters.

12.3 CASE STUDY 1: TOC REDUCTION FOR AEGIS MICROWAVE POWER TUBES

12.3.1 Background

The AEGIS shipbuilding program, arguably one of the largest and most successful acquisition programs in the DoD, has provided the Navy with more than 90 capable surface combatant ships. AEGIS ships now make up the majority of the Navy's destroyer and all of its cruiser fleet. MWTs are a primary component in the radar systems of the AEGIS fleet. Numerous other shipboard systems use MWTs: Radars SPS-48, SPS-49, MK-99, and the Phalanx Close In Weapon System to name just a few. Throughout the world, 57% of MWTs are used in radars; the manufacture of radar MWTs alone is a $280.3M market. Figure 12.3 shows the world market for MWTs by application and type (Dutkowski, 2004–2005).

12.3.1.1 The Story. It was the early 1980s, the height of the Reagan defense buildup, and AEGIS was the centerpiece of the Navy's shipbuilding program. Initial deployment of AEGIS cruisers was completed, and departmental focus was shifting to include life-cycle cost control as well as enhanced system operational availability (A_o). At the same time, the DDG-51 class was in engineering design and development. The weapons systems of the ships were heavily dependent on MWTs with cruisers using 176 and destroyers using 90 per system. Not surprisingly, MWTs became the cost drivers for TOC.

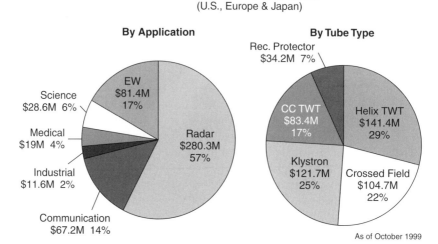

FIGURE 12.3 World market for MWTs by application and type. *Source*: Dutkowski, E.J., Jr. (2004–2005).

In a regular monthly meeting with his staff, the PM for the AEGIS shipbuilding project was briefed on the status of MWTs. These routine meetings were held to review program execution, cost, and schedule issues. Staff noted that the unit cost of a microwave tube had exceeded their expectations and that the mean time between failures of a microwave tube was very low. This situation had existed for some time, and improvement was not expected without additional management attention.

12.3.1.2 The MWTs. The AEGIS weapons system had been in design and development for several years. Program decisions were based on, among other considerations, cost estimates, and maintenance concepts for MWTs. The use of MWTs in combat systems was extensive. The principals in the program knew well the extent of the applications of MWT in AEGIS ships. In addition there were several different MWTs utilized in AEGIS ships: Cross Field Amplifiers in the radar systems and traveling-wave tubes (TWTs) in electronic warfare systems. Although there were numerous applications of MWTs then, now they are used in even more war-fighting applications. Figure 12.4 (Dutkowski, 2004–2005) shows a current example of MWTs' utilization in a CG-47 class ship. As shown, many shipboard systems depend on MWTs. At the time the PM was briefed by his staff, the microwave tube manufacturing industry was in its infancy with little process or configuration control. The tubes were state-of-the-art technology for that era, but they were in an early stage of evolutionary acquisition (Apte, 2005). This situation presented great challenges to the team to put manufacturing controls in place to reduce production risk and increase yield.

Through contract terms that dictated configuration and process controls, investment of government dollars into product improvements and continual

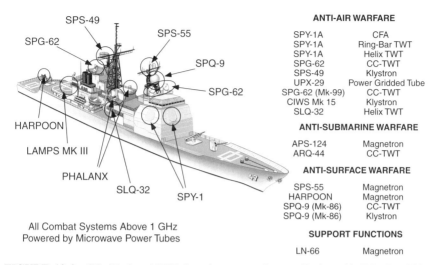

ANTI-AIR WARFARE	
SPY-1A	CFA
SPY-1A	Ring-Bar TWT
SPY-1A	Helix TWT
SPG-62	CC-TWT
SPS-49	Klystron
UPX-29	Power Gridded Tube
SPG-62 (Mk-99)	CC-TWT
CIWS Mk 15	Klystron
SLQ-32	Helix TWT

ANTI-SUBMARINE WARFARE	
APS-124	Magnetron
ARQ-44	CC-TWT

ANTI-SURFACE WARFARE	
SPS-55	Magnetron
HARPOON	Magnetron
SPQ-9 (Mk-86)	CC-TWT
SPQ-9 (Mk-86)	Klystron

SUPPORT FUNCTIONS	
LN-66	Magnetron

FIGURE 12.4 CG-47 class MWTs-based systems. *Source*: Dutkowski, E.J., Jr. (2004–2005).

monitoring of product would enable process control over time. Given the unit cost and MTBF data of the microwave tube, the PM estimated tube replacement costs to be $1M/ship/year. Based on a then-projected AEGIS fleet size of 40 ships, the PM understood the total annual cost for cross field amplifier tube replacement to be $40M—just to keep the AEGIS radar systems operational. Use of CFAs in ships is depicted in Figure 12.5.

12.3.2 The Problem and Solution

The data presented to the PM highlighted that MWTs were costly to replace and costly to produce. With MTBF in the range of 1,300–12,000 hours and a high unit cost, they were a major contributor to TOC. The key players in the program understood the risks involved in the "cottage industry" character of microwave tube production. This set the stage for tasking Warfare Center engineering specialists to craft engineering and logistics solutions. In any case, performance-versus-cost of MWTs was a problem and not using the tubes was not a technical solution.

A parallel and nagging question of the day was that if MWTs are so difficult and expensive to produce and at the same time essential to so many applications of AEGIS, then why not replace the tubes with solid-state components? This issue took considerable engineering expertise and time to answer. Simply put, solid-state devices at that time did not cover the power frequency spectrum provided by vacuum tubes. This remains the case today.

12.3.2.1 The Opportunity.
The PM was facing many issues revolving around the microwave tube components. But an opportunity presented itself. An upcoming conference was to be held at the Naval Postgraduate School (NPS) on MWTs. The

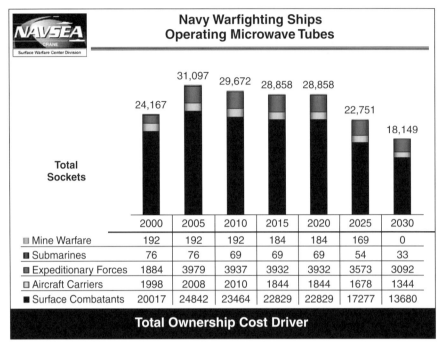

FIGURE 12.5 Current MWT usage in the U.S. Navy. *Source*: Dutkowski, E.J., Jr. (2004–2005).

CEOs of all the companies that provided MWTs to the AEGIS shipbuilding project were attending the conference. The companies would benefit from the success of AEGIS and would suffer if the performance/cost of MWTs remained the same. The PM sensed the potential for turnaround.

12.3.2.2 The Solution. The PM decided that instead of treating adversity as a constraint, he was going to exploit it to solve the problem. He made a pitch specifically to the CEOs of the vendor companies. Focusing on CFAs, he explained to them that the problem he was facing was that CFAs had an operating cost of $1M/ship/year. He conveyed to them the criticality of CFAs to the AEGIS fleet. At the MWT conference, he challenged them, "Propose something to fix this problem. Let us work together. As Project Manager, I promise you full cooperation. Let us collaborate and resolve this issue." At the PM's invitation, the CEOs came to his quarters in Hermann Hall at NPS. A frank dialogue at this meeting laid the groundwork for implementing a successful process that ultimately led to a reduction in TOC and to an increase in MTBF for CFAs. Although limited contractually on the extent he could be involved in internal company affairs, the PM did offer all the CEOs the opportunity to have him speak to each of their workforces to underscore the importance of their efforts in this critical national defense program.

12.3.2.3   The Process. As he had agreed, the PM visited each of the tube production facilities with his staff. This staff included staff engineers, operations managers, and financial experts from the Naval Surface Warfare Center at Crane Indiana, the Navy's In Service Engineering Agent (ISEA) for MWTs. In essence, an IPT was formed with a focus on human systems integration. The PM challenged his staff to devise a solution. He felt confident they could do so because of the technical competence they demonstrated through the years in combat systems engineering and in particular microwave tube technology.

After initial evaluation, staff proposed a two-fold approach to the combined TOC/MTBF issues. The first was to ensure an accurate diagnosis and repair of the tube (Hoffer, 2003); the second was an improved ability to locate and track all tubes that had been produced (Dutkowski, 2004–2005). Initial diagnostic troubleshooting revealed a metallurgical problem. The anode of the CFA was made of high-purity, electrical-grade copper that is quite soft. This copper also has a very low melting point. When a CFA turns on or during the change to a long waveform, the CFA has a tendency to arc between cathode and anode. Excessive arcing leads to premature failure and is detrimental to the overall performance of the amplifier. The PM's team discovered that the anode vanes were melting slightly and progressively because of arcing. The deterioration from arcing was increasing—to the point that the tube would arc at shorter pulse lengths and at lower power levels. This erosion eventually led to tube failure. It was clear that better operating life could be obtained if the anode vanes could be prevented from melting because of arcing. The solution devised was to add a thin layer of molybdenum, which has a higher melting point than copper, at the ends of the anode vanes and to reduce the arcing by better processing of the tubes.

The second issue to be resolved was an inventory control issue. The CFAs were high-value assets and capturing them for repair vice disposal was crucial. This required knowing where each tube was and then providing for their return to the designated repair facility. Additionally, tubes with the new modifications needed to be installed where and when appropriate so they could fit the empty sockets left by the unimproved tubes. Therefore, tracking the CFAs was vital. The team developed a method of serial number tracking of each tube. The principal behind it was the same as that behind the now common barcode and the recently introduced technology of radio frequency identification (RFID). The history of the operating cost of the sockets is given in Figure 12.6.

The Crane team successfully implemented a serial number tracking program. It was successful for two reasons: They had in-house technical skills to develop successful engineering solutions to the arcing problem, and they had the managerial skills necessary to develop an effective inventory control and repair protocol. This process, in addition to configuration control in manufacturing, added the ability to track changes introduced through the evolutionary acquisition process to give visibility to the impact these changes had on performance. In addition to the physical location, tracking offered another opportunity. This was the facility to measure the ability of the shipboard technician who maintained the system containing the MWT. The process provided the mechanism to track the maintenance competence

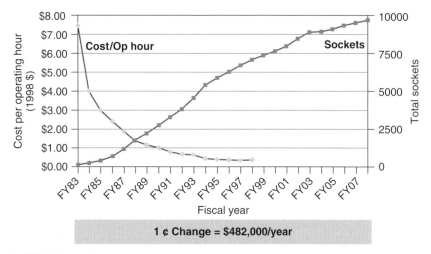

FIGURE 12.6   Cost/operating hour history. *Source*: Dutkowski, E.J., Jr. (2004–2005).

of that technician. The allowed corrective action to be taken was in the form of technical assistance to the ship or technician retraining.

At the next biennial conference, the PM presented the results to industry as a "good improvement." Thanks to the improved tracking and repair processes of CFAs, the MTBF had increased substantially—to approximately 5000 hours (Greene, 2004)—and the operating cost was reduced by a third. However, he challenged his staff and industry to keep building on the three pillars for success of the initiative: accurate tracking of tubes, maintenance of the knowledge base necessary to stay abreast of technical developments in the tube industry and continuous improvement of the tube production process.

12.3.2.4   The Accomplishment.   The success of the initiatives had tremendous impact on the operational availability and cost of the AEGIS system. The solution implemented regarding the CFA components in AEGIS is an excellent example of a successful pursuit of reduction of TOC (Boudreau & Naegle, 2003). The effort initiated in the mid-1980s is still paying off. The arcing that led to the melting of anodes (which, as mentioned, precipitated the failure of CFAs—resulting in a very low MTBF) prompted a series of well-managed steps to continue to improve the tubes. Tracking of the tubes was essential for locating and tracking the result of the changes. This was an extremely valuable initiative because it helped reduce the cycle-time for repair and change insertion by the real-time feedback to the manufacturer. Knowing what caused the failure and where and when the failure occurred was critical to increasing operational availability. The ability to track the CFAs was critical to the success of the program. The V-chart in Figure 12.7 shows that investment in engineering and management initiatives throughout the life cycle of a system can dramatically reduce TOC.

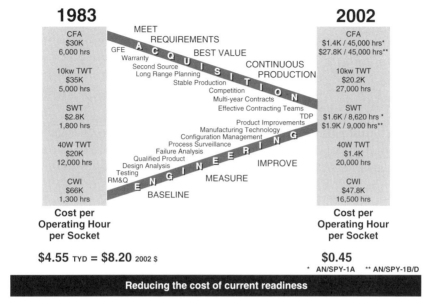

FIGURE 12.7 AWS microwave tube engineering/acquisition program. *Source*: Dutkowski, E.J., Jr. (2004–2005).

12.3.3 The Analysis

Several key factors contributed to the success of this initiative. Of special note is the use of HSI: the program achieved a high level of collaboration among all key parties, including the fleet, field activities, and headquarters.

12.3.3.1 The Champion. For a project to succeed across functional departments such as design, planning, contracting, and engineering, there needs to be a leader who will champion the new initiatives and processes. This leader, along with key team players, has to devise innovative programmatic and contractual provisions. The PM's ability to integrate his team was a key element in that team's success, and he was able to leverage the existing AEGIS team to do so.

12.3.3.2 Integrated Product Team. The concept of an IPT is an initiative common today, but it was not known as such 20 years ago. Likewise, although theoretically the use of an IPT is an effective business tactic, the difficulty in implementation can be overcome only if a clear focus and vision is maintained by the team leader. The success of this endeavor must be attributed to establishment of the IPT and its leader.

12.3.3.3 In-house Knowledge. In addition to the strength of the PM's leadership, none of these successful steps could have been achieved without the in-house (government) technical competence and knowledge of the PM's staff. In-house technical knowledge was essential to ensuring industry collaboration. A knowledgeable (smart) buyer is essential in evaluating industry recommendations for technical and process changes. Consequently, retaining in-house expert technical personnel is critical to improving system operations and costs. These assets may be in the field or at headquarters, or both, but they must exist on the government team.

12.3.3.4   Customization. The previous discussion of the small industrial base and process/material dependent manufacturing of the CFAs suggests that there are very few (sometimes no) manufacturers for customized defense products. Because of this lack of competition, integrity has to be maintained in writing best value contracts. It has been proven in the private sector over and over again that competition breeds both diversity in products and helps the best among them to survive. The absence of competition in the manufacturing of some defense products adds responsibility to the program manager; he/she must ensure the quality of the product and contract. Therefore, evaluating best value contracts is a key in this type of acquisition. Many times the government must buy critical products in a small production lot environment, precluding the ability to have a competitive contract situation. The small production lot environment requires competent government oversight to be effective.

In addition, the production pipeline of such customized defense products has to be smooth to achieve significant cost savings. A disruptive supply will likely stifle MTBF improvements. The stop-and-start of production adds the disadvantage of a high fixed cost. Therefore, continuous production is desirable. This requires a procuring agency that views long-term requirements and vendor loading versus the short-term (meet the current demand) perspective. This continuity is especially necessary when manufacturers are training workers for such tailored products. In the case of the CFAs, the very stable AEGIS shipbuilding project provided just such a stable demand for CFAs.

12.3.3.5   Business Issues. On the surface, this discussion is a simple success story of identifying CFA failures and engineering their prevention by tracking and modifying the microwave tubes, which increased MTBF, and at the same time lowered production costs. But what is impressive and powerful about this accomplishment is the process through which the problem was diagnosed and the cure implemented. "In my wildest dreams, I did not think we could come this far," said the PM about the success of this program (Greene, 2004). This detection, analysis, and solution process encompasses profound technical, managerial, and policy issues. Some of the business issues this story highlights—Human systems integration, evolutionary acquisition, theory of constraints, total quality management or Six Sigma, outsourcing, and process management—are instructive for any program manager.

- **HSI/IPT/HFE:** The dramatic improvement implemented by the PM and his team was and is a result of integration of key players: the AEGIS Program

Office, Communications and Power Industries (CPI, a vendor that was formerly part of Varian and provided the CFAs), Crane Naval Surface Warfare Center, the Navy Man Tech Office, and Raytheon (the prime contractor).

- **Evolutionary Acquisition:** All the CFAs were customized products. Versions were redesigned based on user input so that there was reduction of the flaws and reassessment of risks and assumptions. The evolving production of the CFAs reminds us of the current strategy of evolutionary acquisition (Apte, 2005; Apte & Lewis, 2007). There are similarities between the process through which the CFAs evolved and the current trends of incremental design and production based on the input of war fighters.

- **Theory of Constraints:** Theory of constraints suggests that identified system constraints can be exploited to the advantage of the system. That is precisely what was done here. For example, the PM exploited the high unit production cost and low MTBF by challenging the manufacturers to discover a process that would deliver a quality tube to the program. Another instance is that the lack of communication among the key players was eliminated by the PM by visiting the manufacturing plants.

- **Total Quality Management:** One of the important tools in total quality management is the cause-and-effect or Ishikawa diagram to get to the root of the problem. Brainstorming sessions of groups of personnel involved identify lists of causes of the problem and the relationship between them and the effects that educate everyone involved in the system. This process revealed the melting of the CFAs during arcing by asking questions such as follows: Why did the tube fail? Why did the vanes melt? What was the cause of low melting point? Why cannot the solid state be used? The PM's team was able to expose and correct the root cause of the low MTBF statistics for the CFAs.

- **Outsourcing:** Research in outsourcing indicates that outsourcing products, not services, are more advantageous to most systems, especially where the service involves the defense of a nation. Outsourcing the support of a weapon system eliminates the need for in-house technical knowledge. Absence of such personnel not only constrains evaluating competitive contracts but prohibits creative approaches to system failures. It also inhibits challenges to the industry that support the same system. Decreasing or eliminating certain services may prove to be penny wise and pound foolish.

- **In-House Expertise:** Based on the CFA case study, one can conclude that it is essential that a technically competent in-house workforce both at program inception and during system deployment should be maintained. This role has been historically filled in the Navy by competent ISEAs resident in Naval Warfare Centers, each specializing in particular technologies. It is imperative that this skill set be maintained if the DoD hopes to contain TOC in an evolutionary setting.

- **Stewardship:** The CFA case raises the notion that it is crucial for a DoD entity to play a "stewardship" role when necessary to preserve DoD's ability to obtain an affordable product/process that is critical to national defense needs. Such a role is envisioned when products or processes have limited commercial interest and support, on-shore sources are nonexistent or insufficient, and/or unique military logistics requirements exist. In a stewardship role, a facility such as the Crane Naval Surface Warfare Center would facilitate communication and knowledge sharing among industry, academia, and military users of products and processes; maintain crucial capabilities and knowledge required for test and evaluation, logistics, and for certain manufacturing and repair; serve as an advocate for programs targeted at maintaining the viability of the product or process; and work to maintain a balanced budget strategy to support the critical industrial base.

- **Process Management:** Process management is an ongoing methodology for evaluating, analyzing, and improving the performance quality of a key business process. In the first phase, a process is evaluated by establishing process ownership, determining user requirements, and evaluating and rating the process; in the second phase, it is analyzed by benchmarking, developing and reviewing solutions with participants, and developing improvement plans; and in the last phase, the plan is implemented, results are measured, user feedback is obtained, and at the end, the entire process is repeated. Clearly, without such a continuous process improvement strategy, the CFA program would not have achieved such impressive results.

12.4   CASE STUDY 2: TOC REDUCTION FOR AEGIS RADAR PHASE SHIFTERS

12.4.1   Background

Advances in technology throughout the 1980s made it possible to build an AEGIS system with a smaller ship while maintaining multimission capabilities. The smaller ship was designed using an improved sea-keeping hull form, reduced infrared and radar cross-section, and upgrades to the AEGIS Combat System such as the SPY-1D. The first ship of the DDG-51 class, USS ARLEIGH BURKE, was commissioned on July 4, 1991. The DDG-51 class was named after a living person, the legendary ADM Arleigh Burke, the most famous destroyer man of World War II.

12.4.1.1   Principles of Phased-Array Radar Antennas. There are many benefits to electronically scanned antennas, including fast scanning, the ability to host multiple antenna beams on the same array, and the elimination of mechanical complexity and reliability issues associated with rotating antennas. Because phased-array radar antennas require no physical movement (Figure 12.8), the beam can scan at thousands of degrees per second, fast enough to irradiate and track many individual targets and still run a wide-ranging search periodicity.

A SPY-1 Phased-array Radar Antenna consists of 4,500 elements that are essentially miniature individual antennas. These elements are arrayed in patterns depending on the desired performance characteristics needed by the application,

FIGURE 12.8 Spy-1D phased-array radar antennas (two of four shown). *Source*: After Global Security.org, 2006.

such as operating frequencies, antenna gain, sensitivity, and power requirements. Each of these elements requires a phase shifter.

Beams are formed by shifting the phase of the signal emitted from each radiating element to provide constructive/destructive interference so as to steer the beams in the desired direction. In Figure 12.9a, both radiating elements are fed with the same phase. In Figure 12.9b, both elements are fed with different phases. The signal achieves maximum gain by constructive interference in the main direction. The beam sharpness is improved by the destructive interference.

12.4.1.2 Cost Drivers. Operating and support costs may be dramatically reduced by identifying cost drivers and correcting them—often, but not always, through redesign. The most efficient time to accomplish this is during the pre-acquisition and development phases while the system is only a paper design and may be changed relatively inexpensively. However, acquisition cost drivers that are discovered during the production phase also may lead to redesign or other actions to reduce the APUC or may reduce the cost of manufacturing by improving the process yield to save or avoid future expenditures. The focus in this study will be on design and redesign of the SPY-1 phase shifters, which dramatically improved performance without increasing the APUC and the reduction of costs to manufacture SPY 1-B/D phase shifters by improving the process yield. Additionally, this case will present various process-improvement programs used to reduce "touch-labor" and improvements to programs that affected phase-shifter production either directly or indirectly (i.e., consolidated purchasing, Lean and Six Sigma, productivity improvement projects, etc.).

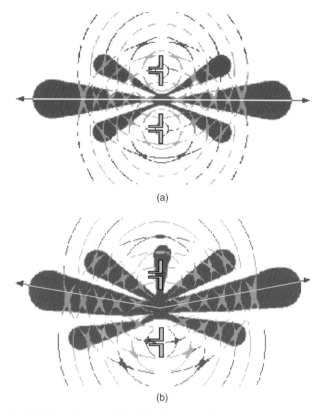

(a)

(b)

FIGURE 12.9   (a) Two elements fed with the same phase. *Source*: From Radar Tutorial, 2006. (b) Two elements fed with different phases. *Source*: From Radar Tutorial, 2006.

12.4.1.3 Lockheed Martin at Moorestown. Lockheed Martin Corporation

(LMCO) is principally engaged in research, development, manufacture, integration, and sustainment of advanced technology systems, products, and services. The corporation serves customers worldwide in defense and commercial markets, with its principal customers being agencies of the U.S. Government. With its corporate headquarters in Bethesda, Maryland, LMCO is organized into five business areas: Aeronautics, Electronic Systems, Information & Technology Services, Integrated Systems & Solutions, and Space Systems. LMCO employs 135,000 personnel at 939 facilities worldwide and achieved $37.2 billion in sales for 2005 (Figure 12.10).

Lockheed Martin Maritime Systems & Sensors (MS2) in Moorestown, New Jersey, is part of the Electronic Systems business area that manages complex programs and provides integrated hardware and software solutions to ensure the mission readiness of armed forces and government agencies worldwide; this facility achieved $10.6 billion in sales for 2005. The MS2 facility was established in 1953 as part of RCA Corporation and later merged with General Electric-Aerospace Group, was sold to Martin Marietta in 1992, and merged with Lockheed in 1995.

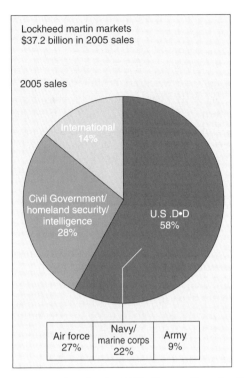

FIGURE 12.10   Lockheed Martin sales for 2005. *Source*: From Lockheed Martin, 2006.

LMCO-Moorestown is the prime contractor for manufacturing and integration of the Aegis Weapons System and Aegis Depot Operations for the Navy. Its successful history in large-scale systems integration, radar technology, software development, microelectronics, lifetime support, vertical launching systems, and fire-control systems enabled the company to establish a solid foundation for creating future innovative solutions.

12.4.1.4 SPY-1A Phase Shifter.

A phase shifter is a two-port device whose basic function is to provide a change in phase of RF signal with minimal attenuation. There are two types of phase shifters: mechanical and electronic. From the late 1940s up to the early 1960s, prior to the development of electronically variable phase shifters, all phase-shifting requirements including those of beam-steering array antennas were mostly met by mechanical phase shifters. In 1957, Reggia and Spencer reported the first electronically variable ferrite phase shifter, which was employed in an operational phased array (Koul & Bhat, 1991). The 1960s saw the emergence of another important type of phase shifter—the semiconductor diode phase shifter. Since then, significant advances have taken place in both ferrite and semiconductor diode phase shifters, resulting in a wide variety of practical devices. Major growth of phase-shifter technology came from its known potential utility in phased arrays.

A typical phased array may have thousands of radiating elements, and with each antenna element connected to an electronically variable phase shifter, the array acquires the basic capability for inertia-less switching or scanning of the radiated beam with minimal time. With this capacity, the array achieves complete flexibility to perform multiple functions in three-dimensional space, interlaced in time and even simultaneously. The evolution of phased-array technology to its present sophisticated form is strongly based on the development of electronically variable phase shifters. In turn, new areas of application have opened up in radar, communication, and civilian sectors, demanding newer techniques and technologies for phase shifters. In addition to ferrite and semiconductor diode phase shifters, several other types have emerged in recent years; these, however, are not the focus of this case and, hence, will not be discussed.

12.4.2   The Big Breakthrough

Because of the lack of electronic media available from the 1970s, and multiple corporate mergers (RCA/GE/Martin Marietta/Lockheed) spanning three decades, detailed engineering/production data of the AEGIS Weapons System transition from the EDM-1 to the SPY-1A is virtually nonexistent or not available. Research relating to this case is based on the recollections of current and retired production engineers and managers from Lockheed Martin at Moorestown.

12.4.2.1   The RCA Role. In the 1960s and 1970s, ferrite phase shifters were preferred for the large phased-array radars. However, they were extremely expensive. The first phase shifters used in EDM-1 were in the neighborhood of $2000 per unit in 1974 dollars. One phased-array radar antenna requires 4,500 phase shifters, and one AEGIS combatant requires four phased arrays, thus totaling approximately $36 million in phase shifters alone. In 2006 dollars, this equates to approximately $148 million, clearly representing a significant cost for a single part in one system on a AEGIS equipped ship. Although AEGIS was a huge leap in national defense capability, RCA knew that ferrite phase shifters would have far-reaching effects on acquisition management, design, and life-cycle support of a modern navy.

In an effort to drive down AEGIS Weapons System costs, RCA embarked in a 2–3-year effort to productize the phase shifter, that is, something that could be practically specified, repeatable, and producible. This product development effort resulted in RCA designing its own version of the ferrite phase shifter for use in the next generation of SPY-1 radars.

RCA was able to meet the cost objective to produce one phase shifter unit for approximately $200—a monumental effort considering it brought down the cost of one ship-set (18,000 units) from approximately $148 million to approximately $15 million (2006 dollars). Although RCA designed the phase shifter for AEGIS, critical materials for the phase shifter were procured from other companies. The garnet material was provided by Trans Tech who has continued to provide all of the garnet material for AEGIS production. RCA's "best value" pitch proved its worth. To date, 23 years and 76 AEGIS capable cruisers and destroyers later, a garnet phase shifter has never been replaced failure (Lockheed Martin, 2006).

Assembly of the phase shifter has always been a significant challenge because of the sensitive nature of the material interactions among the garnet material, the aluminum housing, and the ancillary RF and logic control wire interfaces. A highly skilled assembly team using advanced manufacturing process control techniques has continually managed this process closely to provide the high yields necessary to achieve the demanding cost requirements.

12.4.2.2 The New Requirements. Although the AEGIS SPY-1A radar was a huge success, the Navy continued to push RCA throughout the 1980s to improve phase-shifter insertion loss, bit-phase shifting, and differential phase error— ultimately reducing side lobe levels (Figure 12.11). Low side lobes were among the highest priorities for several reasons: reduction of radar and communications intercept probability, reduction of radar clutter and jammer vulnerability, and increasing spectrum congestion in satellite transmissions (Lockheed Martin, 2006). The big challenge for RCA was how to meet the Navy's new performance requirements and keep down costs.

Differential phase error is the root-mean-square (rms) phase-shift error caused by variations with frequency, phase state, power, and temperature. When considering a large number of phase shifters (4500 in one array), the calculation of this error may include variations from unit to unit. Phase error reduces the antenna gain in a transmitting array and raises side lobes in a receiving array. The rms phase error permissible for the SPY-1A phase shifter was ≤5.8 deg rms (Lockheed Martin, 2006).

The Navy's new differential phase error performance parameter for SPY-1B was ≤2.1 deg rms (Lockheed Martin, 2006). This was a 64% improvement requirement over the SPY-1A. To achieve these numbers, RCA had to make one major modification and one major tradeoff—increase phase-shifter bit capacity and allow more insertion loss.

The SPY-1B phase shifter was as big a breakthrough as was the SPY-1A in that RCA was able to increase phase-shifter performance by leaps and bounds for the

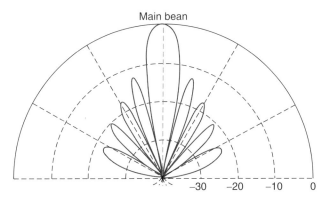

FIGURE 12.11 Depiction of main beam-side lobe relation. *Source*: After Radar Tutorial, 2006.

next generation of radars, yet do it without increasing the average unit procurement costs. Since it is not possible with the passage of time to ascertain detailed quantitative cost data in specific areas that improved process yield and reduced touch labor, this case provides only some very general graphs (Figure 12.12) that, although incomplete, provide a good overarching snapshot of the impact of process yield and touch labor on costs. Figure 12.12 shows how LMCO-Moorestown brought the APUC of a phase shifter from $200 in 1984 down to almost $100 in 2002. The APUC for phase shifters in 2006 is now $80 ($1.44 million per ship-set). This $200 APUC in 1984 is $5.91 million in 2006, which is a substantial reduction in APUC of 76%—a sizable cost reduction considering this is a single part in one system on an AEGIS equipped ship.

12.4.3 Production Improvement Processes

Since 1984, there have been many process improvement initiatives to improve process yield and to reduce touch labor of the SPY-1B/D phase shifter. We now offer a discussion of some important management initiatives and their impact on APUC.

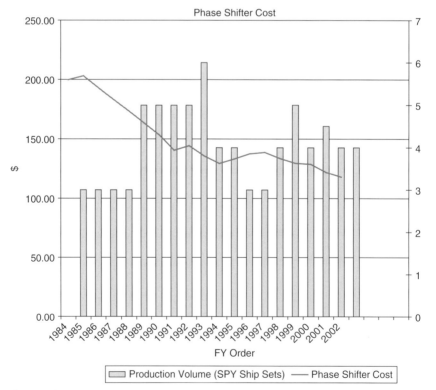

FIGURE 12.12   Phase shifter cost per unit. *Source*: Used with permission from Lockheed Martin, 2006.

12.4.3.1 Defect and Scrap Reduction. Scrap generation and defect production are important conditions when evaluating a company's performance. LMCO-Moorestown established a program in 1991 to both measure and reduce these parameters. For scrap, prior to 1991, there was a limited breakdown of collected data, and the data were not in a format that allowed meaningful analysis (Office of Naval Research, 1995). In addition, little of the analyzing that was conducted by engineering management was relayed to floor personnel where it could produce the greatest impact. In other words, the worker was not allowed to share in the knowledge of the experts. One can clearly see that if an IPT approach was used or HSI were applied here, the program could have benefited.

LMCO-Moorestown has since established multifunctional IPTs in each work center. Each team brainstorms a list of metrics for the work center that are monitored, including defects and scrap. Performance is then measured against the metrics weekly. An important aspect of this effort includes the linking of the company suggestion program to team efforts and performance. Benefits have been demonstrated throughout the company. For example, phase-shifter defect free yield of a hoped-for 80% in the 1970s improved to 99.5% in 2006 (Lockheed Martin, 2006). Scrap costs have been reduced by 60% from 1994 through 2006 (Figure 12.13). In addition, this approach has yielded intangible benefits, such as improved problem solving and corrective action skills, increased sense of ownership by the team, lower costs, higher quality, and a more educated workforce.

12.4.3.2 Touch-Labor. Changes in the defense environment since the mid-1980s affected most government contractors. In 1989, LMCO-Moorestown responded to changes by abolishing thousands of positions. However, the Local 106 union moved to team with LMCO in a partnership as both sides realized they had to work together to remain a viable business. This initiative demonstrated LMCO's determination to maintain a level workforce. LMCO listened to new ideas, facilitated implementation, and opened lines of communication. Aggressive goals were set—and exceeded—such as reducing touch labor by 26%. By

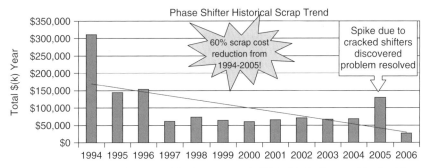

FIGURE 12.13 Phase shifter historical scrap trend 1994–2006. *Source*: Used with permission from Lockheed Martin, 2006.

implementing the initiative, what was scheduled to become additional outsourced work on components for the AEGIS system translated into the retention of 400 labor positions planned for elimination (Office of Naval Research, 1995). Today, touch labor is down by 40% in phase shifters alone (Figure 12.14). The remainder of this chapter discusses some of the larger contributors to this accomplishment.

12.4.3.3   Ultraviolet Acrylic Cure Process. The ultraviolet (UV) acrylic process reduced touch labor. The iris epoxy cure process was labor intensive prior to the introduction of the UV acrylic. The epoxy was a two-part adhesive that required one operator 16 hours a week to mix enough material for one week's production of phase shifters. Additionally, the two-part epoxy was time-sensitive once mixed; so material that did not get used expired rapidly. The UV acrylic is a one-part material that is dispensed directly from the manufacturer's container and has a greater shelf-life.

12.4.3.4   Automation. In 2002, a process improvement introduced automation to the mixing process. A machine mixed the two parts on-demand, resulting in no waste and no shelf-life. Additionally, the material was fresher and more consistent. The new process eliminated the 3-hour mixing operation and cut the dispensing effort by 50%. Before, an operator had to prep the pumping machine with the potting material; now the machine mixes and pumps it directly into the phase shifter.

PHASE SHIFTER

PRODUCT TOUCH LABOUR RATE (Hours per unit) HISTORY

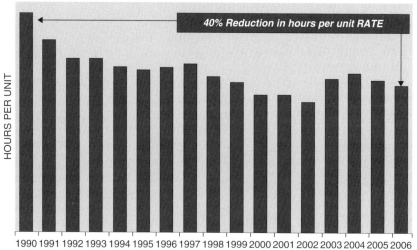

FIGURE 12.14 Phase shifter touch labor rate history 1990–2006. *Source*: Used with permission from Lockheed Martin, 2006.

12.4.4 Business Issues

The following business issues are tools used by a company to survive in the global competitive markets. This is an instance where LMCO-Moorestown implemented concepts from contracting, HSI, and operations management.

- **Consolidated Purchasing:** Each business unit of LMCO-Moorestown maintained and operated a complete and independent purchasing department, resulting in inconsistent sourcing and quality practices. Opportunities for increased buying efficiencies and overall cost effectiveness were often lost. Lockheed Martin resolved this situation by consolidating the business units into three purchasing organizations, one of which is the Material Acquisition Center Mid-Atlantic Region (MAC-MAR). MAC-MAR provides full-service sourcing including direct major subcontracting and indirect buying, supplier management, technology engineering, receiving, supplier quality assurance, inspection, freight management, and cost estimating.

- **Supplier Process Surveillance:** LMCO-Moorestown used traditional supplier product acceptance methods that relied on costly inspections upon receipt or at the supplier's location. MAC-MAR implemented supplier process surveillance (SPS), which shifted the emphasis of quality from inspections to process controls (Office of Naval Research, 2001). Minimum requirements are determined by supplier category (e.g., manufacturer and distributor, manufacturer only, distributor only, and manufacturer of custom parts). Reviews are predetermined by the TDP team.

- **Eight-Step Process Improvement Program:** The eight-step process improvement program follows a detailed process flow that focuses on critical suppliers, materials, and processes; uses analytical tools to identify supplier trends; identifies critical manufacturing and/or part processes; and employs process surveillance to monitor risk areas. Of interest here is the team composed of a lead engineer facilitating a team of three to six people for each supplier, personnel from other business units who have expertise in the products/processes related to the product to be delivered.

- **Lean and Six Sigma:** LMCO identified several key roles in pulling the Lean and Six Sigma methodology together, and it started from the top. A Senior Leadership Team (SLT) of top executives provides visible support through programs and resources to drive overall change throughout the organization. Functional organizations select management points of contact to be the focal point of Lean and Six Sigma (e.g., project measurements, performance, and results) in their areas. The return on investment of Lean and Six Sigma techniques is directly proportional to the commitment of business leadership.

- **Productivity Improvement Projects:** At LMCO, the productivity improvement projects include renewed planning, improved reporting structure, and better capture of improvements versus baseline. The approach engages the company's strong experience base, with the technical support team and operations management initiating and facilitating brainstorming sessions in selected micro businesses. The company also developed a process flow to facilitate the new approach and a comprehensive database to support the entire operation from initiation to benefits tracking. These PIPs take advantage of tools and concepts offered by Lean and Six Sigma.

- **Employee Suggestion Program:** The employee suggestion program features a suggestion tracking system that operates as a comprehensive tool for inputting, storing, evaluating, and communicating suggestions throughout the organization. Every suggestion receives feedback of disposition and an explanation of the evaluation decision.

- **Engineering Change Notice (ECN):** ECNs are formal mechanisms for revising released engineering drawings. Problem sheets are formal mechanisms for documenting issues with engineering or process documentation. The automated ECN/problem sheet system provides Lockheed Martin with an automated tool for creating, processing, and monitoring ECNs and problem sheets in the program management office, engineering, manufacturing, and sourcing departments. Standard and custom review screens give employees the ability to develop meaningful metrics of their processes.

- **Engineering Change Notice Reduction:** Engineers originating an ECN present the root cause and corrective action to the ECN Review Board (ERB). The Board can either approve or disapprove the action. The ECN Review Board assigns individuals to map the processes and to determine the costs associated with implementing the corrective action. To prevent ECNs from recurring, a database tracking process is used. Additional benefits include improved design practices and tools, and a reduction in ECNs, rework, and cycle-time. Since implementing the ERB, Lockheed Martin realized more than one million dollars in cost-avoidance and savings.

12.5 CONCLUSION

The goal of Case Study 1 was to provide an illustrative case study that chronicled the operational and engineering processes used to reduce one aspect of the total ownership cost for the Aegis Project. As the study illustrates, because of these improvement plans in the early 1980s, the MTBF for the MWTs increased substantially—to approximately 5,000 hours (Greene, 2004). The initiatives continue to thrive at Crane Naval Surface Warfare Center. It is still operating with increasingly impressive statistics: MTBF is up to 40,000–45,000 hours, and the cost per operating hour in 2002 dollars has been reduced to $0.45/socket from $8.20/socket (Dutkowski, 2004–2005).

In the early days (circa 1983) of the Aegis Shipbuilding Project, the program office recognized that the numerous microwave tubes used in the Aegis radar system would significantly contribute to operational costs. As a result, an initiative was put in place to focus on substantially reducing these costs. The initiative was eventually applied across the whole spectrum of Navy microwave tubes. This reduction in total ownership cost (R-TOC) effort has been extremely successful as various initiatives have driven down cost metrics (such as dollars/operating hour) while achieving a significant increase in MTBF. The engineering and management processes used to achieve these results are important to understand in light of recent manpower reductions in the services as well as an erosion of the in-house engineering skill base. In particular, this case highlights the role that Naval Surface Warfare Centers can and do play in the acquisition process. This case study validates these successes and identifies the underlying factors such as the IPT that catalyzed them.

The objective of Case Study 2 was to capture the production and design processes and program management solutions used to reduce TOC of AEGIS radar phase shifters. The phase shifter was an AEGIS Weapon System major acquisition cost driver that was reduced to a medium-priced component through design and redesign, various process improvement projects, and other programs that improved phase-shifter production either directly or indirectly. Thereafter, the Navy sought to improve phase-shifter performance to reduce side lobe levels. This was the next big challenge for LMCO because not only did the Navy want to improve performance, but also it had incentivized LMCO to improve performance while concurrently keeping down the APUC. The result was SPY-1B: a radar system that incorporated significant advances over the SPY-1A radar, with improved detection capabilities as well as lower side lobes. LMCO was able to increase phase-shifter performance by leaps and bounds for the next generation of radars without increasing the APUC.

Throughout the 1980s and 1990s, there have been many LMCO process improvement initiatives to improve process yield and reduce touch labor. Through various defect- and scrap-reduction initiatives, LMCO improved defect yield from approximately 80% in the 1970s to 99.5% in 2006. It brought down touch labor by 40% between 1990 and 2006 through the implementation of robotics and other automation processes. The culmination of these process improvements has brought the APUC of a phase shifter from $200 in 1984 down to $80 in 2006—reducing the APUC another $4.47 million (or 76%) in 2006 dollars per ship set—thus, reducing the costs of future acquisitions of AEGIS Weapons Systems. In addition, the establishment of MAC-MAR improved manpower productivity by more than 26% in its first four years and reduced overall procurement costs of the AEGIS program by 32%. This success was achieved by the initiatives that could not have been implemented without a clear focus on considering the human element in business/management decisions—HSI in action.

In conclusion, as a system progresses from early concept through prototyping, into production, and finally reaches the sustainment phase, the opportunities to reduce TOC significantly diminish. This clearly indicates that R-TOC efforts are most effective early in the developmental cycle where changes are least expensive and easiest to implement. However, TOC reductions can be effective throughout the system's life cycle. The balance between capabilities and affordability means that more warfighting assets are available to the warfighter. TOC stakeholders have a vested interest in influencing the system design and development to yield a suitable, effective, and affordable solution. The challenge is how to accomplish this goal.

This challenge becomes greater in today's restructured acquisition environment. A key to success outlined in both the case studies of the Aegis TOC reduction efforts was the single program management office for the entire weapons system throughout its life cycle. As one may envision, because of its size and complexity and long-term life-cycle requirements, separating program management (thus, ownership) of the major weapons acquisition function from shipbuilding may present significant challenges to major TOC reduction efforts for future ships.

An acquisition strategy prevalent in Aegis that enabled increased operational availability with reduction of TOC was the use of IPT and strategic partnering. Both these program management practices represent a long-term, mutually beneficial business relationship containing specific elements unique to the relationship; it is an agreement detailing performance requirements and conditions, structures to promote successful interaction between parties, organizational alignment, clear measures of success, and a high level of mutual commitment. Long-term contracts and collaboration generally foster lower costs because of the greater incentive to make transactional-specific investments, the sharing of information, and value engineering with the resulting enhanced learning curves. These AEGIS cases clearly demonstrate a compelling and undeniable example of this. One of the more intangible benefits of IPTs and strategic partnering worth mentioning is the longevity of both government and contractor employees in a program, and the benefits it lends to program success by way of capturing experience and corporate knowledge.

12.6 ACKNOWLEDGEMENT

I am grateful to RADM Jim Greene for his invaluable insights without which these case studies would not have been possible. As most readers might have realized by now, "the PM" in the first case study was RADM Jim Greene.

REFERENCES

Apte, A. (2005), Spiral development: A perspective. *Acquisition Research Sponsored Report Series*. Monterey, CA: Naval Postgraduate School.

Apte, A., & Dutkowski, E. (2006), Total ownership cost reduction case study: AEGIS microwave power tube. *Acquisition Research Sponsored Report Series*, NPS-AM-06-008, Monterey, CA: Naval Postgraduate School.

Apte, A., & Lewis, I.A. (2007), The logistics impact of evolutionary acquisition. *Acquisition Research Sponsored Report Series*, NPS-LM-07-046, Monterey, CA: Naval Postgraduate School.

Boudreau, M.W., & Naegle, B.R. (2003). Reduction of total ownership cost. *Acquisition Research Sponsored Report Series*, NPS-AM-03-004, Monterey, CA: Naval Postgraduate School.

Bridger, R., & Ruiz, M. (2006). Total ownership cost reduction case study: AEGIS radar phase shifters. *Acquisition Research Sponsored Report Series*, NPS-AM-06-050, Monterey, CA: Naval Postgraduate School.

Dutkowski, E.J., Jr. (2004-2005). [Electronic communication and phone interviews with Aruna Apte].

Greene, J.B., Jr. (2004-2005). [Interviews and electronic communication with Aruna Apte].

Hoffer, K. (2003). [Electronic communication with Aruna Apte].

Jane's.com. (2006). *AEGIS Weapon System MK-7*. http://www.janes.com/defence/naval_forces/news/misc/AEGIS01 0425.shtml.

Koul, S.K., & Bhat, B. (1991). *Microwave and Millimeter Wave Phase Shifters*. (Volume 1). Norwood, MA: Artech House.

Lockheed M. (2006). [Interviews and electronic communication with LCDR Wray W. Bridger & Capt. Mark D. Ruiz].

Office of Naval Research. (1995). *Best Manufacturing Practices*. Report of Survey conducted at Lockheed Martin Government Electronic Systems, Moorestown, NJ. College Park, MD: BMP.

Office of Naval Research. (2001). *Best Manufacturing Practices*. Report of Survey conducted at Lockheed Martin Naval Electronics & Surveillance Systems-Surface Systems, Moorestown, NJ. College Park, MD: BMP.

Radar Tutorial. (2009) *Radar Principles*. http://www.radartutorial.eu/index.en.html#this.

Truver, S. (2002). Unpublished Report on AEGIS.

USS NORTON SOUND (2006). *History of USS NORTON SOUND*. http://www.ussnortonsound.com/.

Chapter **13**

The Economic Impact of Integrating Ergonomics within an Automotive Production Facility

W. Gary Allread and William S. Marras

13.1 INTRODUCTION

Since Henry Ford perfected the assembly line in the early 20th century, automotive production work has been associated with repetitive activities. That is, an assembly worker performs the same or similar activities over and over throughout his or her shift. In addition, advances in technology and process control throughout the past several decades have shortened cycle times, further increasing efficiency and productivity. From a strict industrial engineering perspective, these improvements in manufacturing help to streamline production and use of resources.

Unfortunately, repetitive movements and rapid work can negatively impact one valuable production resource—the employee. The National Research Council and the Institute of Medicine (2001) found that increased reports of low back pain were related to frequent bending and twisting as well as to load moment (i.e., the combination of an object's weight and the distance from the spine it is held) and heavy physical work. Musculoskeletal disorders (MSDs) of the upper extremities also were clearly associated with repetitive work, along with high-force exertions and exposure to vibrating surfaces.

The Economics of Human Systems Integration: Valuation of Investments in People's Training and Education, Safety and Health, and Work Productivity. Edited By William B. Rouse
Copyright © 2010 John Wiley & Sons, Inc.

Recent statistics suggest that, when employees produce goods using repetitive and rapid movements, their injury risk increases. In the United States, the manufacturing sector comprises just 13% of all goods-producing industries; however, it accounted for the most injuries (20.1%) and illnesses (36.0%) in 2006 (Bureau of Labor Statistics, U.S. Department of Labor, 2008). More specifically, those working for U.S. motor vehicle manufacturers developed an average of 11.4 work-related injuries or illnesses per 100 full-time employees. Sprains, strains, or tears of the body's soft tissues accounted for 43.4% of all injury types in motor vehicle manufacturing, whereas overexertion was the event associated with 21.4% of all injuries or illnesses (Bureau of Labor Statistics, U.S. Department of Labor, 2009a).

Health-care costs associated with injuries in U.S. motor vehicle production continue to rise. For example, Downey (2004) reported that Daimler Chrysler paid approximately $1,300 per vehicle to cover employee health-care costs. In 2008, the *Wall Street Journal* found that General Motors spent roughly $4.8 billion on employee health care, which added about $1,500 to the cost of every car and truck it produced (Boudette, 2009).

Clearly, better human-systems approaches are needed to improve the interface between people and their work. Many vehicle manufacturers have found that applying the principles of ergonomics is a valid approach to do this. *Ergonomics* is the science of adapting work environments, products, and machines to the capabilities and limitations of people. Inherent in this multidisciplinary field is its integration with the occupational biomechanics, cognitive engineering, medical, psychological, human performance, and systems safety knowledge bases.

The objective of this chapter is to present details of various case studies in which ergonomics principles were used to evaluate and modify repetitive assembly processes in three automobile manufacturing plants.

13.2 ERGONOMICS CASE STUDIES

In 1982, Honda of America Manufacturing constructed the first Japanese-owned auto plant in the United States. Located in Marysville, Ohio, the 3.6 million-square-feet facility employs more than 5,000 associates. It has the capacity to produce 440,000 vehicles (i.e., Accord, Acura RDX, and Acura TL models) annually. Honda's Alliston, Ontario facility opened in 1986 as the first Japanese-owned automotive manufacturing plant built in Canada. It currently employs more than 5,000 associates, who make the Honda Ridgeline pickup and Civic as well as Acura's MDX and (Canadian-exclusive) CSX. Honda opened a second U.S. assembly facility in 1989. This 1.9 million-square-feet plant, in East Liberty, Ohio, produces the Element, CR-V, and (until 2008) Civic models. Its nearly 2,500 associates can assemble up to 240,000 vehicles each year. The primary operations of these facilities include: metal stamping, welding, painting, plastic injection molding, and final assembly.

These Honda facilities have teams of ergonomists and engineers who study human–machine interaction issues and develop solutions aimed at improving associate safety, vehicle quality, and process efficiency. These individuals are guided by corporate-level ergonomists as well as by safety and health professionals, who are charged with understanding production issues across all of Honda's North American facilities and with developing a systematic approach to address and solve ergonomics-related concerns.

For more in-depth ergonomics analyses and those geared toward specific issues, Honda worked with faculty and researchers at the Institute for Ergonomics, a center on The Ohio State University's main campus in Columbus. Ohio State is home to several ergonomics laboratories. For example, in the Biodynamics Laboratory, occupational joint loading is studied under dynamic conditions. Its goal is to better understand occupational ergonomics through quantification and analysis of data gathered directly from industrial settings. The mission of the Orthopaedic Ergonomics Laboratory is to improve the interactions among employees, their jobs, and their work environments. This is conducted by studying how the musculoskeletal system responds to a variety of occupational tasks and potential workplace interventions.

In 2007, the Center for Occupational Health in Automotive Manufacturing (COHAM) opened at Ohio State. In this facility, production environments are simulated to study high-tech manufacturing technology using state-of-the-art occupational health risk assessment techniques. COHAM features a variety of new production equipment, including overhead car carriers and adjustable-height skillet systems. These were developed with the aim of reducing musculoskeletal stresses associated with automotive assembly by orienting a vehicle around the production employee. These technology systems are studied at COHAM, along with other assistive devices used in manufacturing, including tools, rail systems, balancers, and carts. Researchers at COHAM work with automobile manufacturers and their parts suppliers to test these types of manufacturing technology and determine how it can best be used to improve employee health. COHAM is an interdisciplinary partnership funded by several OSU departments, automobile manufacturers (including Honda), and automobile suppliers. Engineers and ergonomists at Honda have used COHAM to study possible equipment and production improvements with their decisions based on scientific evidence regarding employee health and safety.

Because of its research capabilities and geographical proximity to several Honda manufacturing facilities, Ohio State and Honda formed a University–Industry partnership. This unique collaboration allows for the study of design processes aimed to optimize new vehicle production while minimizing occupational health risk. This partnership provides educational and research opportunities for Ohio State students and faculty. It also benefits Honda, by advancing their ability to solve complex production issues and produce higher-quality products.

13.2.1 Job-Specific Case Study

In early 2004, Welding Department managers and those trained in ergonomics at Honda's Marysville facility became especially concerned with the numbers of MSDs occurring among its associates. These were injuries primarily to individuals' tendons, ligaments, and muscles. Upon analysis of the department's previous

3 years of Occupational Safety and Health Administration (OSHA)-recordable injury data, the job with the highest MSD rate was the Door Line process. It required associates to assemble various door parts manually, which were then welded together. Injuries to associates had occurred across multiple body parts (i.e., hands/wrists, shoulders, abdomen, low back, and elbows).

Associates working on the Door Line rotated every 2 hours among three subsequent steps in this welding process. These steps and their descriptions are:

1. Door Skin Process. The outer door skin (weighing between 13 and 32 lbs, depending on the number of skins lifted at any one time) was lifted from a cart (Figure 13.1) and placed on a nearby table. A tool was used to apply sealer to specific areas. The skin then was placed onto a set fixture and clamped in place for welding. The associate finished this step by activating the machine's palm buttons.

2. Door Panel Process. The 7-pound door sash was removed from a parts basket and loaded onto a second welder, as were two 9-pound subcomponents (Figure 13.2). The associate then loaded the door panel (11 lbs to 14 lbs in weight) onto the fixture and triggered the machine.

3. Door Finishing Process. The completed door was first visually inspected. It was then removed from the set fixture (Figure 13.3) and placed on a table for inspection (door weights varied; sedan front doors weighed 35 lbs, sedan rear door were 27 lbs, and coupe doors weighed 42 lbs). Following this check, it was manually moved from the table and placed on a drop-lift hanger.

Across all three steps in this process, each associate working at the Door Line process performed approximately 170 separate materials-handling tasks per hour.

FIGURE 13.1   Associate transferring a part for the Door Skin process of the Door Line job.

FIGURE 13.2   Associate lifting a part for the Door Panel process of the Door Line job.

FIGURE 13.3   Associate moving a completed part for the Door Finishing process of the Door Line job.

Together, these processes required associates to reach for and to place parts at different locations, lift parts having a wide weight range, and move and twist their backs. Thus, relevant ergonomics assessments were used to evaluate the demands of these processes and their subtasks. They were:

- Honda Guidelines for Vertical and Horizontal Reaching. These guidelines were developed by Honda's ergonomists, from anthropometric data contained in Pheasant (1996), PeopleSize Professional software, version 2.05 (Open Ergonomics, Ltd., Kent, England), and *Ergonomics Design Guidelines* (Auburn Engineers, Auburn, AL). From these sources, the ergonomists determined three color-coded categories for both occasional and repetitive reaches that were considered to be either within their goal of keeping associates injury-free (green), "acceptable upon [further] review" of the task (yellow), or "not acceptable" and, as a result, more likely to cause an injury (red).

- ACGIH Threshold Limit Values (TLVs) for Lifting. These values (American Conference of Industrial Hygienists, 2002) provide limits for what is considered to be safe levels of lifting. That is, they are working conditions under which it is believed most employees can work repeatedly, across time, without developing disorders to the low back or shoulders. They are based on various workplace factors found to be associated with MSDs, including lifting frequency, duration of the activity, and the vertical and horizontal locations of the object being handled. These lifting TLVs were based, in part, on research conducted at the Ohio State University and the University of Waterloo and on equations developed by the National Institute for Occupational Safety and Health and the Washington State Department of Labor & Industries.

- Lumbar Motion Monitor. The Lumbar Motion Monitor (LMM) is a patented tri-axial electrogoniometer (Marras et al., 1992). It acts essentially as a lightweight exoskeleton of the lumbar spine and was developed in response to the need for a practical method to directly measure dynamic trunk motions in occupational settings. It is worn on the back of an individual, directly in line with the spine, and is attached using a waist belt at the pelvis and a harness worn over the shoulders. Data from the LMM (positions, velocities, and accelerations of the trunk in its three planes of motion) are input into a validated risk model (Marras et al., 1993, 2000), which determines the probability that the measured activity is similar in nature to previous materials-handling jobs with high low-back injury rates. It also identifies those workplace factors most responsible for the measured level of risk.

The primary findings from these ergonomics analyses are summarized as follows:

- The reach required to remove door skins from their supply carts was deemed "not acceptable," according to Honda's vertical reach guidelines.

- Access to the gun for applying sealer to the door skins as well as clamping this part into place in the welding machine was a cautionary "yellow zone" activity, per Honda's horizontal reach guidelines.

- Repetitively reaching into parts baskets for door panel subcomponents triggered a red "not acceptable" rating, assessed using Honda's guidelines for vertical reach.

- Because of the location and lifting frequency of finished doors, sedan front doors, and coupe doors exceeded the ACGIH TLV for safe amounts of lifting;

- Assessment of the entire Door Line job (comprising the three assembly steps) using the LMM's low-back disorder risk model was determined to be "moderately risky" for sedan doors as well as "high risk" for coupe doors.

- Workplace factors that contributed most to these risk levels were spine moment (i.e., a multiple of the part's weight and the distance held from the spine) when handling finished doors, high levels of trunk twisting when transferring door parts between workstations, and extreme amounts of forward bending when reaching into baskets.

Because of the numerous injury risks that were present, several possible workstation improvements were considered by the Honda and Ohio State teams. Each was aimed at reducing associates' exposures to the various workplace demands. Those changes decided upon focused on improving the efficiency of the Door Line process. This involved relocating the area's existing materials-handling equipment that assisted with lifting the completed doors as well as modifying guards that surrounded the door drop-lifts, so they could be loaded more easily.

The project costs involved in making improvements to the Door Line process, as well as the documented and projected savings across a 5-year period, are listed in Table 13.1. The upper portion of this table shows that there was a cost to conduct the ergonomics evaluation itself. Its findings guided the decision to streamline the production process and create a more efficient workplace layout. The total project costs were $89,000. As Table 13.1 shows, there are no additional costs projected for this process change throughout the subsequent 4 years.

The improvements made to the Door Line process resulted in both manpower and injury reductions. The more-efficient workplace arrangement required three fewer associates to perform the same work, at the same production capacity (Note: these individuals were moved to other production processes.) The lower portion of Table 13.1 shows that the increase in work efficiency accounted for a large portion of the first year's project savings. However, additional savings also were realized through a reduced number of injuries that occurred within this redesigned process. Company records showed that approximately 2.45 work-related MSDs are estimated to be reduced each year since the modifications were made for a savings of more than $100,000. In total, the first-year's estimated savings was more than $362,000. In relation to this project's costs, the payback period for these changes was slightly less than 3 months.

Table 13.1 also projects cost savings for the Door Line process during the 4 years following modifications. It was assumed there would be no change in the number of

TABLE 13.1   Door Line Process Improvement Costs and Projected Savings Across a 5-Year Period

Project Costs	Year					5-Year Total
	1	2	3	4	5	
Ergonomics Assessment	$ 34,000	$ —	$ —	$ —	$ —	$ 34,000
Workstation Modifications	$ 55,000	$ —	$ —	$ —	$ —	$ 55,000
Total Cost per Year	$ 89,000	$ —	$ —	$ —	$ —	$ 89,000
Project Savings						
Increased Work Efficiency	$ 252,000	$ 259,560	$ 267,347	$ 275,367	$ 283,628	$ 292,137
Reduced Injury Costs	$ 110,250	$ 110,250	$ 110,250	$ 110,250	$ 110,250	$ 551,250
Total Savings per Year	$ 362,250	$ 369,810	$ 377,597	$ 385,617	$ 393,878	$ 843,387

associates needed for the process during this time period. Thus, the annual savings of salary and benefits costs following the workstation changes would continue. National compensation statistics show the annual salary increase per motor vehicle production employee to be 4.1% (Bureau of Labor Statistics, U.S. Department of Labor, 2009b). However, Honda reports that its annual rate for associates is approximately 3.0%, so this number was used in Table 13.1 instead. Savings also would continue from the reduced numbers of injuries on this process. Although conservative, an assumption was made that these costs would remain the same for subsequent years.

Even though the payback period for this job modification was short, Table 13.1 shows even greater potential savings over an extended period. It was estimated that, from the initial $89,000 investment in analyses and modifications to the Door Line process, the company would realize a savings of nearly $850,000 after 5 years.

13.2.2   Department-Specific Case Study

The aforementioned case study detailed one of several assembly processes in the Welding Department that was evaluated and modified based on ergonomics principles. Other processes, identified from medical records as having high numbers of MSDs associated with them, are described as follows.

- Fender Installation. This process, shown in Figure 13.4, included a sub-assembly of the vehicle fender and its subsequent installation on the vehicle. Subassembly required bolts to be attached at several difficult-to-reach locations, and awkward upper body postures were used to attach it to the vehicle frame. Improvements to these tasks involved greater use of counter-balanced

FIGURE 13.4 Task required in the fender subassembly and installation process.

tools and a rebalance of the line to allow some tasks to be performed in less awkward postures.

- Bumper Beam Installation. Associates working on this process lifted front and rear beams (weighing approximately 15 lbs) from baskets and placed them onto the vehicle (Figure 13.5). This task produced high amounts of trunk twisting. The beams then were attached with bolts, using an impact gun. For many associates, this was done with the arms at or above shoulder level. Modifications to this process included reorienting parts baskets and training

FIGURE 13.5 Installing a bumper beam on the vehicle.

FIGURE 13.6   Attaching a gusset to the vehicle's rear bulkhead.

associates to time their work so the vehicle was at its lowest level on the conveyor before beginning work.

- Rear Gusset Installation. This process (Figure 13.6) required mounting the part to the bulkhead inside the vehicle and securing it with five bolts using an air tool. Its location caused associates to bend, twist, and reach to do the task. Also, the number of bolts needed to fasten the part made this a highly repetitive activity, done as many items as were held in the hands. Ergonomics-based solutions were to have the task done at a point in the assembly process where the vehicle had fewer obstructions as well as providing associates with a pouch to hold bolts.

- Spatter Cut. In this process, various parts of the vehicle (e.g., pillars, roof flanges, and door openings) were visually checked for weld spatter. A powered grinding tool was used to grind down these areas as needed (see Figure 13.7). These spatter locations often required associates to maintain awkward shoulder, neck, back, and wrist postures. A systems solution was found for this ergonomics issue. The robots causing the spatter were adjusted, serviced, and replaced (as needed). Robots also were installed to cut some of the spatter. Although these modifications did not reduce the extent of awkward body positions that took place, they did greatly lower the amount of grinding required.

- SR Station. This quality control function involved visual checks of vehicle bodies for incorrectly set parts, misaligned flanges, and weld problems (Figure 13.8). Weld integrity also was checked, using hammers and screwdrivers, and spot-welding was done on any needed areas. Like the Spatter Cut process (described above), associates worked in awkward postures. Also, like

FIGURE 13.7 Removing weld spatter using a grinder.

FIGURE 13.8 Checking weld integrity at the SR Station.

Spatter Cut, the demands of this job were reduced by correcting problems upstream in the welding process.

- Mig Welding. A large group of associates rotated among numerous manual welding processes that could not be done by robots (Figure 13.9). This included the front and rear door and fender areas, the wheel wells, and inside the trunk. Static postures in awkward positions were found to occur mostly in the lateral and transverse planes of the trunk. Improvements to these jobs

FIGURE 13.9   Manual welding of parts to the vehicle.

focused on administrative practices, particularly limiting a single associate's exposure to those specific areas of the vehicle where static postures were the most awkward.

- Moving Carts and Baskets. Across the plant, and especially in the Weld Department, parts contained on carts and in baskets were manually pushed and pulled into and out of the work area (Figure 13.10). Carts vary both in size

FIGURE 13.10   Typical cart design and work position.

and in the amount of weight they contain. Also, the speed at which associates must move these carts is dependent on production demands. Laboratory-based testing of these activities found that:

- Handling medium and large carts produced much higher levels of loads on the spine than did small carts.
- Carts more heavily loaded increased both spine loading and use of shoulder muscles.
- There was a trade-off between spine loading and shoulder muscle activity when pulling carts compared with pushing.
- Handling carts at a more "hurried" pace increased both spine loading and shoulder muscle activity.
- Position of the hands when moving carts affected loading on both the spine and the shoulder.

Improvements to this activity included changes in both cart designs and work practices. Carts were outfitted with handles at locations that lowered stress on the body, and worn wheels were replaced with those that reduced push/pull forces. Also, associates were advised to seek assistance from coworkers when maneuvering large or heavy carts.

- Final Body Adjustment. This process involved a fit-and-finish check of door, hood, and trunk with the vehicle body. As depicted in Figure 13.11, tasks involved inspecting and evening gaps between parts (as needed) and ensuring they are flush with one another. These activities required the use of rubber mallets and other tools to make these corrections and to adjust hinges and grommets. There were several improvements made to this activity, including

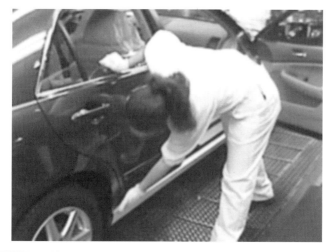

FIGURE 13.11 Performing final body checks on assembled vehicle.

working upstream in the assembly process to eliminate or reduce the need to make these adjustments at all, changing rotation schemes to reduce exposure to physical demands on associates' upper extremities, and educating associates who work on initial weld processes about the impact of their work quality downstream.

Three full years of production have occurred since these interventions were fully implemented. Figure 13.12 shows how total OSHA-recordable cases and injury severity have changed in the Marysville Welding Department as a result. As compared with the total number of OSHA-recordable injuries in 2005, injury counts have dropped by nearly 20% after 3 years. More notably, the severity of these injuries (defined as lost workdays) has been reduced by nearly 50% over this same 3-year period.

Also of interest was how injuries resulting specifically from ergonomics issues (e.g., cumulative trauma) compared with those injuries that were more acute in nature (e.g., lacerations, contusions). Figure 13.13 shows these trends for 3 years (2006 through 2008) following the full implementation of ergonomics-based improvements. In 2005, MSDs represented 60% of all OSHA-recordable cases in the Welding Department. However, in 2008, MSDs represented only 43% of all injuries to associates. The reduction in injury severity was even more striking. Data from 2005 found that four of every five lost workdays in this department were from MSDs. However, after 3 full years of production using the improved processes, MSDs represented only half of the lost workdays recorded.

13.2.3 Facility-Wide Case Studies

The aforementioned case studies focused on ergonomics modifications to a single welding process and to a specific production department. However, these were

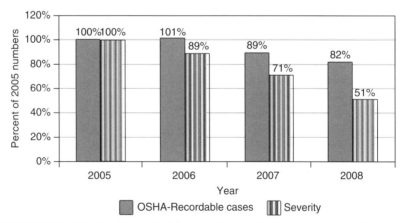

FIGURE 13.12 Relative change in injury numbers and injury severity following implementation of numerous ergonomics-related modifications in the Weld Department of Honda's Marysville, Ohio facility.

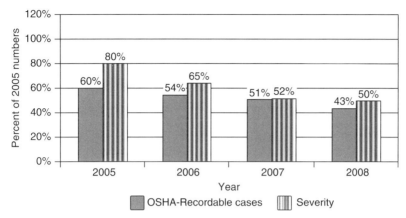

FIGURE 13.13 Changes in the percentage of injuries and lost workdays attributed to MSDs, compared with acute injuries, in Honda's Marysville Weld Department.

part of dozens of similar efforts that have occurred recently across Honda's North American facilities. Many of these were made in conjunction with the launch of new vehicle models. This section will not detail the specific methods used to evaluate and modify all of these existing assembly processes. Instead, it will illustrate the collective results of the comprehensive approach used to evaluate and modify assembly systems in which humans are integrally involved.

For several years, plants in the United States (East Liberty, Ohio) and Canada (Alliston, Ontario) have manufactured the Honda Civic. It was redesigned for the 2006 model year. As with any new-model changeover, many assembly tasks had to be modified. This provided company engineers and ergonomists with an opportunity to also study and improve the specific assembly processes linked to high numbers of MSDs reported by associates.

Ergonomics assessments conducted in both Civic assembly plants found there to be MSD risk factors present in specific assembly processes. These generally included: exerting high levels of force to install parts, reaching overhead, moving repetitively, and working in awkward postures. As a result, several improvements were made to these tasks. These included:

- Changing to grommets and fasteners that require less insertion force.
- Reducing the number of couplers required to install parts.
- Creating subassembly areas, so less work was done inside the vehicle.
- Providing larger access areas throughout the vehicle.
- Adding assist arms for parts and tools that exceeded recommended weight guidelines.

The results from these various process changes were substantial (Figure 13.14). Compared with the number of injuries occurring annually to associates in the East

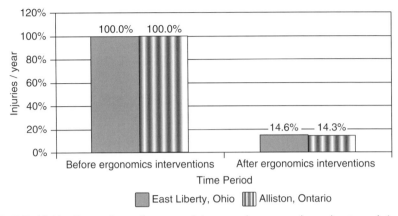

FIGURE 13.14 Comparison of average injury numbers occurring prior to and 1 year following the implementation of ergonomics interventions at two Honda Civic production facilities.

Liberty facility before the ergonomics interventions were implemented, the number of injuries dropped by more than 85% just 1 year after the new model's production. Figure 13.14 shows that similar results were found in the Canadian plant as well.

The direct costs incurred by Honda for associates who develop MSDs are proprietary. However, national cost data are available to provide an estimate of the savings possible from implementing a facility-wide ergonomics initiative. The National Safety Council (2007) reported that the average direct cost of a single work-related strain was $18,600. If one were to apply this estimate to a manufacturing facility in which 100 MSDs were reported annually by employees, the average direct costs would be $1,860,000. As reported in Figure 13.14, Honda used an ergonomics process to reduce their injuries by more than 85%. In this theoretical example, a reduction of this magnitude would save a company more than $1.5 million annually for every 100 injuries it had previously incurred.

The Honda facility in Marysville, Ohio manufactures Accord sedans and coupes. As with the Civic model change 2 years earlier, the Accord underwent a major redesign in 2008. Prior to this, several job evaluations were conducted by company ergonomists within the production facility. Specific processes also were studied by researchers at Ohio State's COHAM facility. Before the redesign, these efforts found that risk factors for Accord production were similar to those with the Civic. That is, for some assembly processes, high forces were needed to install grommets, clips, and springs; components had to be attached in hard-to-access areas; and parts orientations for installation required associates to work in awkward postures.

Creative improvements were made to the 2008 Accord assembly process; many of these were implemented to address the ergonomics issues. Changes involved:

- Modifying the material properties of parts to require less insertion force.
- Making subassembly lines height-adjustable, so components would automatically adjust to the appropriate vertical level for parts installation.

- Altering parts, such as wiring harness clips, to reduce time-consuming manipulation during installation.
- Reworking process steps, to eliminate unnecessary motions.
- Moving problematic assembly processes to subassembly areas.

In total, these improvements vastly lowered injuries (Figure 13.15). By comparing the average number of injuries resulting from Accord production each year prior to 2008, the injury total 1 year following these changes dropped by 93%. Although Honda does not disclose its injury costs, one can approximate the financial savings from such a reduction. Again using the National Safety Council's (2007) statistics ($18,600 for a single work-related strain), a reduction in injuries of this magnitude could save a company $1.7 million in direct medical costs for every 100 incidents it previously incurred.

In addition to injury reduction, analysis of the *severity* of those injuries that do occur is another method to gauge the impact of ergonomics-related changes. Honda used the number of days that associates were away from their jobs following an injury as an indication of severity. These were expressed as a rate, that is, the number of days lost because of an injury based on 100 associates working full-time per year. This severity trend is shown in Figure 13.16. Honda determined that, after 6 years of systematically making ergonomics modifications across their North American assembly plants, their injury severity rate dropped by nearly 88%.

A tremendous amount of savings in direct medical costs can be realized from reduced injury rates that follow ergonomics interventions (Figure 13.14 and Figure 13.15). However, companies can also benefit financially from reduced injury severity rates. When employees miss work because of an MSD, the employer must pay wages for a temporary replacement. These individuals often have less work experience, which can impact both work quality and productivity and, thus, operational costs. The same can be said for situations in which the injured employee is still on the job but working at a restricted capacity. Thus, ergonomics improvements can produce savings at many different levels.

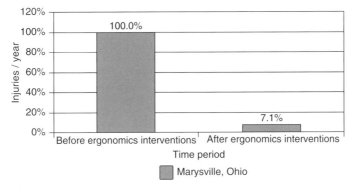

FIGURE 13.15 Comparison of the average numbers of injuries occurring prior to and 1 year following the implementation of ergonomics interventions at Honda's North American Accord production facility.

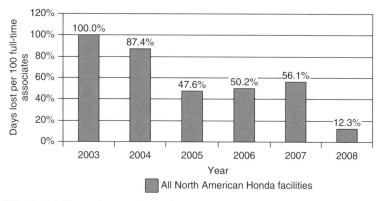

FIGURE 13.16   Change in annual severity rates (i.e., days lost because of injury) following a systematic integration of ergonomics improvements across Honda's North American production plants, compared with 2003.

13.3   CONCLUSIONS

These case studies illustrate the substantial benefits that can be realized through the improvement of manual, repetitive vehicle assembly processes, which were based on ergonomics principles. They show that sizable reductions in both injuries and their severity can be realized. Just as important, they are yet additional examples of the considerable benefits to be gained through more efficient, and safer, human-systems interactions.

Although most privately held companies do not make public the costs of injuries that occur to their employees, these results imply that companies can reap enormous amounts of direct and indirect savings through the design of systems that maximize both productivity *and* safety. In our increasingly global marketplace, the integration of ergonomics into production systems can make U.S. companies more competitive, stable, and better able to withstand economic downturns and market-driven changes.

13.4   ACKNOWLEDGMENTS

The authors wish to acknowledge the valuable assistance of Honda Associates José Banaag, Todd Gordon, and Jim Wolever. Without their expertise, these case studies could not have been developed.

REFERENCES

American Conference of Industrial Hygienists (ACGIH). (2002). *2002 Threshold Limit Values for Chemical Substances and Physical Agents & Biological Exposure Indices.* Cincinnati, OH: ACGIH.

Boudette, N. (2009). Ten hard questions facing the "Car Czar." *Wall Street Journal*. January 22.

Bureau of Labor Statistics, U.S. Department of Labor. (2008). *Incidence Rates of Nonfatal Occupational Injuries and Illnesses by Industry and Case Types, 2007*. Washington, DC: U.S. Department of Labor.

Bureau of Labor Statistics, U.S. Department of Labor. (2009a). *Survey of Occupational Injuries and Illnesses in Cooperation with Participating State Agencies*. Washington, DC: U.S. Department of Labor.

Bureau of Labor Statistics, U.S. Department of Labor. (2009b). *Employment, Hours, and Earnings from the Current Employment Statistics Survey (National)—Motor Vehicles*. Washington, DC: U.S. Department of Labor.

Downey, K.A. (2004). A heftier dose to swallow: Rising cost of health care in U.S. gives other developed countries an edge in keeping jobs. *Washington Post*. March 6.

Marras, W.S., Allread, W.G., Burr, D.L., & Fathallah, F.A. (2000). Prospective validation of a low-back disorder risk model and assessment of ergonomic interventions associated with manual materials handling tasks. *Ergonomics*, *43*(11), 1866–1886.

Marras, W.S., Fathallah, F., Miller, R.J., Davis, S.W., & Mirka, G.A. (1992). Accuracy of a three dimensional lumbar motion monitor for recording dynamic trunk motion characteristics. *International Journal of Industrial Ergonomics*, *9*(1), 75–87.

Marras, W.S., Lavender, S.A., Leurgans, S., Rajulu, S., Allread, W.G., Fathallah, F.A., & Ferguson, S.A. (1993). The role of dynamic three dimensional trunk motion in occupationally-related low back disorders: The effects of workplace factors, trunk position and trunk motion characteristics on injury. *Spine*, *18*(5), 617–628.

National Research Council and the Institute of Medicine. (2001). *Musculoskeletal Disorders and the Workplace: Low Back and Upper Extremities*. Washington, DC: National Academy Press.

Pheasant, S. (1996). *Bodyspace, Anthropometry, Ergonomics and the Design of Work* (2nd Edition). New York: Taylor & Francis.

National Safety Council. (2007). *Injury Facts*. Itasca, IL: National Safety Council.

Chapter 14

How Behavioral and Biometric Health Risk Factors Can Predict Medical and Productivity Costs for Employers

RON Z. GOETZEL, ENID CHUNG ROEMER, MARYAM TABRIZI, RIVKA LISS-LEVINSON, AND DANIEL K. SAMOLY

14.1 INTRODUCTION

Just as businesses provide regular maintenance for equipment to ensure that machinery is tuned and running at optimal levels, investing in the health, well-being, and safety of workers, -an organization's human capital, is just as vital to the success of the enterprise. One way for businesses to invest in the health, well-being, and safety of employees is to offer multicomponent worksite health-promotion and disease-prevention programs that address individual and environmental health and safety risks. Businesses that provide worksite health-promotion and disease-prevention programs to their employees may accrue benefits on two fronts by helping employees better manage high blood pressure and cholesterol levels, eat a healthier diet, increase their physical activity, manage weight gain, reduce excess alcohol consumption, and adopt safe motor vehicle habits such as wearing seatbelts, many debilitating diseases and injuries may be avoided. There is now compelling evidence that well-crafted programs that are based in behavioral and social-ecological theory can improve health and reduce health risks in employees that adopt positive lifestyle habits (Task Force Community Preventive Services, 2007).

The Economics of Human Systems Integration: Valuation of Investments in People's Training and Education, Safety and Health, and Work Productivity. Edited By William B. Rouse
Copyright © 2010 John Wiley & Sons, Inc.

Second, studies have shown that worksite health-promotion programs can produce economic benefits for employers in the form of reduced health-care expenditures and improved worker productivity, in particular, lower worker absenteeism (Chapman, 2005). Studies spanning three decades that evaluated the financial impacts of health-promotion initiatives at Johnson & Johnson (Ozminkowski et al., 2002), The Dow Chemical Company (Goetzel et al., 2005), Citibank (Ozminkowski et al., 2000), Procter and Gamble (Goetzel et al., 1998c), Chevron (Goetzel et al., 1998b), and Highmark (Naydeck et al., 2008) have demonstrated that appropriately resourced programs achieve cost savings in various categories and may even produce savings that exceed program expenses. Thus, with the potential of demonstrating a positive return-on-investment (ROI), there is a business case to be made for employers' adoption of worksite health-promotion programs.

This chapter reviews a sampling of economic models that project cost savings from successful worksite programs. These models, built on research linking health risk and cost data for employees are used to support a business case for employer adoption of health-promotion and disease-prevention programs. Additionally, organizations utilize these models to project cost savings from alternative risk reduction scenarios (i.e., how much an employer can save in medical and absenteeism expenditures costs by reducing employees' health risks). The chapter begins with a synopsis of the business case for worksite health-promotion programs and ends with ROI case studies.

14.2   THE BUSINESS CASE FOR WORKSITE HEALTH PROMOTION AND DISEASE PREVENTION PROGRAMS

14.2.1   Health Care is Expensive

The United States spends more on health-care services than any other country in the world. In 2007, health-care costs in the United States totaled $2.25 trillion (Hartman et al., 2009), but only a small portion (estimated at 1–3%) was spent on prevention and health promotion (Woolf, 2008). At the same time, compliance with recommended clinical preventive services has been dismal, with only 56% of patients receiving the services recommended by the U.S. Preventive Services Task Force (McGlynn et al., 2003). Furthermore, costs associated with the provision of health-care are expected to rise an average of 6.6% annually through the year 2015, and the proportion of the gross domestic product (GDP) attributed to medical expenses is projected to increase from approximately 16% today to 25% in 2030 (Poisal et al., 2007).

Although some cost increases are from advances in expensive medical treatments or lower thresholds for treating certain conditions such as high blood pressure, high cholesterol, and metabolic syndrome, much of this steep rise in spending is from an increase in the prevalence of chronic disease. Today, approximately three fourths of all health-care spending is directed at the treatment of persons with one or more chronic health conditions (Thorpe, 2005). These conditions often are caused by modifiable lifestyle habits such as smoking, poor eating habits, and lack of exercise (Mokdad et al., 2004). One major contributor to chronic disease burden over the past 15 years has been the epidemic rise in obesity rates in the United States, which has contributed 27% to the overall increase in medical care costs for the period of 1987—2002 (Thorpe, 2004).

14.2.2   Healthy Workers Cost Less and are More Productive

For employers, poor employee health is costly. Not only does it affect direct medical costs, but also it produces a spillover effect on worker productivity. In an analysis of medical claims and productivity data for approximately 375,000 workers, Goetzel et al. (2003) found that many of the most costly medical conditions paid for by employers were from preventable health risks, (Figure 14.1). The analysis considered not just direct medical costs but also indirect costs such as employee absenteeism, disability, and presenteeism (on-the-job productivity losses associated with having a chronic health condition).

The National Business Group on Health (NBGH), an organization dedicated to finding ways to help employers reduce health-related expenses and maintain high-quality and affordable health services, periodically releases issue briefs listing steps employers can take to help control health-care spending. Recommendations are formulated based on discussions with and surveys of their large employer members. Close to the top of the NBGH list of employer strategies is adoption of health-improvement programs to address health risks and costs "upstream" before they become a major liability for the business.

The logic flow for such a recommendation can be summarized as follows. A large proportion of high-cost chronic diseases result from modifiable health risk factors such as smoking, overweight and obesity, low levels of physical activity, and poor eating habits (Amler & Dull, 1987; Breslow & Breslow, 1993; Healthy People, 2010; McGinnis & Foege, 1993; Mokdad, et al., 2004). In employed populations, these risk factors are associated with increased medical and indirect costs. The reduced productivity resulting from these risk factors occurs within a relatively short time window (Anderson, et al., 2000; Bertera, 1991; Goetzel, et al., 1998a; Brink, 1987; Pronk, 1999; Yen, et al., 1992).

Several health risks have been shown to predict cost. For example, Goetzel et al. (1998a) found that the additional costs of employees at high risk can be significant (Figure 14.2). An analysis of medical claims, employee health risk behaviors, and biometric laboratory values for more than 46,000 workers in public- and private-sector organizations determined that individual health risks can cost employers up to an additional 70% more annually in medical costs, controlling for demographics and other co-occurring health risks. For example, employees at high risk for depression had approximately 70% higher medical expenditures compared with those with a lower depression risk, and employees at high risk for stress had approximately 46% higher costs compared with employees at low risk.

In addition to medical care expenditures, employee health risks have been associated with productivity losses. A recent study for Novartis Pharmaceuticals estimated how health risks can impact medical care *and* productivity costs in a population

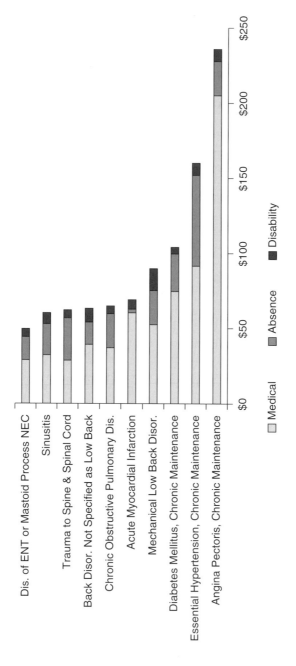

Source: Goetzel, Hawkins, Ozminkowski, Wang, JOEM 45:1, 5–14, January 2003.

FIGURE 14.1 Top 10 physical health conditions: medical, absence, and disability expenditures (1999 annual $ per eligible).

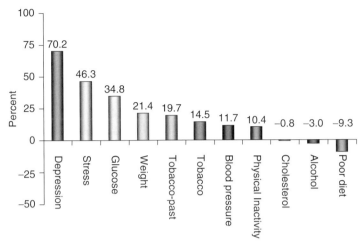

FIGURE 14.2   Percent Difference in Medical Expenditures for High-Risk versus Low-Risk Employees.

of approximately 6,000 employees (Goetzel et al., 2009). Factor analysis methods were used to identify relationships among individual health risks (tobacco, alcohol, weight, emotional health, exercise, triglycerides, cholesterol, blood sugar, and blood pressure), and their combination into the following broad risk factor groupings: (1) high biometric laboratory values, (2) tobacco and alcohol use, and (3) poor emotional health. The relationships among these combined risk factors and costs were estimated using multiple regression analyses.

Annual medical care costs were significantly impacted by two of the three factors. Costs were approximately 13% higher in females and 22% higher in males with high biometric laboratory values and poor emotional health. The percentage increase in the annual number of unproductive days was also significantly higher in females and males for all three factors, ranging from 116% to 238% higher costs for those at higher risk. Female workers' annual absenteeism costs were also significantly greater for the three factors (ranging from 31% to 38% in higher absenteeism costs). This study with Novartis employees illustrated the potential for medical and productivity savings for employers able to reduce health risks among workers through health-promotion programs.

Employers are well positioned to address health risks among workers. Growing evidence supports the effectiveness of workplace health-promotion and disease-prevention programs (Community Guide Task Force, 2007; Heaney & Goetzel, 1997; Pelletier, 1999; Wilson, 1996). These programs also hold open the option of achieving cost savings from risk-reduction programs, an attractive option for employers, especially in light of some evidence that well-designed interventions can produce a positive ROI. (Bertera, 1990; Edington, 2001; Fries et al., 1993, 1994; Goetzel et al., 1998b, 1998c, 1999; Ozminkowski et al., 1999, 2002).

14.2.3   Estimating ROI

Beyond cost savings, decision makers often also require ROI projections for various corporate investments, including those focused on employee health improvement. ROI analyses can provide justification to company officials for investing in health-promotion programs that achieve health benefits for the organization (Loeppke et al., 2008). If savings exceed investments, then the health-promotion program can be viewed as cost-beneficial and as a wise business decision. Health-promotion programs also can be deemed successful when savings equal investments, thus achieving cost neutrality if the health of the workforce is improved and reductions in health risk are achieved.

Although there is controversy regarding the specific methods of determining program-related ROI (Goetzel et al., 1999; Ozminkowski & Goetzel 2001; Sexner et al., 2001), the basic approach calls for the calculation of program savings achievable through a reduction in the health risk profile of an employee population contrasted with the amount that needs to be spent to achieve risk reduction. The difference between the costs of the intervention and the projected savings from the program is determined to be program savings, and these savings are compared with program investments. ROI projections often are calculated for various "scenarios" of program impact throughout time, allowing for shifts in worker demographics, changing prevalence of health risk factors, and subsequent changes in medical and productivity-related expenditures. Using the preceding research as a foundation, economic models can be developed to predict cost impacts from risk-reduction programs. (see Figure 14.3).

14.2.4   Predictive Modeling Tools

Using research from Goetzel et al. (1998a) and Burton et al. (2005) on the relationships between employee health risks and medical/productivity expenditures as a foundation (see Figure 14.2), a predictive modeling tool was developed to help forecast the ROI of health-promotion and risk-reduction programs. The tool projects a financial return based on the information entered regarding participants' health

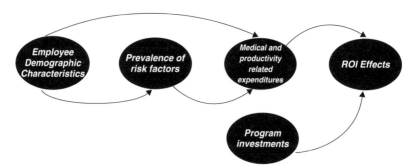

FIGURE 14.3   Basic framework for prospective ROI calculation.

risks, demographic characteristics, the amount spent on the program, and the degree of risk reduction expected among program participants. For example, individuals who eat right, exercise, and do not smoke, incur lower medical expenses and have greater productivity. This information is then used to project organizational costs associated with a healthier employee risk profile. The tool is ideal for organizations that wish to determine the potential program impact on expenses as well as projected cost savings expected from risk reduction achieved in an employee population following the introduction of health-promotion programs. Appendix A provides more detailed information on the methods employed for this type of predictive modeling tool.

14.3 CASE STUDIES

14.3.1 Case Study 1: The Dow Chemical Company

An example of an ROI simulation conducted for a large employer was completed on behalf of The Dow Chemical Company (Dow) and reported by Goetzel et al. (2005). The analysis focused on a projection of medical cost savings from three risk-reduction scenarios contemplated by the company. The first scenario (the base or reference case) assumed that no health-promotion program would be put in place, and increases in health risks alone would drive health-care costs driving increased costs.

Scenario two assumed the company would pay for and implement a health-promotion program that would produce a modest impact on the risk profile of the population by achieving a net reduction of 0.1 percentage points for ten common modifiable risk factors during a 10-year period. At the end of 10 years, the company would achieve a net reduction in all health risks of 1 percentage point.

Scenario three assumed that the health-promotion program would produce a much larger impact and that all ten modifiable risks would be reduced, on a net basis, by a 1.0 percentage point per year, or by 10 percentage points in 10 years. Dow intended to spend approximately $15.5 million across 10 years on its health-promotion program and wished to determine the ROI from such an investment as well as the break-even point.

Risk probabilities were based on two sets of inputs. The first was the demographic profile of Dow employees in 2001, the base year for the study, and projected forward to 2011, assuming an inflow and outflow of employees into the company with varying demographic characteristics (see Table 14.1). Dow's historical employment experience was used to generate predicted changes in workforce size and composition during the following decade. The second set of inputs were derived from employees' responses to the Dow Health Risk Appraisal (HRA), a self-report questionnaire and biometric screening administered on a 3-year cycle to employees with high participation rates in each cycle (75–90%). The HRA asked employees to report their health behaviors and practices related to issues such as nutrition, physical activity, stress, alcohol consumption, tobacco use, and depression. Biometric screenings collected data on workers' height and weight, blood pressure,

**TABLE 14.1 Demographic Profile
of Dow Chemical Company
Employees, 2001**

⇒	Population: 25,828 employees
⇒	Mean Age: 43
⇒	Male: 75%
⇒	White: 82%
⇒	College Educated: 46%
⇒	Professional/Managerial: 44%

total cholesterol, and blood glucose levels. These data were compiled and formed the foundation for a baseline analysis of employee health risks.

Multivariate regression studies were then run to estimate future health risks and costs by risk factor. Independent variables consisted of Dow employee demographic characteristics and risk factors profiles of workers, dichotomized into low- and high-risk groups. Ultimately, these were combined to predict medical expenditures using methods described in Leutzinger et al. (2000) and Ozminkowski et al. (2004). Table 14.2 presents the predicted changes in health risks during the 10-year period.

Health-care expenditures across a 10-year period for Dow were then projected, assuming no net reductions in risk. The inflation-adjusted costs reflect medical expenditures during 10 years associated with a *status quo* or reference condition in which employees aged and their risks generally worsened.

The combination of health risk and demographic data were used to estimate medical care costs for each of the following simulation scenarios: no reduction in

TABLE 14.2 Estimated Health Risk Profile: Dow Employees, 2001–2011

	Summary of Adjusted Probabilities of Being at High Risk Over Time					
Variable	2001 Risk	2003 Risk	2005 Risk	2007 Risk	2009 Risk	2011 Risk
Poor Exercise Habits	23%	24%	25%	26%	27%	28%
Poor Eating Habits	20%	17%	16%	15%	14%	14%
Obesity	40%	41%	42%	43%	44%	45%
Current Smoker	19%	19%	19%	19%	19%	19%
Former Smoker	31%	31%	31%	31%	31%	31%
High Cholesterol	14%	15%	17%	18%	20%	21%
High Blood Glucose	7%	8%	9%	11%	12%	14%
High Blood Pressure	2%	2%	3%	3%	3%	4%
High Stress	7%	7%	7%	7%	7%	7%
Depression	5%	5%	5%	5%	5%	5%
Heavy Alcohol Use	4%	3%	3%	3%	3%	2%

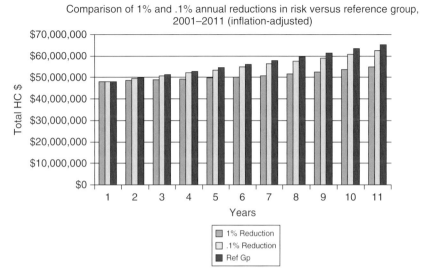

FIGURE 14.4 Simulating the effects of risk reduction under three scenarios, Dow Chemical Company, 2001–2011.

any risk factor, a 1-percentage point net reduction in all risks during 10 years, and a 10-percentage point net reduction in all risks during 10 years. The simulation results are illustrated in Figure 14.4. As shown, there was a clear dose-response relationship between program impact and medical expenditures in which the lowest costs were associated with the high-impact program compared with the base-case scenario. Furthermore, the difference between scenario costs became more pronounced across time.

Results of the simulation study, shown in Table 14.3, demonstrated that to "break even," Dow's program would need to achieve an annual risk reduction of 0.17%. A moderate risk-reduction program achieving 0.1% in net improvements would only return $0.76 for every dollar invested, whereas an aggressive program that achieved a 1.0% risk reduction would save $3.21 for every dollar invested (see Table 14.3).

14.3.2 Case Study 2: Novartis Pharmaceuticals

Using the results of the study referenced previously (Goetzel et al., 2005), Novartis Pharmaceuticals (Novartis) sought to develop a cost calculator that would support its health risk management efforts. The calculator needed to be based on actual company data and allow Novartis management to produce a summary "health index" score that connected employees' health risks to company costs, and then estimate potential savings from risk reduction programs. Furthermore, the calculator needed to quantify the potential savings from changes in employees' health risk on the company's medical care, short-term disability, absenteeism, and presenteeism

TABLE 14.3 Results from Simulating the Impact of Three Health Risk-Reduction Scenarios on Health-Related Expenditures

Year	Reference Case: Total Expenditures with demographics and risk shifting as forecasted (i.e., preexisting trends remain)	Scenario 2: Total Expenditures with 10% decrease in risk across 10 years (1% per year) and demographics change as forecasted	Scenario 3: Total Expenditures with 1% decrease in risk across 10 years (0.1% per year) and demographics change as forecasted	Scenario 4: Break-Even (Reduce risks by 0.17% per year)
Increase in Expenditures From 2001–2011	$17,094,174.26	$6,608,877.16	$14,324,879.51	$13,434,028.14
Percent change between first and last years	35.48	13.72	29.73	27.88
Sum of Total Expend.	$617,074,003.89	$556,469,544.50	$602,640,734.47	$598,059,428.40
Potential Benefits of Risk Management (with a 3% discount rate)	Not applicable— base case	$49,512,590.66	$11,705,745.61	$15,426,727.88
Dow investment (also with a 3% discount rate)		$15,426,671.88	$15,426,671.88	$15,426,671.88
Return on Investment		$3.21	$0.76	$1.00

Return on Investment is calculated relative to scenario in which demographics and risk shift as according to preexisting trends.

Dow investment based on $70.02 per person per year for 10 years, all in 2001 Year Dollar Equivalents, then discounted by 3% per year to adjust for the changing value of money across time.

costs. Novartis wanted a simple graphical display of its Health Index that could be used to visually summarize the data, allowing senior managers to quickly grasp the relationship between employee health risks and financial metrics relevant to business operations.

An overview of the calculator design is presented in Figure 14.5. Constructed using Microsoft Excel software, the user inputs the demographic characteristics of the target population for specific health interventions and the proportion of employees at high risk at baseline for ten modifiable risk factors. User inputs also

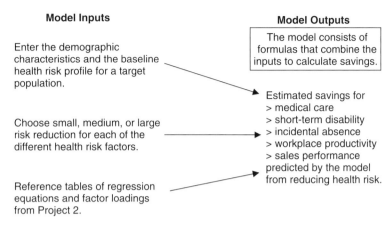

Model Inputs

Enter the demographic characteristics and the baseline health risk profile for a target population.

Choose small, medium, or large risk reduction for each of the different health risk factors.

Reference tables of regression equations and factor loadings from Project 2.

Model Outputs

The model consists of formulas that combine the inputs to calculate savings.

Estimated savings for
> medical care
> short-term disability
> incidental absence
> workplace productivity
> sales performance
predicted by the model
from reducing health risk.

FIGURE 14.5 Application of data—building the Novartis Health Index.

include employees' average daily wage and the percentage of employees expected to participate in the intervention programs.

The user then selects the expected magnitude of changes in risk achieved from the health-promotion interventions for each cluster of major risk categories (high biometric laboratory values, smoking–drinking behavior, and poor emotional health). The user decides whether changes are expected to be small (a 1% annual decrease in risk), moderate (a 5% annual decrease in risk), or large (a 10% annual decrease in risk). The calculator determines the baseline costs for the population with the given set of risks as well as the anticipated changes in costs, assuming the alternative risk reduction scenarios proposed, and reports results in current dollars with the option to input adjustments to inflate dollars to a different year.

Figure 14.6 displays the costs and "Health Index" scores for Novartis employees under two risk reduction scenarios. Scenario A assumes a 1% annual reduction in alcohol, tobacco, and emotional health risk resulting in a 0.1-point increase in the health index score and a reduction in costs of approximately $15 per employee per year. Scenario B assumes a more dramatic annual reduction of 10% in alcohol, tobacco, and emotional health risk that results in saving $153 per employee per year. Furthermore, the second scenario would also increase the health index score by 1.5 points.

14.4 CONCLUSION

Given the high and ever increasing costs of providing health care to their workers, employers are increasingly turning to workplace health-promotion and disease-prevention programs as a strategic tool for helping workers stay healthy and productive. With an increasing national focus on prevention and the growing evidence base, worksite health-promotion and risk-reduction programs are gaining prominence as a means of averting unnecessary costs. In 2009, a Hewitt Associates

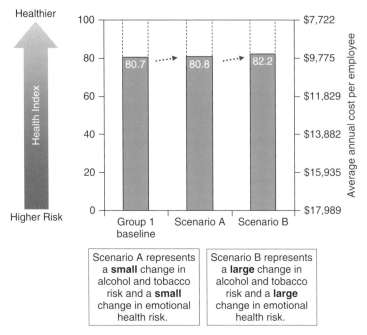

FIGURE 14.6   Novartis Health Index: Change in estimated baseline costs and score.

survey found that, of 343 executives whose companies employ more than 5 million workers across a broad range of industries, almost half offered health and productivity management programs aimed at reducing employees' health risks (Hewitt Associates, 2009). To support significant investment in these programs, employers have spurred researchers to develop predictive models that estimate cost savings from population health improvement initiatives. This chapter has reviewed two such models developed for The Dow Chemical Company and for Novartis Pharmaceuticals.

When successful, health-promotion and disease-prevention programs produce positive health benefits for employees (and in some cases, also dependents) and financial returns for employers in the form of lower medical care expenditures and lower rates of absenteeism, short- and long-term disability, presenteeism, and workers' compensation claims.

Predictive models are supported by a growing body of evidence that improving the health of workers will save businesses money by reducing health-related losses and limiting absence and disability. Employers' investment in human capital can also achieve secondary gains such as improved employee morale, lower turnover rates, and improved safety performance. Taken together, evidence-based health-promotion and risk-reduction programs can improve the financial performance of organizations instituting these programs while improving the health and well-being of workers.

14.5 ACKNOWLEDGMENTS

No external funding was provided for the preparation of this chapter. The opinions expressed in this chapter are the authors' and do not necessarily represent the opinions of Thomson Reuters or Emory University.

REFERENCES

Amler, R.W., & Dull, H.B. (1987). *Closing the gap: The burden of unnecessary illness*. Oxford University Press, New York.

Anderson, G.F., Hurst, J., Hussey, P.S., & Jee-Hughes, M. (2000). Health spending and outcomes: Trends in OECD countries, 1960–1998. *Health Affairs, 19*(3), 150–157.

Bertera, R.L. (1991). The effects of behavioral risks on absenteeism and health-care costs in the workplace. *Journal of Occupational Medicine, 33*, 1119–1123.

Bertera, R.L. (1990). The effects of workplace health promotion on absenteeism and employment costs in a large industrial population. *American Journal of Public Health, 80*, 1101–1105.

Breslow, L., & Breslow, N. (1993). Health practices and disability: some evidence from Alameda County. *Preventive Medicine, 22*(1), 86–95.

Brink, S. (1987). Health risks and behavior: The impact on medical costs. Brookfield, Wi Milliman & Robertson Inc.

Burton, W.N., Chen, C., Conti, D.J., Schultz, A.B., Pransky, G.P., & Edington, D.W. (2005). The association of health risks with on-the-job productivity. *Journal of Occupational and Environmental Medicine, 47*(6), 769–777.

Chapman, L.S. (2005). Meta-evaluation of worksite health promotion economic return studies: 2005 update. *American Journal of Health Promotion, 19*, 1–10.

Edington, D.W. (2001). Emerging research: A view from one research center. *American Journal of Health Promotion, 15*(5), 341–349.

Fries, J.F., Bloch, D.A., Harrington, H., Richardson, N., & Beck, R. (1993). Two-year results of a randomized controlled trial of a health promotion program in a retiree population: The Bank of America study. *American Journal of Medicine, 94*, 455–62.

Fries, J.F., Harrington, H., Edwards, R., Kent, L.A., & Richardson, N. (1994). Randomized controlled trial of cost reductions from a health education program: The California Public Employees' Retirement System (PERS) study. *American Journal of Health Promotion, 8*, 216–223.

Goetzel, R.Z., Anderson, D.R., Whitmer, R.W., Ozminkowski, R.J., Dunn, R.L., Wasserman, J., & HERO Research Committee. (1998a). The relationship between modifiable health risks and health care expenditures. *Journal of Occupational Medicine, 40*, 843–854.

Goetzel, R.Z., Carls, G.S., Wang, S., Kelly, E., Mauceri, E., Columbus, D., & Cavuoti, A. (2009). The relationship between modifiable health risk factors and medical expenditures, absenteeism, short-term disability, and presenteeism among employees at Novartis. *Journal of Occupational and Environmental Medicine, 51*(4), 487–499.

Goetzel, R.Z., Dunn, R.L., Ozminkowski, R.J., Satin, K., Whitehead, D., & Cahill, K. (1998b). Differences between descriptive and multivariate estimates of the impact of

Chevron Corporation's Health Quest Program on medical expenditures. *Journal of Occupational and Environmental Medicine*, *40*, 538–545.

Goetzel, R.Z., Hawkins, K., Ozminkowski, R.J., & Wang, S. (2003). The health and productivity cost burden of the "top 10" physical and mental health conditions affecting six large U.S. employers in 1999. *Journal of Occupational and Environmental Medicine*, *45*(1), 5–14.

Goetzel, R.Z., Jacobson, B.H., Aldana, S.G., Vardell, K., & Yee, L. (1998c). Health care costs of worksite health promotion participants and nonparticipants. *Journal of Occupational and Environmental Medicine*, *40*, 341–346.

Goetzel, R.Z., Ozminkowski, R.J., Baase, C.M., & Billotti, G.M. (2005). Estimating the return-on-investment from changes in employee health risks on the Dow Chemical Company's health care costs. *Journal of Occupational and Environmental Medicine*, *47*(8), 759–768.

Goetzel, R.Z., Juday, T.R., & Ozminkowski, R.J. (1999). What's the ROI? A systematic review of return on investment (ROI) studies of corporate health and productivity management initiatives. *AWHP's Worksite Health*. Summer.

Hartman, M., Martin, A., McDonnell, P., & Catlin, A. (2009). National health spending in 2007: Slower drug spending contributes to lowest rate of overall growth since 1998. *Health Affairs*, *28*(1), 246–261.

Heaney, C.A., & Goetzel, R.Z. (1997). A review of health-related outcomes of multicomponent worksite health promotion programs. *American Journal of Health Promotion*, *11*(4), 290–307.

Hewitt Associates (2009). *Challenges for health care in uncertain times. Survey findings: Hewitt's 10th Annual Health Care Report*. Hewitt Associates, LLC.

Loeppke, R., Nicholson, S., Taitel, M., Sweeney, M., Haufle, V., & Kessler, R.C. (2008). The impact of an integrated population health enhancement and disease management program on employee health risk, health conditions, and productivity. *Population Health Management*, *11*(6), 287–296.

Leutzinger, J.A., Ozminkowski, R.J., Dunn, R.L., Goetzel, R.Z., Richling, D.E., Stewart, M., & Whitmer, R.W. (2000). Projecting future medical care costs using four scenarios of lifestyle risk rates. *American Journal of Health Promotion*, *15*(1), 35–44.

McGinnis, J.M., & Foege, W.H. (1993). Actual causes of death in the United States. *Journal of the American Medical Association*, *270*(18), 2207–2212.

McGlynn, E.A., Asch, S.M., Adams, J., Keesey, J., Hicks, J., DeCristofaro, A., & Kerr, E.A. (2003). The quality of health care delivered to adults in the United States. *New England Journal of Medicine*, Jun 26; *348*(26), 2635–45.

Mokdad, A.H., Marks, J.S., Stroup, D.F., & Gerberding, J.L. (2004). Actual causes of death in the United States, 2000. *Journal of the American Medical Association*, *291*(10), 1238–1245.

Naydeck, B.L., Pearson, J.A., Ozminkowski, R.J., Day, B.T., & Goetzel, R.Z. (2008). The impact of the Highmark employee wellness programs on 4-year healthcare costs. *Journal of Occupational and Environmental Medicine*, *50*(2), 146–156.

Ozminkowski, R.J., Dunn, R.L., Goetzel, R.Z., Cantor, R.I., Murnane, J., & Harrison, M. (1999). A return on investment evaluation of the Citibank, N.A. health management program. *American Journal of Health Promotion*, *14*, 31–43.

Ozminkowski, R.J., & Goetzel, R.Z. (2001). Getting closer to the truth: Overcoming research challenges when estimating the financial impact of worksite health promotion programs. *American Journal of Health Promotion*, *15*, 289–295.

Ozminkowski, R.J., Goetzel, R.Z., Santoro, J., Saenz, B.J., Eley, C., & Gorsky, B. (2004). Estimating risk reduction required to break even in a health promotion program. *American Journal of Health Promotion*, *18*(4), 316–325.

Ozminkowski, R.J., Goetzel, R.Z., Smith, M.W., Cantor, R.I., Shaughnessy, A., & Harrison, M. (2000). The impact of the Citibank, NA, health management program on changes in employee health risks over time. *Journal of Occupational and Environmental Medicine*, *42*(5), 502–11.

Ozminkowski, R.J., Ling, D., Goetzel, R.Z., Bruno, J.A., Rutter, K.R., Isaac, F., & Wang, S. (2002). Long-term impact of Johnson & Johnson's Health & Wellness Program on health care utilization and expenditures. *Journal of Occupational and Environmental Medicine*, *44*(1), 21–29.

Pelletier, K.R. (1999). A review and analysis of the clinical and cost-effectiveness studies of comprehensive health promotion and disease management programs at the worksite: 1995–1998 update (IV). *American Journal of Health Promotion*, *13*(6), 333–345, iii.

Poisal, J.A., Truffer, C., Smith, S., Sisko, A., Cowan, C., Keehan, S., & Dickensheets, B. (2007). Health spending projections through 2016: Modest changes obscure part D's impact. *Health Affairs*, *26*(2), 242–253.

Pronk, N.P., Goodman, M.J., O'Connor, P.J., & Martinson, B.C. (1999). Relationship between modifiable health risks and short-term health care charges. *Journal of the American Medical Association*, *282*(23), 2235–2239.

Sexner, S., Gold, D., Anderson, D., & Williams, D. (2001). The impact of a worksite health promotion program on short-term disability usage. *Journal of Occupational and Environmental Medicine*, *43*(1), 25–29.

Task Force Community Preventive Services. 2007. *Proceedings of the Task Force Meeting: Worksite Reviews*. Atlanta, GA: Centers for Disease Control and Prevention.

Thorpe, K.E. (2004). The impact of obesity on rising medical spending. *Health Affairs*, *23*(6), 480–486.

Thorpe, K.E. (2005). The rise in healthcare spending and what to do about it. *Health Affairs*, *24*(6), 1436–1445.

U.S. Department of Health and Human Services (2000). *Healthy People 2010: Understanding and Improving Health*. Washington, DC, U.S. Government Printing Office. 2nd ed.

Wilson, M.G. (1996). A comprehensive review of the effects of worksite health promotion on health-related outcomes: an update. *American Journal of Health Promotion*, *11*(2), 107–108.

Woolf, S.H. (2008). The power of prevention and what it requires. *Journal of the American Medical Association*, *299*(20), 2437–2439.

Yen, L.T., Edington, D.W. & Witting, P. (1992). Prediction of prospective medical claims and absenteeism costs for 1284 hourly workers from a manufacturing company. *Journal of Occupational Medicine*, *34*, 428–435.

Appendix

Return on Investment (ROI) Model—Sample Data Entry Materials

INTRODUCTION

The Health and productivity management (HPM) return on investment (ROI) tool is based on research showing that health care costs and workplace productivity are influenced by individual health risks (as well as by demographic characteristics). For example, individuals who eat right, exercise, and do not smoke cost less. By reducing these risks, you can thereby expect to see cost reductions. But you also need to consider how much money is invested to achieve these reductions. The tool predicts this financial return based on the information you enter regarding participants' health risks, demographic characteristics, the amount spent on the HPM program, and the degree of risk reduction among program participants.

Information provided in the data entry sheet below can help forecast an ROI, or desirable program cost levels, for many different HPM interventions. Once you complete the data entry sheet, information provided will be run through the ROI tool, and results will be provided to you. All you have to do is enter statistics for as many scenarios as you want to investigate. If you want to leave some or many rows blank, then that's fine too, and the model will then apply the default values listed. The default values come either from U.S. Census information or from the Centers for Disease Control and Prevention's Behavioral Risk Factor Surveillance Survey, for people aged 18–64 in the United States.

The Economics of Human Systems Integration: Valuation of Investments in People's Training and Education, Safety and Health, and Work Productivity. Edited By William B. Rouse
Copyright © 2010 John Wiley & Sons, Inc.

STEPS FOR COMPLETING THE DATA SHEET TO RUN THE ROI MODEL

Step 1. Open Attached Excel Data Entry Sheet (example is provided below) Open the attached excel data entry sheet named: "ROI Model Data Entry Sheet.xls"

Step 2. Enter Program Participation and Cost Information This section asks you to supply values for cost and several related items that influence the cost of HPM programs, such as the number of participants, cost per participant, etc.

Step 3. Enter Demographic Information Please enter demographic information such as age, gender, race, and job characteristics for the people who participate in the HPM program and how their demographics may change across time.

Step 4. Enter Health Risk Information These sections ask you to enter information regarding the health risks of participants at the start of the program such as the percentage of participants with various health habits and hypothetical health risk reductions achieved by the HPM program. See "Assessing Health Risks" for additional information on how health risks are assessed.

Step 5. Enter Productivity Assumptions This section asks whether you would like to assume an aggressive (large) amount of risk change if a good HPM program is put in place to improve productivity at work (true or false) as opposed to a more conservative (small) change. The table below shows the number of hours of lost productivity, for people with each risk factor. These numbers are based on a review of the literature (see the Bibliography page for the list of studies that were reviewed).

Estimated Productive Hours Lost Annually		
Health Risk Factor	Absenteeism	Presenteeism
Poor exercise habits	7.8	43.1
Poor eating habits	10.7	49.1
Overweight or obese	17.2	33.8
Current smoker	14.6	41.4
Former smoker	0.0	0.0
High cholesterol	11.7	8.1
High blood glucose	71.7	58.5
High blood pressure	12.5	36.0
High stress	9.8	107.0
Depression	41.9	165.9
Heavy alcohol use	3.9	15.1

Step 6. Save Excel File Once you have filled in the data for each scenario you wish to run through the model, save the Excel file.

Data sheet for defining scenarios	For each scenario (column) enter as many nondefault values as you like. Leave cells blank if you want to keep the default value. Enter as many scenarios as you like.	For percentages, enter the decimal value or type a % sign.	
Model Section	**Input Parameter of Model**	**Default Value**	**Scenario # 1**
Program participation and cost	Number of eligible employees enrolled in the base year?	10,000	
	Annual change in the number of eligible employees?	0%	
	Participation rate in the program?	45%	
	Program cost per eligible employee in the base year?	$130.00	
	Medical payment per eligible employee in the base year?	$2,379	
	Discount rate applied for ROI calculation?	5%	
	Time horizon (1–10 years)?	10	
	Number of years until program levels off?	10	
	Please specify an average daily wage.	$185.20	
Demographics in the base year	Average age in years	41.0	
	Female (%)	44.7%	
	African American (%)	10.3%	
	Hispanic (%)	11.5%	
	Other nonwhite (%)	5.0%	
	Sales job (%)	11.2%	
	Professional job (%)	38.0%	
Expected annual change in demographics	Average age in years	1.0	
	Female (%)	0.0%	
	African American (%)	0.0%	
	Hispanic (%)	0.0%	
	Other nonwhite (%)	0.0%	

Model Section	Input Parameter of Model	Default Value	Scenario # 1
	Sales job (%)	0.0%	
	Professional job (%)	0.0%	
Percentage at high risk in the base year	Poor exercise habits	51.2%	
	Poor eating habits	43.2%	
	Overweight	45.1%	
	Current smoker	24.8%	
	Former smoker	21.2%	
	High cholesterol	23.4%	
	High blood glucose	5.7%	
	High blood pressure	20.1%	
	High stress	9.4%	
	Depression	8.7%	
	Heavy alcohol use	6.5%	
Expected annual change in risk factors with no program	Poor exercise habits	1.1%	
	Poor eating habits	−0.9%	
	Overweight	0.5%	
	Current smoker	0.0%	
	Former smoker	0.7%	
	High cholesterol	1.1%	
	High blood glucose	0.4%	
	High blood pressure	1.2%	
	High stress	0.0%	
	Depression	0.0%	
	Heavy alcohol use	−0.3%	
Expected annual change in risk factors with user-specified program	Poor exercise habits	−2.8%	
	Poor eating habits	−7.1%	
	Overweight	1.9%	
	Current smoker	−1.8%	
	Former smoker	1.8%	
	High cholesterol	3.2%	
	High blood glucose	2.0%	
	High blood pressure	−1.5%	
	High stress	−5.2%	
	Depression	−0.4%	
	Heavy alcohol use	−3.2%	
Productivity	Use aggressive assumptions (TRUE or FALSE)?	FALSE	

Assessing Health Risks

Health risks among employees are generally assessed using a health risk appraisal (HRA) instrument. Because different HRA tools employ various definitions of high risk, we offer operational definitions for each of the risk categories in the ROI model to guide your assessment of the prevalence of high risk in your employee population.

Health Risks	BRFSS Definition of High Risk	HERO Definition of High Risk
Poor exercise habits	No moderate or vigorous physical activity or insufficient exercise to meet at least moderate level, and not pregnant.	Did not exercise vigorously at all during a typical week.[1,2,3]
Poor eating habits	Consumes fewer than four fruit/vegetable servings per day.	Defined based on total fat and saturated fat intake; consumption of fruits, vegetables, and other complex carbohydrates; salt intake; use of low-fat dairy products; and consumption of lean meat.[4]
Overweight or Obese	Body mass index (BMI) $30\text{kg}/\text{m}^2$ or more and not pregnant.	Weight was either 30% or more above or 20% or more below the midpoint of their frame-adjusted desirable weight range for their height.[5]
Current smoker	Current smoker.	Current smoker of pipe, cigar, snuff, or smokeless tobacco.
Former smoker	Former smoker.	Former smoker.
High cholesterol	Doctor or health professional told cholesterol high and cholesterol checked within past year.	Total cholesterol level greater than or equal to 240 mg/dl.
High blood glucose	Diagnosed with diabetes and not pregnant.	Blood glucose level greater than 115 mg/liter (as currently recommended by the American Diabetes Association).[6]
High blood pressure	Doctor or health professional told blood pressure high and not pregnant.	Systolic blood pressure greater than or equal to 160 mg Hg or diastolic blood pressure greater than or equal to 100 mg Hg.
High stress	Felt sad, blue, or depressed 14–30 days out of past 30 days.	"Almost always" troubled by stress and did not handle stress well.[7]

Health Risks	BRFSS Definition of High Risk	HERO Definition of High Risk
Depression	Mental health not good 14–30 days out of past 30 days.	"Almost always" were depressed.[8,9]
Heavy alcohol use	Male age younger than 50: more than 2 drinks per day or more than 14 drinks per week; Male age 50 and older: more than 1 drink per day or more than 7 drinks per week; Female: more than 1 drink per day or more than 7 drinks per week.	Consuming 5 or more drinks per day on 2 or more days per week.[10]

[1] American College of Sports Medicine. (1990). Position stand on the recommended quantity and quality of exercise for developing and maintaining cardiorespiratory and muscular fitness in healthy adults. *Medicine & Science in Sports & Exercise, 22*, 265–274.

[2] American Heart Association. (1983). *Recommendations of the Nutrition Committee*. Dallas, TX: American Heart Association.

[3] Pate, R.R., Pratt, M., Blair, S.N., Haskell, W.L., Macera, C.A., Bouchard, C., Buchner, D., Etinger, W., Heath, G.W., King, A.C., et al. (1995). Physical activity and public health: A recommendation from the Centers for Disease Control and Prevention and the American College of Sports Medicine. *Journal of the American Medical Association, 273*, 402–407.

[4] U.S. Department of Agriculture. (1990). *Nutrition and Your Health: Dietary Guidelines for Americans, Third Edition*. Home and Garden Bulletin No. 232. Washington, DC: U.S. Department of Agriculture.

[5] Metropolitan Life Insurance Company. (1983). Metropolitan height and weight tables. Statistical 10. *Bulletin of the Metropolitan Life Insurance Company, 64*, 3–.

[6] American Diabetes Association. (1997). Screening for Diabetes. *Diabetes Care, 20* (1), 522–523.

[7] Lazarus, R.S. (1996). *Psychological Stress and the Coping Process*. New York: McGraw-Hill.

[8] Burnam, M.A., Wells, K.B., Leake, B., & Landsverk, J. (1988). Development of a brief screening instrument for detecting depressive disorders. *Medical Care, 26*, 775–789.

[9] Stoudemire, A., Frank, R., Kamlet, M., & Hedemark, N. (1987). Depression. In: R.W. Amler, & H.B. Dull, Eds., *Closing the Gap: The Burden of Unnecessary Illness*. New York: Oxford University Press.

[10] National Institute on Alcohol Abuse and Alcoholism. (1992). *Alcohol Alert: Moderate Drinking*. Washington, DC: U.S. Department of Health and Human Services. Publication No. 16, PH 315.

REFERENCES

References Describing the Structure and Content of the ROI Model

Goetzel, R.Z., Anderson, D.W., Whitmer, R.W., Ozminkowski, R.J., Dunn, R.L., Wasserman, J., & the HERO Research Committee (1998). The relationship between modifiable health risks and health care expenditures: An analysis of the multi-employer HERO health risk and cost database. *Journal of Occupational and Environmental Medicine, 4*, 843–857.

Goetzel, R.Z., Ozminkowski, R.J., Baase, C., & Billotti, G.M. (2005) Estimating the return-on-investment from changes in employee health risks on The Dow Chemical Company's health care costs. *Journal of Occupational and Environmental Medicine, 47*(8), 759–768.

Leutzinger, J.A., Ozminkowski, R.J., Dunn, R.L., Goetzel, R.Z., Richling, D.E., Stewart, M., & Whitmer, R.W. (2000). Projecting future medical care costs using four scenarios of lifestyle risk rates. *American Journal of Health Promotion, 15*, 35–44.

Ozminkowski, R.J., Goetzel, R.Z., Santoro, J., Saenz, B.J., Eley, C., & Gorsky, B. (2004) Estimating risk reduction required to break-even in a health promotion program. *American Journal of Health Promotion, 18*(4), 316–325.

References Pertaining to Specific Health Risks or HPM Program Impact

Aldana, S.G., Merrill, R.M., Price, K., Hardy, A., & Hager, R. (2005). Financial impact of a comprehensive multisite workplace health promotion program. *Preventive Medicine, 40*, 131–137.

Alderman, M.H., & Schoenbaum, E.E. (1975). Detection and treatment of hypertension at the work site. *The New England Journal of Medicine, 2*(293), 65–68.

American Cancer Society. Facts and Figures. (1994). *Economic Impact of Smoking in the Workplace*.

American College of Sports Medicine. (1990). Position stand on the recommended quantity and quality of exercise for developing and maintaining cardiorespiratory and muscular fitness in healthy adults. *Medicine Science in Sports & Exercise, 22*, 265–274.

American Diabetes Association. (1997). Screening for Diabetes. *Diabetes Care, 20*(1), 522–523.

American Heart Association. (1983). *Recommendations of the Nutrition Committee*.

Bauer, J.E., Hyland, A., Li, Q. Steger, C., & Cummings, K.M. (2005). A longitudinal assessment of the impact of smoke-free worksite policies on tobacco use. *American Journal of Public Health, 95*(6), 1024–1029.

Bauer, R.L., Heller, R.F., & Challah, S. (1985). United Kingdom Heart Disease Prevention Project: 12 year follow-up of risk factors. *American Journal of Health Promotion, 121*, 563–569.

National Center for Chronic Disease Prevention and Health Promotion. Behavioral Risk Factor Surveillance Survey, 2002 & 2003. Atlanta, GA: Centers for Disease Control & Prevention.

Bureau of Labor Statistics. (2005). *Employer Costs for Employee Compensation*. Washington, DC: Bureau of Labor.

Burnam, M.A., Wells, K.B., Leake, B., & Landsverk, J. (1988). Development of a brief screening instrument for detecting depressive disorders. *Medical Care, 26*, 775–789.

Burton, W.N., Chen, C., Conti, D.J., Schultz, A.B., Pransky, G.P., & Edington, D.W. (2005). The association of health risks with on-the-job productivity. *Journal of Occupational and Environmental Medicine*, *47*(6), 769–777.

Burton, W.N., & Conti, D.J. (2000). Disability management: Corporate medical department management of employee health and productivity. *Journal of Occupational and Environmental Medicine*, *42*, 1006–1012.

Druss, B.G., Rosenheck, R.A., & Sledge, W.H. (2000). Health and disability cost of depressive illness in a major U.S. corporation. *American Journal of Psychiatry*, *157*(8), 1274-8.

Duan, N. (1983). Smearing estimate: A nonparametric retransformation method. *Journal of the American Statistics Association*, *78*, 605–611.

Edington, D.W., Yen, L.T., & Witting, P. (1997). The financial impact of changes in personal health practices. *Journal of Occupational and Environmental Medicine*, *39*(11), 1037–1046.

Elixhauser, A. (1990). The costs of smoking and the cost effectiveness of smoking cessation programs. *Journal of Public Health Policy*, *11*(2), 218–237.

Erfurt, J.C., Foot, A., & Heirich, M.A. (1991). Worksite wellness programs: incremental comparison of screening and referral along, health education, follow-up counseling, and plant organization. *American Journal of Health Promotion*, *5*(6), 438–448.

Fires, J.F., & McShane, D. (1998). Reducing the need and demand for medical services in high risk persons: A health education approach. *Western Journal of Medicine.*, *169*, 201–207.

Fries, J.F., Harrington, H., Edwards, R., Kent, L.A. & Richardson, N. (1994). Randomized controlled trial of cost reductions from a health education program: The California Public Employees' Retirement System (PERS) Study. *American Journal of Health Promotion*, *8*(3), 216–223

Fries, J.F., Bloch, D.A., Harrington, H., Richardson, N., & Beck, R. (1993). Two-year results of a randomized controlled trial of health promotion program in a retiree population: The Bank of America Study. *American Journal of Medicine*, *94*(5), 455–462.

Fries, J.F., Fries, S.T., Parcell, C.L., & Harrington, H. (1992). Health risk changes with a low-cost individualized health promotion program: Effects at up to 30 months. *American Journal of Health Promotion*, *6*(5), 364–371

Goetzel, R.Z., Ozminkowski, R.J., Bruno, J.A., Rutter, K.R., Isaac, F., & Wang, S. (2002). Long-term impact of Johnson & Johnson's Health & Wellness Program on employee health risks. *Journal of Occupational and Environmental Medicine*, *44*(5), 417–424.

Halpern, M.T., Shikiar, R., Rentz, A.M. & Khan, Z.M. (2001). Impact of smoking status on workplace absenteeism and productivity. *Tobacco Control*, *10*, 233–238.

Haynes, G., & Dunnagan, T. (2002). Comparing changes in health risk factors and medical costs over time. *American Journal of Health Promotion*, *17*(2), 112–121.

Heaney, C.A., & Goetzel, R.Z. (1997). A review of health-related outcomes of multicomponent worksite health promotion programs. *American Journal of Health Promotion*, *11*, 290–307.

Hennrikus, D.J., Jeffrey, R.W., Lando, H.A., Murray, D.M., Brelje, K., Davidann, B., Baxter, J.S., Thai, D., Vessey, J., & Liu, J. (2002). The SUCCESS Project: The Effect of Program Format and Incentives on Participation and Cessation in Worksite Smoking Cessation Programs. *American Journal of Public Health*, *92*, 274–279.

Johansson, M., Partanen, T. (2002). Role of trade unions in workplace health promotion. *International Journal of Health Services*, *32*(1), 179–193.

Koopman, C., Pelletier, K.R., Murray, J.F., Sharda, C.E., Berger, M.L., Turpin, R.S., Hackleman, P., Gibson, P., Holmes, D.M., & Bendel, T. (2002). Stanford Presenteeism Scale: Health status and employee productivity. *Journal of Occupational and Environmental Medicine*, *44*, 14–20.

Kristein, M.M. (1983). How much can business expect to profit from smoking cessation? *Preventative Medicine*, *12*, 358–381.

Lazarus, R.S. (1996). *Psychological Stress and the Coping Process*. New York: McGraw-Hill.

MacKenzie, T.D. Bartecchi, C.E., & Schrier, R.W. (1994). The human costs of tobacco use-second of two parts. *New England Journal of Medicine*, *330*, 975–980.

Max, W. (2001). The financial impact of smoking on health-related costs: A review of the literature. *American Journal of Health Promotion*, *15*(5), 321–331.

McMahon, A., Kelleher, C.C., Helly, G., Duffy, E. (2002). Evaluation of a workplace cardiovascular health promotion programme in the Republic of Ireland. *Health Promotion International*, *17*(4), 297–308.

Metropolitan Life Insurance Company. (1983) Metropolitan height and weight tables. *Statistical 10 Bulletin of the Metropolitan Life Insurance Company*, *64*, 3.

Musich, S., Napier, D., & Edington, D.W. (2001). The association of health risks with workers' compensation costs, *Journal of Occupational and Environmental Medicine*, *43*, 534–541.

Muto, T., & Yamauchi, K. (2001). Evaluation of a multicomponent workplace health promotion program conducted in Japan for improving employees' cardiovascular disease risk factors. *Preventive Medicine*, *33*(6), 571–577.

National Institute on Alcohol Abuse and Alcoholism. (1992). *Alcohol Alert: Moderate Drinking*. Washington, DC: U.S. Department of Health and Human Services.

Ozminkowski, R.J., Goetzel, R.Z., Smith, M.W., Cantor, R.I., Shaughnessy, A., & Harrison, M. (2000). The impact of the Citibank, NA, health management program on changes in employee health risks over time. *JOEM*, *42*(5), 502–511.

Ozminkowski, R.J., Ling, D., Goetzel, R.Z., Bruno, J.A., Rutter, K.R., Isaac, F., & Wang, S. (2002). Long-term impact of Johnson & Johnson's Health & Wellness Program on health care utilization and expenditures. *Journal of Occupational and Environmental Medicine*, *44*(1), 21–29.

Pate, R.R., Pratt, M., Blair SN, Haskell, W.L., Macera, C.A., Bouchard, C., Buchner, D., Ettinger, W., Heath, G.W., & King, A.C. (1995). Physical activity and public health: A recommendation from the Centers for Disease Control and Prevention and the American College of Sports Medicine. *JAMA*, *273*, 402–407.

Pelletier, K. A review and analysis of the clinical and cost-effectiveness studies of comprehensive health promotion and disease management programs at the worksite: 1995–1995 update (IV). *Am J Health Promot*, 1999, *13*(6), 333–345.

Pelletier, K. A review and analysis of the clinical and cost-effectiveness studies of comprehensive health promotion and disease management programs at the worksite: 1998–2000 update. *Am J Health Promot*, 2001, *16*(2), 107–116.

Pelletier, K. A review and analysis of the health and cost-effective outcome studies of comprehensive health promotion and disease prevention programs at the worksite: 1991–1993 update. *Am J Health Promot*, 1993; *8*, 43–49.

Pelletier, K. A review and analysis of the health and cost-effective outcome studies of comprehensive health promotion and disease prevention programs at the worksite: 1993–1995 update. *Am J Health Promot*, 1996; *10*, 380–8.

Pelletier, K. A review and analysis of the health and cost-effective outcomes studies of comprehensive health promotion and disease prevention programs. Am J Health Promot, 1991; *5*, 311–315.

Pescatello, LS, Murphy, D, Vollono, J, Lynch, E, Bernene, J, Costanzo, D. The Cardiovascular Health Impact of an Incentive Worksite Health Promotion Program. Am J Health Prom. Sept/Oct, 2001 *16*(1), 16–20.

Prochaska, J.O., Norcross, J., DiClemente, C. (1995) Changing for Good, Avon Publishing.

Poole, K, Kumpfer, K, Pett, M. The Impact of an Incentive-based Worksite Health Promotion Program on Modifiable Health Risk Factors. Am J Health Prom. Sept/Oct, 2001 *16*(1), 21–26.

Ramsey, S, Summers, KH, Leong, SA, Birnbaum, HG, Kemner, JE, Greenberg, P. Productivity and medical costs of diabetes in a large employer population. Diabetes Care, 2002 Jan; *25*(1), 23–9.

Schultz, AB, Lu, CF, Barnett, TE, et al. Influence of participation in a worksite health-promotion program on disability days. J Occup Environ Med. Aug, 2002 *44*(8), 776–780.

Serxner, S., Gold, D., Grossmeier, J., and Anderson, D., (2003) The relationship between health promotion program participation and medical costs: A dose response, Journal of Occupational and Environmental Medicine, February, *45*(11), 1196–1200.

Sexner, S, Gold, D, Anderson, D, Williams, D. The Impact of a worksite health promotion program on short-disability usage. J Occup Environ Med, 2001; *43*(1) 25–29

Sherman, B. Worksite health promotion—A critical investment. Disease Management & Health Outcomes, 2002 *10*(2), 101–108.

Sidorov, J, Shull, R, Tomcavage, J, Girolami, S, Lawton, N, Harris, R. Does diabetes disease management save money and improve outcomes? A report of simultaneous short-term savings and quality improvement associated with a health maintenance organization-sponsored disease management program among patients fulfilling health employer data and information set criteria. Diabetes Care, 2002 Apr; *25*(4), 684–9.

Stave, G., Muchmore, L., Gardner, H. (2003) Quantifiable impact of the Contract for Health and Wellness: Health behaviors, health care costs, disability, and workers' compensation, Journal of Occupational and Environmental Medicine, February, *45*(2), 109–117.

Stein, A., Shakour, S., and Zuidema, R., (2000) Financial incentives, participation in employer-sponsored health promotion, and changes in employee health and productivity: HealthPlus Health Quotient Program, Journal of Occupational and Environmental Medicine, December, *42*(12), 1148–1155.

Stewart, W.F., Ricci, J.A., Chee, E. & Morganstein, D. (2003). Lost productive work time costs from health conditions in the United States: results from the American Productivity Audit. Journal of Occupational and Environmental Medicine, *45*(12), 1234–1246.

Stokols, D, Pelletier, K, Fieldings, J. Integration of medical care and worksite health promotion. JAMA, 1995; *273*, 1136–42.

Stoudemire, A, et al. Depression. In: Amler, RW, Dull, HB, eds., Closing the Gap: The Burden of Unnecessary Illness. New York, NY: Oxford University Press, 1987.

Tsai, S.P., Wendt, J.K., Ahmed, F.S., Donnelly, R.P., & Strawmyer, T.R. (2005). Illness absence patterns among employees in a petrochemical facility: impact of selected health risk factors. Journal of Occupational and Environmental Medicine, *47*(8), 838–846.

Tucker, LA, Clegg, AG, Differences in Health Care Costs and Utilization Among Adults With Selected Lifestyle-Related Risk Factors. Am J Health Prom. March/Apr, 2002; 211–213.

Umans-Eckenhausen, MA, Defesche, JC, van Dam, MJ, Kastelein, JJ. Long-term compliance with lipid-lowering medication after genetic screening for familial hypercholesterolemia. Arch Intern Med, 2003 Jan 13; *163*(1), 65–8

U.S. Department of Health and Human Services (1990). The health benefits of smoking cessation. A report to the Surgeon General, Public Health Service, Centers for Disease Control, 90–8416.

Wang, F, Schultz, AB, Musich, S, McDonald, T, Hirschland, D. The Relationship Between National Heart, Lung and Blood Institute Weight Guidelines and Concurrent Medical Costs in a Manufacturing Population. Am J of Health Prom, 2003 *17*(3), 183–189.

Warner, K.E., Smith, R.J., Smith, D.G, & Fries, B.E. (1996). Health and economic implications of a work-site smoking cessation program: a simulation analysis. Journal of Occupational Medicine, *38*(10), 981–992.

Wasserman, J, Whitmer, RW, Bazarre, RL, Kennedy, ST, Merrick, N, Goetzel, RZ, Dunn, RL, Ozminkowski, R.J., and Anderson, D.R. Gender-specific effects of modifiable health risk factors on coronary heart disease and related expenditures. J Occup Env Med, 2000; *42*, 1060–1069.

Weiss, S.J., Jurs, S., Lesage, J.P., & Iverson, D.C. (1984). A cost-benefit analysis of a smoking cessation program. Evaluation and Program Planning, *7*, 337–346.

Whitmer, R.W., Pelletier, K.R., Anderson, Dr, Basse, C.M. and Frost, G.J. (2003). A wake-up call for corporate America. J Occup Environ Med, *45*(9), 916–925.

Wright, D.W., Beard, M.J., Edington, D.W. Association of Health Risks With the Cost of Time Away from Work. J Occup Environ Med, 2002; *44*, 1126–1134.

U.S. Department of Agriculture. Nutrition and Your Health: Dietary Guidelines for Americans, Third Edition. Home and Garden Bulletin No. 232, 1990.

Zhu, S.H., Anderson, C.M., Johnson, C.E., Tedeschi, G., & Roeseler (2000). A centralized telephone service for tobacco cessation: The California experience. Tobacco Control, *9*(suppl 2:II), 48–55.

Zhu, S.H., Anderson, C.M., Tedeschi, G.J., Rosbrook, B., Johnson, C.E., Byrd, M., et.al. (2002). Evidence of Real-World Effectiveness of a Telephone Quitline for Smokers, New England Journal of Medicine, *347*, 1087–1093.

Zhu, S.H., Rosbrook, B., Anderson, C., Gilpin, E., Sadler, G., & Pirece, J. (1995). The demographics of help-seeking for smoking cessation in California and the role of the California Smokers' Helpline. Tobacco Control, *4*(suppl 1), 9–15.

Zhu, S.H., Stretch, V., Balabanis, M., Rosbrook, B., Sadler, G., & Pierce, J.P. (1996). Telephone counseling for smoking cessation: Effects of single-session and multiple-session intervention. Journal of Counseling and Clinical Psychology, *64*(1), 202–211.

Zhu, S.H., Tedeschi, G., Anderson, C.M., Rosbrook, B., Byrd, M., Johnson, C.E., et.al. (2000) Telephone counseling as adjuvant treatment for nicotine replacement therapy in a "real world" setting. Preventive Medicine, *31* (4), 357–363.

Zhu, S.H., Tedeschi, G.J., Anderson, C.M., & Pierce, J.P. (1996). Telephone counseling for smoking cessation: What's in a call? Journal of Counseling Development, *75*, 93–102.

Zhu, S.M., & Pierce, J.P. (1995). A new scheduling method for time-limited counseling. Professional Psychology: Research and Practice, *26* (6), 624–625.

Chapter 15

Options for Surveillance and Reconnaissance

WILLIAM B. ROUSE

15.1 INTRODUCTION

The case study in this chapter emerged from a keen interest in "the value of defense" or, put differently, the question, "What is defense worth?" Of course, considerable attention has been paid to the cost of defense. Ideally, one would like the cost to be significantly less than the value. That makes defense a good investment in terms of the concepts, principles, models, methods, and tools discussed in Chapters 7 and 10.

We have applied this approach to a number of case studies in the private sector in industries ranging from aerospace, automotive, and electronics, to semiconductors, computers, and pharmaceuticals (Rouse & Boff, 2004 Rouse et al., 2000). In all these applications, an upfront investment secured the possibility of future free cash flow. The difficulty with defense investments is identifying the free cash flow.

It would seem that money always flows from taxpayers through Congress to the Department of Defense and then to defense contractors. Cash does not seem to flow the other way. However, reduced expenditures in the future, relative to what they would have been without investment, can be characterized as cash flows. Thus, we should, in principle at least, be able to estimate the cash flow returns from defense investments.

There is one significant difficulty, however. What is the value of the performance impacts of the systems acquired through defense expenditures? This leads to questions of the relative value of an aircraft sortie compared with a ship patrol.

The Economics of Human Systems Integration: Valuation of Investments in People's Training and Education, Safety and Health, and Work Productivity. Edited By William B. Rouse
Copyright © 2010 John Wiley & Sons, Inc.

One quickly gets to the point of having to compare the value of things that serve completely different purposes.

We finessed this difficulty by taking performance requirements as a given. Thus, all alternatives are characterized in ways that meet performance requirements. As this case study illustrates, assuring parity in this way can take some creativity. However, once this is accomplished, all comparisons are primarily economic. We can, for example, view current investments in research and development (R&D) as providing options for future reductions in operating costs. Framed in this way, the models, methods, and tools in Chapters 7 and 10 can be used to perform economic valuations.

The case study in this chapter was conducted for the Singapore Ministry of Defense (MINDEF). The chapter first addresses framing the investment decisions of interest. Alternative investments are then considered. Investment valuations are then discussed from both net present value and net option value perspectives. The economic valuation results are then integrated with a broader multistakeholder, multiattribute analysis. Finally, the recommended investment strategy is summarized, and the resulting decisions are discussed.

15.2 FRAMING THE INVESTMENT

Investment analyses should always begin with consideration of the goals of the investment. It is certainly the case that all investors desire "returns" on their investments. In many cases, the returns sought are purely financial. However, for public sector investments, "returns" can be much more multidimensional.

15.2.1 Effects and Capabilities

For defense investments, the goal is national defense. Recent thinking has emphasized characterizing the goals of defense investments in terms of the effects sought, rather than platforms acquired (Rouse & Boff, 2001). In this study, the effects sought were characterized as deterrence and superiority. The first goal concerns inhibiting aggression. The second goal involves competitive advantage in the face of aggression.

Effects are achieved by capabilities that may be provided in various ways. Three capabilities were sought from the investment under consideration:

- Pervasive battlespace awareness
- Rapid response
- Dominant battlespace presence

Succinctly, MINDEF wanted to be fully aware of the current and emerging states of the environment of interest, be able to respond to these states quickly, and be dominant in their response. There are many alternative platforms that have the potential to provide these capabilities. The objective of the case study in this chapter was to assess the economic value of these alternatives.

15.2.2   Mission Scenarios

It is helpful to think about effects and capabilities in the contexts of specific missions. Three scenarios were developed to help define more specific capabilities.

- Battlefield
  - Detect, identify, and monitor forces and their movements
  - Recognize and classify patterns of movements, intent, and so on.
- Period of tension
  - Detect (anticipate) transition to conflict
  - Demonstrate presence to deter transition
- Terrorist
  - Detect violation of geographical boundaries
  - Monitor ongoing status of activities
  - Search and identify targets with known characteristics
  - Monitor transactions (financial and supply chains)

As will be discussed, these three scenarios served as surrogates for the public's interests in the investments being considered. This is a useful approach to answer the question, "What does the taxpayer want?" The answer is that the public wants the military to be able to perform successfully the missions for which it is responsible.

15.2.3   Sensing Requirements

Analysis of the scenarios led to the determination of sensing requirements for the capabilities sought from the investment of interest. The quantitative nature of these requirements cannot be reported here.

- Physical location
- Physical size and shape
- Physical dispersion
- Physical weight and density
- Physical movement
- Communications
- Electrical activity
- Temperature

- Light
- Rates of change

15.3 ALTERNATIVE INVESTMENTS

The framing of the investment decision led to consideration of alternative means for providing the desired capabilities. Seven alternatives were defined:

- Manned Aircraft
- GEO Satellite
- LEO Satellite
- Micro Satellite (MICRO)
- Predator
- Global Hawk
- LALEE

MICRO and LALEE were the alternatives being pursued by internal R&D at MIN-DEF. MICRO is a very small satellite. LALEE is a very large, unmanned air vehicle. An overarching question was which of these two should be pursued further or whether both should be abandoned for capabilities procured externally.

Manned aircraft provide a traditional approach to surveillance and reconnaissance. Geosynchronous and low-earth-orbiting satellites can be almost purchased off the shelf. Predator and Global Hawk are U.S. unmanned air vehicles, at that time not available to Singapore but, nevertheless, interesting alternatives.

Substantial discussion was devoted to determining how each of these alternatives could provide the desired capabilities. A central question was how many units would be needed to satisfy the sensing requirements and support typical duty cycles. For example, it was determined that they would need one GEO or many LEOs. Numbers for all the alternatives cannot be reported here.

It was very important that all of the alternatives be capable of the same sensing performance. As noted earlier, this eliminated the need to address cost versus performance tradeoffs. Because the payloads of the alternatives varied significantly, this strongly affected the number of units needed to achieve performance parity. Although the parity achieved was not perfect, the key stakeholders in this case study felt that it was close enough to focus solely on economic valuation, at least initially.

15.3.1 System Schedules

The study team then shifted its attention to the timelines for development, acquisition, and deployment of the alternatives. MICRO and LALEE were the only alternatives that needed development. The other five alternatives could, in principle, be acquired and deployed. Consequently, the timelines of the seven alternatives varied greatly.

TABLE 15.1   Development, Acquisition, and Deployment Schedules

System	Development Duration	Acquisition Duration	Deployment Date	Comments
Aircraft				
GEO				
LEO				
Micro				
Predator				
Hawk				
LALEE				

Table 15.1 shows the format developed for compiling the timeline information. The contents of this table cannot be provided here. However, the most important point is that considerable effort was invested in reaching agreement on the contents of this table.

15.3.2   System Costs

One of the most time-consuming aspects of this case study was compiling cost information in the format shown in Table 15.2. Although Table 15.1 defined the starting year in each row of Table 15.2 for each alternative; developing the cost estimates for subsequent cells required some sleuthing.

The data compilation process employed involved having representatives of all key stakeholders around a large table in a conference room. Each person's laptop was connected to the MINDEF network. Each person also had a cell phone. Over the course of a couple of days, many "data calls" were made either via email or cell phone. Of particular importance was the estimation of operating costs for each alternative. Acquisition costs were more readily available. Development costs came from internal proposals for MICRO and LALEE.

The cost estimates developed in this way were vetted following these intense meetings. Interestingly, few changes resulted. It is, of course, very important to have realistic and credible cost estimates. Such estimates can also include uncertainties that can be employed in Monte Carlo simulations to yield probability distributions of economic values rather than just point values. Such distributions can also provide the means for estimating risks as well as returns, perhaps expressed as the probability that the return exceeds zero.

TABLE 15.2   Cost Projections Developed for Each System Alternative

Costs	Year 1	Year 2	Year 3	. . .	Year 20
Development					
Acquisition					
Operating					
Total					

15.4   INVESTMENT VALUATIONS

The data from Table 15.2 was first used to calculate the net present value for each alternative (see Chapter 7 for how this calculation is done). Figure 15.1 shows the results, stated in terms of net present cost, so the numbers are positive. Note that the cost of the aircraft is dominated by operating costs, whereas the cost of LEO is dominated by the number of units that must be deployed to satisfy sensing requirements.

At this point, LALEE looked most attractive followed by GEO and then Global Hawk. This might suggest that MINDEF should select LALEE and move on. However, this choice would result in not having the desirable capabilities for many years. The only way to gain these capabilities relatively quickly was to employ manned aircraft. (Recall that Predator and Global Hawk were not really available to Singapore.)

15.4.1   Investment Options

Discussion of this situation led to framing two options. Both involved purchasing and deploying the aircraft as soon as possible. The options were defined as follows:

- Replace Aircraft With MICRO → Deploy aircraft in Year X, replace with MICRO in Year Y with operating costs reduced $A per year.
- Replace Aircraft With LALEE → Deploy aircraft in Year X, replace with LALEE in Year Z with operating costs reduced $B per year.

Note that the years and amounts indicated cannot be provided in this exposition.

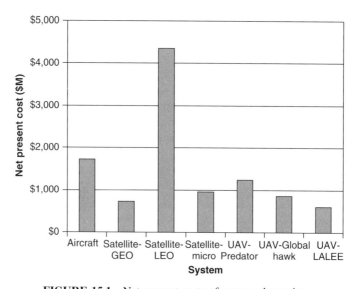

FIGURE 15.1   Net present costs of system alternatives.

Using the real options model introduced in Chapter 7 and discussed in more depth in Chapter 10, the net option value was calculated for each of these options using the ***Technology Investment Advisor*** (ESS, 2000a; Rouse, 2001, 2007; Rouse et al., 2000). The results are shown in Figure 15.2. The black bars on the left represent the initial results. Both MICRO and LALEE are very unattractive. This is because of the acquisition costs of the aircraft. In other words, the purchase price of the option—that is, the costs of the aircraft plus the costs of R&D for MICRO or LALEE—are too high considering the likely downstream returns.

This led to an intense discussion of what to do with the aircraft. Selling it on the open market was a possibility. However, it was decided that it could be "sold" to a different mission. An equivalent "purchase price" was determined and represented as a positive cash flow in the year of its "sale." The bar to the right of the black bars in Figure 15.2 shows the result.

Therefore, at this point, both MICRO and LALEE looked economically attractive. The decision makers had two concerns at this point, both of which involved possible variability in actual costs versus those projected in Table 15.2.

- What if R&D costs exceed projections?
- What if operating savings are overestimated?

The impact of these possibilities was assessed using Monte Carlo analysis assuming the mean R&D costs were 110% of projections with a 10% standard deviation.

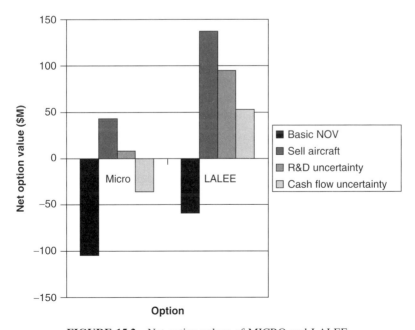

FIGURE 15.2   Net option values of MICRO and LALEE.

Mean operating savings were assumed to be 90% of projections also with a 10% standard deviation. The two rightmost bars in Figure 2 show the results of this analysis. The attractiveness of MICRO is at risk with these new assumptions, whereas LALEE remains attractive despite these possibilities.

15.5   MULTIATTRIBUTE ANALYSIS

Attention now shifted to noneconomic attributes of these investments. The team developed a list of 22 attributes, five of which were economic. The types of attributes are shown as follows:

- Sensor coverage
  - Geographical
  - Spectral
  - Temporal
- Benefits and costs
  - Opportunity benefits
  - Development costs
  - Acquisition costs
  - Operating costs
- Other factors
  - Intrusiveness
  - Development time
  - Development risks

The attribute "intrusiveness" merits discussion. MINDEF decision makers were concerned with the public's perception of the alternatives. How would people feel about a large unmanned aircraft circulating the city? Possible negative perceptions were carefully considered and included in the multiattribute analysis.

Another consideration was how to represent the public's interest in the capabilities provided by the alternatives. As indicated earlier in this chapter, it was concluded, after much discussion, that the public is primarily interested that the capabilities being procured can successfully perform the missions for which they were intended. Thus, the public as a stakeholder entered the multiattribute analysis via preferences for mission success.

A multistakeholder, multiattribute analysis, as described in Chapter 7, was performed using the ***Product Planning Advisor*** (ESS, 2000b; Rouse, 2001, 2007). A qualitative summary of the results is shown in Table 15.3. Dark grey indicates areas where alternatives are strong, and light grey indicates potential weaknesses.

The overall analysis showed that LALEE had the greatest expected utility across stakeholders and attributes. The manned aircraft had strengths but was hurt by high operating costs, which far exceed acquisition costs. In fact, the manned aircraft would be unattractive even if the acquisition costs were zero!

TABLE 15.3   Relative Strengths of Alternatives

	Aircraft	GEO	Micro	LALEE
Sensor Coverage	███			███
Benefits and Costs			███	
Other Factors	███			

The analysis also addressed where increased investment would best pay off. Not surprisingly, the technologists associated with MICRO and LALEE had many ideas for increased functionality. However, this analysis unequivocally showed that the most attractive incremental investments should focus on decreasing time until deployment and decreasing development risks, not increasing system functionality.

The importance of executing faster and managing risks is often underestimated. Economic analyses that address the time value of money and explicitly consider the impacts of uncertainties can enable attaching value to these process-oriented investments. A recent study of success and failure in the automobile industry highlights the importance of such investments (Hanawalt & Rouse, 2007).

15.6   INVESTMENT STRATEGY

The results of the investment analyses provided the foundation for a recommended investment strategy are

- Proceed with aircraft
- Replace with MICRO or LALEE
- Invest in options for MICRO and LALEE
  - Invest in R&D for these development efforts
  - Consider other means to gain options in the future
- Use R&D to deploy faster as well as reduce risks and uncertainties

This proposed strategy, along with the supporting analyses, was presented to the Singapore Secretary of Defense along with heads of other ministries who were interested in the overall approach to investment analysis. The outcome of this presentation was an immediate decision to pursue the recommended strategy.

It is useful to reflect on what was decided. Quite simply, the Secretary of Defense committed to the R&D funding for MICRO and LALEE for the next year. In each subsequent year, this decision would be revisited. Considering the time value of money, these investments should become more attractive, unless something changes. This possibility motivated the "consider other means" phrase included in the strategy.

Options-based valuations and, more importantly, options-based thinking provide important management flexibility to commit incrementally and only to continue commitments that still make economic sense. In this way, investments are buying options that may or may not be exercised but, nonetheless, have value as hedges against future contingencies.

15.7 CONCLUSIONS

This chapter has summarized a case study of economic valuation of defense investments using the real options methodology discussed in Chapters 7 and 10. Particular emphasis was placed on framing investment decisions and on considering a range of capabilities for achieving the defense effects sought. The surveillance and reconnaissance context provides a rich set of alternatives and a good illustration of the steps of economic valuation.

This case study focused on alternative systems for providing desired capabilities. The real options methodology also can be applied to process improvements. Pennock and his colleagues considered improvements in the process of acquiring naval ships. As indicated in Chapter 7, this study showed that somewhat risky process improvements, when framed as multistage options, can have substantial investment value (i.e., large net option value) even though the traditional net present value is negative (Pennock et al., 2007).

This chapter has demonstrated the ability to assess the economic value of defense investments. We can go beyond the question, "What will it cost?" to address the question, "What will it be worth?" This is critical for investment decision making. It does require, however, for those considering funding initiatives of interest to see themselves as investors, not just as "bill payers."

Unfortunately, the common view of defense expenditures tends to not be willing to spend significantly more now to substantially decrease costs later—to invest now for future returns. Consequently, out-year life-cycle costs are greatly increased and future taxpayers find more of their resources committed to yesterday's lack of perspective. This chapter has, hopefully, shown that there is a better way to think about defense investments.

REFERENCES

ESS. (2000a). *Technology Investment Advisor*. Atlanta, GA: Enterprise Support Systems. http://www.ess-advisors.com/software.htm.

ESS. (2000b). *Product Planning Advisor*. Atlanta, GA: Enterprise Support Systems. http://ess-advisors.com/software.htm.

Hanawalt, E., & Rouse, W.B. (2007). *Car Wars: Factors Underlying the Success or Failure of New Automobiles*. Atlanta, GA: Tennenbaum Institute, Georgia Institute of Technology.

Pennock, M.J., Rouse, W.B., & Kollar, D.L. (2007). Transforming the acquisition enterprise: A framework for analysis and a case study of ship acquisition. *Systems Engineering*, *10*(2), 99–117.

Rouse, W.B. (2001). *Essential Challenges of Strategic Management*. New York: Wiley.

Rouse, W.B. (2007). *People and Organizations: Explorations of Human-Centered Design*. New York: Wiley.

Rouse, W.B., & Boff, K.R. (2001). Impacts of next-generation concepts of military operations on human effectiveness. *Information • Knowledge • Systems Management*, *2*(4), 347–357.

Rouse, W.B., & Boff, K.R. (2004). Value-centered R&D organizations: Ten principles for characterizing, assessing & managing value. *Systems Engineering*, *7*(2), 167–185.

Rouse, W.B., Howard, C.H., Carns, W., & Prendergast, J. (2000). An options-based approach to technology strategy. *Information • Knowledge • Systems Management*, *2*(1).

Chapter 16

Governing Opportunism in International Armaments Collaboration: The Role of Trust

Ethan B. Kapstein

16.1 INTRODUCTION

International joint ventures suffer high failure rates, and academic research has placed much of the blame on ungovernable problems of opportunism (Oxley, 1999). Not only do partners sometimes shirk their responsibilities and hold up a given venture—for example, by failing to deliver quality products on time and within budget—but they also engage in "technology poaching" or illicit efforts to procure proprietary knowledge from the other firm(s) (Clemons & Hitt, 2004). Although these problems are difficult enough to manage within a purely domestic setting, they become much more intractable when it comes to operating across borders, where laws and cultural norms may differ between the partner companies, often rendering contracts inefficient.

Central to a successful cross-border alliance (in particular, one that operates in countries where legal regimes are weak or where violations of intellectual property arrangements are difficult to prosecute in court, which is the case with many "secret" defense technologies), is the establishment of *trust* among the partners. But how can trust be "engineered" into complex systems like international, high-technology alliances? This chapter examines that question, drawing insights primarily from the field of behavioral economics.

We should note at the outset that it is *not* our contention that trust is the only or even necessarily the most important key to successful alliance-building. In their

The Economics of Human Systems Integration: Valuation of Investments in People's Training and Education, Safety and Health, and Work Productivity. Edited By William B. Rouse
Copyright © 2010 John Wiley & Sons, Inc.

study of commonplace failures in the building of complex systems, for example, Sosa et al. found that engineers in different departments had often shared only incomplete information with one another, creating any number of "missed inter-faces" (Sosa et al., 2007). This was not necessarily from mistrust. Rather, each engineer was focused on his or her piece of the puzzle with little concern for how all the pieces fit together, and management did not grasp how this could lead to systemic errors. Where trust is missing, however, it is probably safe to assert that engineers will have even less motivation to communicate across their "territorial" boundaries.

A particularly "hard case" for cross-border management and governance, mean-ing one in which opportunism is likely to be especially rife and difficult to monitor and control, is provided by the case of international armaments cooperation—the focus of this chapter. The argument we make is that if the partners in an inter-national armaments project can structure their relationship in such a way as to codevelop a complex weapons system that meets the requirements of their respec-tive governments, then this holds promise for suggesting valuable lessons regarding the governance and management of cross-border joint ventures in high technology more generally.

The chapter is in three sections. After this introduction, we briefly address the question of *why* international armaments collaboration takes place in the first place. Next, based on interviews with public officials and managers who have been involved with some 20 collaborative arrangements between U.S. firms and their foreign (mainly European) counterparts (see the Appendix for a listing), along with a review of the relevant literature, we analyze some major problems these programs have faced and how executives have sought to resolve them. Finally, we seek to address the broader lessons from this case study, along with suggestions for further research.

16.2   WHY COLLABORATE?

Just as there are myriad reasons for forming international joint ventures in the commercial realm—including market access, risk-sharing, and the acquisition of new technology—so too governments and firms have made differing arguments throughout time for building weapons on a collaborative basis. According to the U.S. Government's official "handbook" on armaments cooperation, for example, "The goals or objectives of our major arms cooperative efforts ... can be stated succinctly as a need to achieve the following:

- Deployment and support of common—or at least interoperable—equipment with allies.
- Incentives for the allies to make greater investment in modern conventional military equipment.
- Economies of scale afforded by coordinated research, development, produc-tion, and logistic support programs.

- Department of Defense (DoD) access to, use of, and protection of the best technology developed by our allies, and comparable allied access to, use of, and protection of the best U.S. technology, thereby avoiding unnecessary duplication of developments' (DoD, 1987).

In the early postwar years, say until 1960, U.S. allies were heavily dependent upon imports of U.S. military equipment because they did not possess the capital and technology to build complex weapons on their own. But nations are wont to depend upon others for their security and will pay a premium to retain some autonomous capability. Thus, as nations recovered from World War II, they invested in the rebuilding of their domestic arms industries as well as in civilian industries (like aerospace) with military applications (Kapstein, 1991). Even in Japan, which of course had adopted a "peace constitution" under U.S. occupation, rearmament became a priority with the onset of the Korean War and the perception of growing Chinese and Russian threats to Pacific security (Dower, 1999).

As the allies' dependence on Washington, DC lessened with the renewal of their domestic armaments industries, the number of differing weapons systems deployed by North Atlantic Treaty Organization (NATO) troops multiplied, reducing the interoperability of the armed forces in the event of conflict. Furthermore, with each country investing independently in military research and development (R&D), scarce resources were being squandered. This provided Washington, DC, with both a strategic and fiscal rationale to promote collaborative programs.

Perhaps of equal importance, as countries purchased fewer U.S. weapons, U.S. defense contractors became increasingly dependent upon the Pentagon's annual procurement budget to maintain existing production facilities. But, unfortunately for these contractors, Congressional priorities could shift from one year to the next, making it difficult to plan long production runs or rely upon a stable flow of federal funds. Seeking diversification, these firms wanted to maintain access to foreign markets.

To win sales in an increasingly competitive arms market, U.S. firms began to engage in coproduction of advanced weapons systems. Normally, this took the form of U.S. technology transfers, mainly to European allies, who would then produce U.S. equipment locally or with slight modifications to meet domestic military requirements. The high-water mark of this approach to cooperation occurred in 1975 with the so-called "deal of the century" in which General Dynamics cooperated with four European states—Belgium, Denmark, the Netherlands, and Norway—to produce F-16 fighters under license, an arrangement that was later extended to Turkey among other nations (the F-16 is now built by Lockheed Martin). Under this arrangement, the United States provided the allies with the blueprints to build F-16s in local assembly lines, in a sense giving them the best of both worlds: cutting-edge technology that was still being produced "at home" by local engineers and factory workers.

During this period, the Europeans themselves engaged in any number of collaborative ventures with varying success. On the military side, the Europeans developed a strong market niche in helicopters, among other technologies, with the creation of the Eurocopter joint-venture, and of course, on the civilian side Airbus emerged as a global challenger to Boeing. By the 1980s, then, Europeans were becoming increasingly unwilling to buy U.S. defense equipment "off the shelf."

Since that time, the technological capacity of U.S. allies has continued to improve, although the price tag associated with new weaponry has spiralled ever upward, shifting the name of the game in armaments collaboration from coproduction to codevelopment. In a word, even the United States now finds it financially and technologically challenging to build weapons on its own. The program that best defines this new world order is the F-35 Joint Strike Fighter (JSF), the most expensive defense procurement in history, which the United States is developing and building in collaboration with the United Kingdom (the main foreign partner), along with Italy, the Netherlands, Turkey, Australia, Norway, Denmark, and Canada.

The F-35 is the first cutting-edge weapons platform ever procured by the Pentagon that relies on significant foreign participation in every aspect of the program, including financing, design, and project management. The aircraft incorporates American stealth technology with significance reliance on British vertical takeoff and landing expertise, and its international design team is using collaborative software developed in Denmark among other countries. At a total estimated program cost of some $300 billion for the development and acquisition of an initial 2,443 aircraft, governments engaged in a collaborative effort because none of them—again, including the United States—could build this multirole aircraft on its own (Bolkom, 2002).

In sum, then, international armaments collaboration has a long history among the Western allies. But the nature of that collaboration has changed with shifts in the global political and economic environments. The end of the Cold War, the rising technological capacity of Europe and Asia, and the chronic inflation of weapons costs have all influenced the nature of the collaborative process. Today, arms collaboration has come to mean codevelopment of a complex platform in which each country brings its best technology to the drawing board and, ultimately, to the battlefield. In this world of codevelopment, however, the risks of opportunism—of failures to deliver on military specifications along with unwanted technological leakage—are greater than ever as are the potential security costs associated with such opportunistic behavior.

16.3 OPPORTUNISM IN ARMAMENT COLLABORATION

Interviews in the United States and Western Europe with more than 20 public officials in governments and international organizations (e.g., NATO); defense industry executives and consultants; and trade association representatives have all pointed to opportunism as a major problem in the governance of armaments collaboration.[1]

---

[1]Because of the sensitive nature of this topic, the interviewees requested to remain anonymous.

But what problems of opportunism do these experts have in mind? Three were signaled out by almost every subject in open-ended questioning (e.g., "What are the major problems you confront in managing collaborative weapons programs?"), two of which are well known to students of international joint-ventures, but the third of which has surprisingly received less attention.

Anyone who has examined cross-border alliances from the perspective of transaction cost economics would readily predict the presence of shirking and holdup problems in collective ventures (Das & Rahman, 2002). Within the defense sector, experts equated shirking with the failure of some subcontractors to meet the quality requirements—or military specifications (so-called "milspecs")—written into production agreements. These subcontractors were reported in interviews to have "cut corners" in their efforts to maximize their profits by reducing their costs. Given the difficulty of monitoring the activities of each subcontractor—an arms collaboration project may draw upon more than 1,000 subcontractors scattered around the world (1,200 in the case of the F-35 Joint Strike Fighter)—the problems of shirking in this sector are especially acute.

Shirking problems—and problems of opportunism more generally—are rendered particularly difficult to govern in the arms collaboration space because of the way in which these projects are organized. Because governments are the sole buyers of complex weapons systems, their purchasing decisions can "make or break" a prime defense contractor. This gives them tremendous leverage in setting the terms of defense acquisition agreements. If the Netherlands, for example, agrees to purchase the JSF aircraft, then it will do so only with the "understanding" that its firms will get a substantial share of the work involved in developing, producing, and/or maintaining the aircraft. In Europe, this policy has gone to an extreme called "juste retour," or a guarantee of such work-share as a consequence of purchasing decisions.

This means that a prime contractor—like Lockheed Martin in the case of the F-35—must seek subcontractors in each and every country in which it hopes to sell the plane. But naturally, not every country possesses subcontractors who are capable of working on the world's most technologically sophisticated weaponry. Although corporate executives in the prime contractors stated that they usually attempted to "work with the subcontractors" in bringing them "up to speed," not all firms have proved able or willing to respond. Executives seemed puzzled by firms that were not motivated to meet "milspecs" in such a way as to remain within a program—with the cash flow that is promised over the long-term—but again noted that at least some subcontracting decisions were ultimately made for "political" reasons.

The second problem confronting managers of arms collaboration projects is that of holdup. By this, interviewees (including government officials) meant the *inability* of prime or subcontractors to meet a given set of technological requirements within the cost and time parameters specified in a contract. *It is important to note that in very few cases did interviewees report that such hold-up problems were the result of strategic actions "with guile," to use Williamson's famous phrase* (Williamson, 1985). Thus, contractors did not purposely withhold needed goods and services in an effort to win higher prices or more orders. Instead, holdups were most often from technological complexity and the changing demands of governments (e.g., to add a given set of capabilities to a weapons system).

In addition to these two "usual suspects," the partners in an arms collaboration project face a third challenge: that of illicit technology transfer or technology "poaching." The problem in this sphere is particularly acute because it may not just be firms that are seeking to acquire technology from their partners but the other governments as well. The case of F-35 Joint Strike Fighter reveals these problems in sharp relief.

As previously noted, the JSF will incorporate U.S. leading-edge "stealth" technology, which renders aircraft invisible to radar along with a number of other top-of-the-line avionics and weapons systems. Already, the Pentagon's Inspector General has alleged "that the advanced aviation and weapons technology for the JSF program may have been compromised," by security lapses at the major subcontractor, BAE (formerly British Aerospace) (Shachtman, 2008). The proliferation of such technology to nations or organizations hostile to the United States and its allies obviously poses a security problem of the first degree.

It is crucial to note that, as a general proposition, the partners in international joint ventures seek to protect their proprietary information through patents and other legal devices. That route is not available when it comes to military secrets, which by definition, are never published or made publicly available. Even taking perpetrators to court (say an engineer who acts as a foreign spy) for stealing state or corporate secrets is not straightforward, as governments and firms may not want the technology that has been taken revealed in the courtroom (and they may not wish to have their inability to control that technology made public, either).

If arms collaboration programs face such difficult challenges of shirking, holdup, and poaching—challenges rendered all the more difficult in that they take place across borders, among countries with different cultures and legal standards—then how can they be structured in such a way as to produce complex weapons systems that meet the requirements of the several governments that will buy the end product? An analysis of these collaborative arrangements demonstrate that several structural mechanisms have been used, irrespective of specific governance or control type (e.g., dominant partner with contracts; shared management with equity holdings; and so forth). Most of these will be familiar to students of joint ventures, but some have received less attention in the academic literature. In particular, in addition to engaging in strategic behavior to minimize the risk of opportunism, managers have also sought to build trust within the project team to create a sense of "collective ownership." Often at an unconscious level, managers have drawn from behavioral economics in seeking to develop incentives in which participants seek to contribute to a common good.

In thinking about the management of a major defense program, it is critical to note that what governments do is to break these programs up into several discrete parts or phases, usually as follows:

- Phase I: System Development and Demonstration (at this stage, the technological feasibility of the weapons program is demonstrated by the contractor).
- Phase II: Engineering Management and Development (at this stage, the contractor "gets the kinks out" of the system and demonstrates to the government its ability to build the weapon in quantity within a given set of cost parameters).
- Phase III: Full-Scale Production.

Within each phase, one finds several "milestones" that contractors might meet, and the government increasingly makes use of "award fee" contracts that provide incentives to firms to meet a given set of cost and performance targets. Furthermore, contracting is done on an annual as opposed to a multiyear base during the first two phases; only when the system enters full-scale production does the government provide multiyear contracts. Annual contracting, it is believed, places more pressure on firms to perform for fear of losing the project to a competitor.

What this structure suggests is that, from a broad strategic perspective, collaborative programs are designed in such a way as to incorporate both short-term penalties for "bad behavior" with long-term incentives to perform. Specifically, by breaking programs up into small pieces that are funded annually, the government seeks to send a message to the prime contractor (and perhaps, even more so, the prime to its subs) that it can be eliminated from a project at any time (how realistic this threat is in practice is another matter, albeit one of considerable importance to actual program management. If the threat is weak, then it helps to explain why weapons programs are chronically late and over budget). The prize for good performance, however, is participation in a long-term project (the Joint Strike Fighter program could run for 40 years) with the financial rewards that are associated with such contracts.

Although this structure may help to limit the "classic" problems of opportunism—shirking and holdup—it is less obvious that it can contain technological poaching (although again, the threat of elimination from a project may keep managers vigilant). And poaching is especially complex in arms because governments may join their firms in engaging in industrial espionage. After all, governments tend to value highly their autonomy in the security realm and will often seek to develop or acquire the very best technology that is available.

One way that companies have dealt with "weak" intellectual property rights protection, for example, in the context of emerging market economies, is to "internalize" the research and development process to the greatest extent possible (Zhao, 2006). Something similar may be at work in arms collaboration, to the extent that projects are "modular" in construction, consisting of numerous "black boxes" that control, for example, the avionics and weapons systems. In such a model, each partner pursues its part of the project in relative isolation from the others, whereas the prime contractor puts the pieces together. Note that this may not be the most "efficient" way of building, say, a jet aircraft, but it reconciles the need to produce the weapon collaboratively with the desire to maintain tight control over military secrets.

Strategic approaches to the control of opportunism reflect in a fairly straight-forward way the lessons of transaction-cost economics. But they also reflect its limitations! In particular, transaction-cost economics tends to assume the existence of a legal environment in which firms interact. Firms that are faced with costly opportunism can always sue an unreliable supplier or, instead, make the decision to "buy" that supplier and thus make the goods internally. In building the 787 Dreamliner, for example, Boeing has already been forced to buy some of its leading suppliers given ongoing quality problems.

In the international realm, however, legal safeguards may be weak and firms may simply be prevented by foreign governments from purchasing particular suppliers. If Lockheed Martin is having trouble, say, with an Italian subcontractor, then it may find that suing the subcontractor is too costly and that buying it is impossible given Italian government restrictions. How, then, does one operate in such an environment?

It is in answer to this specific question that managers kept repeating the need to develop "trust mechanisms." As noted previously, trust is a central concept in behavioral economics, which concerns itself with how individuals and societies engage in cooperative behavior. As we will see, trust is also crucial to successful international joint-ventures (Girmscheid & Brockman, 2008). Although behavioral economics remains something of a theoretical grab-bag, it does find that trust is a crucial element in building durable societies; of course, how trust results in the first place remains an elusive question. What is interesting for our purposes is that executives seem to be well aware of the main lessons of this literature even if they have not read it! But what do executives mean by trust building in terms of specific operational measures? The experience of Lockheed Martin with the F-35 is particularly revealing in this respect.

The F-35 team aimed to build trust at several different levels. First, at the governmental level, a joint office was established with representatives from each participating country. This structure was mirrored at the managerial level, where Lockheed Martin would create an oversight group comprising representatives of all the major subcontractors. This was a crucial step because at first Lockheed Martin played its assigned role of prime contractor with a "heavy hand." What this meant was that Lockheed Martin managers were wary of "shared management" structures and wanted to develop a clear chain of command. The company soon found, however, that this hierarchical framework would not work effectively for the F-35 program. As already noted, participating governments had a fair amount of structural power in regard to the program, meaning that they could pressure for their subcontractors to win a piece of the F-35 business. Furthermore, these governments wanted their say in the design process. Like it or not, Lockheed Martin would have to find a way of exercising managerial control while giving each major partner a voice in the F-35 program. And by giving each partner a voice, the program came to have a collective life with each agent recognizing that it was contributing to a common effort.

At the team level, trust building was greatly aided by new forms of collaborative technology, such as work spaces. These enabled engineers to work together across time zones and continents. Again, the sharing of technical knowledge on a *common* technological platform played a key role in creating a team environment.

Finally, at the individual level, every manager on the F-35 project, no matter where they were based, *signed identical nondisclosure agreements*. Lockheed Martin executives believe that this played a key role in building a trust environment because it helped break-down any "insiders"–"outsiders" distinction. Every manager was now bound by the same contract, again contributing to a team spirit and the commitment to a common cause.

The F-35 project is still young, and it remains to be seen how successful it will be. But this experience to date, coupled with lessons from other collaborative programs, already suggests some important lessons. In particular, it is important to recognize upfront that opportunism is likely to be a major problem facing every international joint venture. It is equally important to recognize that strategic actions, although important, will only overcome some of the risks associated with that behavior. A crucial compliment (but not necessarily a substitute) for strategic action is found in trust building *at the different "levels of analysis,"* which means, in the case of weapons programs, at the level of governments, firms, teams, and among individuals.

In sum, arms collaboration represents a "hard case" for students of governance in international joint-ventures because opportunism is likely to be acute and to take some especially pernicious forms. Although it may be possible to devise organizational structures and incentive systems that help limit shirking and holdup, these rationalistic or strategic actions are likely to overcome all the managerial challenges on their own. Mechanisms for trust building will prove a necessary compliment, as they create an environment in which each member believes it is contributing to a common good. The take-away message is that engagement in an international joint venture requires not just clever approaches to management and contracting, but to trust-building as well.

16.4 CONCLUSIONS AND LESSONS FOR FUTURE RESEARCH

During the past 30 years, the United States and its allies have engaged in upward of 90 collaborative weapons programs, and within Europe, a number of projects have also been executed. Today, in light of the rising costs of weaponry and the globalization of the technology base, international collaboration is more in demand than at any time in history. Given the huge budgetary resources that go into these programs, analysis of their management and governance is important in its own right, irrespective of the broader lessons that might exist for managers of international joint ventures and strategic alliances in other sectors. Students of business strategy could play a useful role in public policy debates by paying closer examination to the weapons acquisition process and to the relationship between governments and defense contractors.

Yet the arms collaboration case suggests at least one overarching lesson of more general import: if international collaboration can "succeed" in weaponry, then it can succeed in most other domains as well, despite the high failure rate associated with cross-border joint ventures to date. Furthermore, success has come in this arena despite the lack of strong intellectual property rights or other legal remedies and in the presence of significant problems of opportunism. One could almost argue that international arms collaboration represents cross-border activities in the "state of nature," where no legal framework exists to bind the parties together.

How is this outcome possible? Our analysis suggests that a combination of factors has made it feasible for governments and firms to engage in effective collaborative weapons programs. Each is worthy of close, in-depth examination by students of other sectors as well.

First, these projects have been structured in such a way as to balance short-term penalties for bad behavior against the long-term prize of multiyear production contracts. Thus, breaking complex products up into their constituent parts or phases and placing "real options" on each phase seems like it should be part of a joint venture's winning strategy.

Second, as the list in the Appendix reveals, successful participation in one program seems to lead firms to participate in other projects. Thus, a firm's reputation for getting a job done, on time and on budget, matters greatly to its long-run prospects. Much of transaction cost economics makes it seem like firms are engaged in "one-shot" deals (although scholars naturally recognize the significance of repeated contracts), but of course, that is often not the case in reality. Firms have broad and deep relations, and that is increasingly the case as global production networks advance.

Third, the partners in international joint ventures must recognize head-on the problem of illicit technology transfer and understand that it may not be easy to rely on patents or property rights to secure one's "stuff." This may lead to innovative approaches for placing technologies in "black boxes" or making them modular. The additional costs of technological security represent the price to be paid for international joint ventures in many settings, especially those where property rights are not respected.

Fourth, and finally, managers of successful programs combine strategic behavior with trust building at different levels in an effort to create a sense of collective ownership of a given project.

In this chapter, we have thus argued that the governance and organization of complex, cross-border projects makes a difference to their ultimate success. Through careful structural arrangements that provide the right incentives, through innovative management of technology, and through trust-building, joint ventures can prove beneficial to all parties even in light of the ongoing risks of opportunistic behavior. Rather than view strategic actions and trust building as different realms of managerial activity—the one "hard" and reserved for engineers, the other "soft" and reserved for human resource specialists—executives must learn to blend these two approaches as they develop their international joint ventures.

REFERENCES

Bolkcom, C. (2002). *Joint Strike Fighter: Background, Status, and Issues*. Washington, DC: Congressional Research Service.

Clemons, E., & Hitt, L. (2004). Poaching and the misappropriation of information: Transaction risks of information exchange. *Journal of Management Information Systems*, *21*(2) 87–108.

Das, T.K., & N. Rahman, N. (2002). Opportunism dynamics in strategic alliances. In F.J. Contractor & P. Lorange, Eds., *Cooperative Strategies and Alliances*. Oxford, UK: Elsevier.

DoD. (1987). *Guide for the Management of Multinational Programs*. Fort Belvoir, VA: U.S. Defense Systems Management College.

Dower, J. (1999). *Embracing Defeat*. New York: Norton.

Girmscheid, G., & Brockmann, C. (2008) *Trust as a Success Factor in International Joint Ventures*. Zurich, Switzerland: ETH Zurich.

Kapstein, E.B. (1991). International collaboration in armaments production: A second-best solution. *Political Science Quarterly*, *106*(4), 657–675.

Oxley, J.E. (1999). Institutional environment and the mechanisms of governance: The impact of intellectual property protection on the structure of inter-firm alliances. *Journal of Economic Behavior and Organization*, *38*, 283–309.

Shachtman, N. (2008). Stealth fighter security may have been compromised. *Wired*.

Sosa, M.E., Eppinger, S., & Rowles, C. (2007). Are your engineers talking to one another when they should? *Harvard Business Review*, 1–8.

Williamson, O. (1985). *The Economic Institutions of Capitalism*. New York: Free Press.

Zhao, M. (2006). Conducting R&D in countries with weak intellectual property rights protection. *Management Science*, *52*(8), 1185–1199.

Appendix

List of International Arms Collaboration Programs Studied, by U.S. Firm

(Participating countries and major foreign corporate partners in parentheses.)

Boeing
U.S. Army Common Missile (US/UK; BAE)
Meteor Missile (US/UK/FR/GE/SP/IT/SE; MBDA)
Brimstone Missile (US/UK/IT; MBDA)
AV-8B Harrier Fighter Jet (US/UK; BAE)
T-45 Goshawk Training Jet (US/UK; BAE)
Precision Guided Munitions (US/UK; MBDA, Insys)
NATO Awacs Surveillance Aircraft (NATO; BAE)
F/A-18 Hornet Fighter Aircraft (Finland; Patria Finavitec)
Ballistic Missile Defense (NATO; BAE, EADS, Finmeccanica)
Arrow Missile (US/Israel; IAI)

Lockheed Martin Corp.
MLRS Rocket System (US/GE/IT/FR/UK; Diehl, BPD, Matra, Insys)
Patriot Missile System (US/GE; EADS)
Euro-Art Radar (US/GE; Siemens, Thales)
F-35 Joint Strike Fighter (UK/US/Neth/Nor/Can/It/Tur/Den/Aus; BAE, NorthropGrumman)
MEADS Missile System (NATO; EADS, Alenia)
Advanced Frigate (US/Nor/Sp; IZAR, Kongsberg; Aerospace AS; Mjellum & Karlsen; BIW)

The Economics of Human Systems Integration: Valuation of Investments in People's Training and Education, Safety and Health, and Work Productivity. Edited By William B. Rouse
Copyright © 2010 John Wiley & Sons, Inc.

F-16 Fighter Aircraft (Bel/Neth/Nor/Kor/Tur; SABCA; Fabrique; Stork; Nordisk Aluminum; Kongsberg; KAI)

NATO Theater Ballistic Missile Defense (NATO; TRW, Matra, Alenia, Astrium, BAE, EADS, Fokker, LFK)

Popeye Missiles (US/Israel; Rafael Armaments)

F-2 Support Fighter (US/Japan; MHI)

Index

The Economics of Human Systems Integration: Valuation of Investments in People's Training and Education, Safety and Health, and Work Productivity. Edited By William B. Rouse
Copyright © 2010 John Wiley & Sons, Inc.

WILEY SERIES IN SYSTEMS ENGINEERING AND MANAGEMENT

Andrew P. Sage, Editor

YACOV Y. HAIMES
Risk Modeling, Assessment, and Management, Third Edition

DENNIS M. BUEDE
The Engineering Design of Systems: Models and Methods, Second Edition

ANDREW P. SAGE and JAMES E. ARMSTRONG, Jr.
Introduction to Systems Engineering

WILLIAM B. ROUSE
Essential Challenges of Strategic Management

YEFIM FASSER and DONALD BRETTNER
Management for Quality in High-Technology Enterprises

THOMAS B. SHERIDAN
Humans and Automation: System Design and Research Issues

ALEXANDER KOSSIAKOFF and WILLIAM N. SWEET
Systems Engineering Principles and Practice

HAROLD R. BOOHER
Handbook of Human Systems Integration

JEFFREY T. POLLOCK AND RALPH HODGSON
Adaptive Information: Improving Business Through Semantic Interoperability, Grid Computing, and Enterprise Integration

ALAN L. PORTER AND SCOTT W. CUNNINGHAM
Tech Mining: Exploiting New Technologies for Competitive Advantage

REX BROWN
Rational Choice and Judgment: Decision Analysis for the Decider

WILLIAM B. ROUSE AND KENNETH R. BOFF (editors)
Organizational Simulation

HOWARD EISNER
Managing Complex Systems: Thinking Outside the Box

STEVE BELL
Lean Enterprise Systems: Using IT for Continuous Improvement

J. JERRY KAUFMAN AND ROY WOODHEAD
Stimulating Innovation in Products and Services: With Function Analysis and Mapping

WILLIAM B. ROUSE
Enterprise Tranformation: Understanding and Enabling Fundamental Change

JOHN E. GIBSON, WILLIAM T. SCHERER, AND WILLAM F. GIBSON
How to Do Systems Analysis